Pleiotropic Action of Selenium in the Prevention and Treatment of Cancer, and Related Diseases

Pleiotropic Action of Selenium in the Prevention and Treatment of Cancer, and Related Diseases

Special Issue Editor

Youcef M. Rustum

MDPI • Basel • Beijing • Wuhan • Barcelona • Belgrade

MDPI

Special Issue Editors
Youcef M. Rustum
Roswell Park Cancer Institute
Elm and Carlton Streets
USA

University of Iowa
Iowa city, Iowa
USA

Editorial Office
MDPI
St. Alban-Anlage 66
4052 Basel, Switzerland

This is a reprint of articles from the Special Issue published online in the open access journal *International Journal of Molecular Sciences* (ISSN 1422-0067) from 2018 to 2019 (available at: https://www.mdpi.com/journal/ijms/special_issues/selenium_cancer)

For citation purposes, cite each article independently as indicated on the article page online and as indicated below:

LastName, A.A.; LastName, B.B.; LastName, C.C. Article Title. *Journal Name* **Year**, *Article Number*, Page Range.

ISBN 978-3-03897-692-9 (Pbk)
ISBN 978-3-03897-693-6 (PDF)

Contents

About the Special Issue Editors

Youcef M. Rustum, emeritus professor of oncology. Research professor of University of Iowa. The aim and focus of my research for the last few years has been to develop more effective and selective therapies for cancer patients based on data generated in preclinical models and translated from bench to bed side and vice versa. In the 1990s, my laboratory, in collaboration with clinical colleagues at RPCI, developed the weekly schedule of 5Fluorouracil in combination with Leucovorin as a standard therapy for the treatment of colorectal cancer. Recently, this laboratory was the first to demonstrate that therapeutic doses and schedule of methyl selenocysteine and seleno-L-methionine are potent inhibitors of hypoxia and constitutively expressed HIF-1α and HIF-2α and associated specific types of miRNAs and VEGF expressed in tumor cells. Pronounced and sustained inhibition of these specific biomarkers was associated with therapeutic synergy with anticancer drugs. On the basis of the data generated, Phase 1 and 2 clinical trials of seleno L- methionine in sequential combination with axitinib are under clinical evaluation in patients with advanced clear cell renal cell carcinoma. On the basis of the data generated, Phase 1 and 2 clinical trials of seleno methionine in sequential combination with axitinib are under clinical evaluation in patients with advanced clear cell renal cell carcinoma.

Preface to "Pleiotropic Action of Selenium in the Prevention and Treatment of Cancer, and Related Diseases"

In recent years, proof of concept of the efficacy of biologically targeted therapies has been documented in several cancer types. Treatment-associated innate and/or acquired resistance, and dose-limiting toxicity and cost for chemo, radio, and target-directed therapies continue to represent major clinical challenges. Our advanced understanding of the molecular, immunological, and biological heterogeneity of tumor cells and their adjacent microenvironment has provided the opportunity for the development of new molecules that target commonly altered biomarkers that control tumor growth and angiogenesis. Selenium is a cheap natural product that exists in multiple forms and is expressed in differential metabolic actions, playing an important role in healthcare and maintenance. Dependent on the selenium types, dose, and schedule, selenium modulates target biomarkers associated with angiogenesis, drug resistance, and immune responses. Various forms of selenium-containing molecules have been extensively evaluated in clinical prevention trials. The potential use of specific types, dose, and schedule of selenium-containing molecules as a selective modulator of in vivo drug response in preclinical and clinical models has been introduced in recent years. The preclinical data generated indicate that to achieve an optimal therapeutic benefit, selenium must be sequentially combined with chemo and biologically targeted therapies. Thus, the dose and schedule of a drug modulator such as selenium in sequential combination with cytotoxic and biologic therapies should be optimized in several well-defined and clinically relevant preclinical models. The optimal therapeutic conditions documented in preclinical models should provide the basis for the design of future clinical trials. With the knowledge that hypoxia-inducible factors and specific microRNAs are altered in tumor cells and that their adjacent microenvironment are selenium targets, proof of concept should be carried out in tumors to significantly express the designated targets and to verify that the modulation of these targets would indeed predict treatment outcome.

It has been a privilege for me to assemble contributions from outstanding authors highlighting the therapeutic potential of selenium in cancer therapy. I must thank my collaborators, who were instrumental in the preclinical and clinical development of selenium, as without them, the progress made would have not been achieved.

Youcef M. Rustum
Special Issue Editor

International Journal of
Molecular Sciences

MDPI

Article

Novel Methylselenoesters Induce Programed Cell Death via Entosis in Pancreatic Cancer Cells

Prajakta Khalkar [1,*,†], Nuria Díaz-Argelich [1,2,3,†], Juan Antonio Palop [2,3], Carmen Sanmartín [2,3] and Aristi P. Fernandes [1]

[1] Division of Biochemistry, Department of Medical Biochemistry and Biophysics (MBB), Karolinska Institutet, SE-171 77 Stockholm, Sweden; ndiaz@alumni.unav.es (N.D.-A.); Aristi.Fernandes@ki.se (A.P.F.)
[2] Department of Organic and Pharmaceutical Chemistry, Faculty of Pharmacy and Nutrition, University of Navarra, Irunlarrea 1, E-31008 Pamplona, Spain; japcubillo@gmail.com (J.A.P.); sanmartin@unav.es (C.S.)
[3] IdiSNA, Navarra Institute for Health Research, Irunlarrea 3, E-31008 Pamplona, Spain
*   Correspondence: prajakta.khalkar@ki.se; Tel.: +46-8-52486990
†   These authors contributed equally to this work.

Received: 27 August 2018; Accepted: 18 September 2018; Published: 20 September 2018

Abstract: Redox active selenium (Se) compounds have gained substantial attention in the last decade as potential cancer therapeutic agents. Several Se compounds have shown high selectivity and sensitivity against malignant cells. The cytotoxic effects are exerted by their biologically active metabolites, with methylselenol (CH_3SeH) being one of the key executors. In search of novel CH_3SeH precursors, we previously synthesized a series of methylselenoesters that were active ($GI_{50} < 10\ \mu M$ at 72 h) against a panel of cancer cell lines. Herein, we refined the mechanism of action of the two lead compounds with the additional synthesis of new analogs (ethyl, pentyl, and benzyl derivatives). A novel mechanism for the programmed cell death mechanism for Se-compounds was identified. Both methylseleninic acid and the novel CH_3SeH precursors induced entosis by cell detachment through downregulation of cell division control protein 42 homolog (CDC42) and its downstream effector β1-integrin (CD29). To our knowledge, this is the first time that Se compounds have been reported to induce this type of cell death and is of importance in the characterization of the anticancerogenic properties of these compounds.

Keywords: selenium; methylselenoesters; entosis; anticancer agent

1. Introduction

Pancreatic ductal adenocarcinoma is an extremely aggressive neoplasm and one of the cancers with the poorest prognosis, with a five-year survival of only 8% [1]. In addition, it is predicted to become the second leading cause of cancer-related death by 2030 [2]. Late diagnosis in advanced cancer stages due to a lack of prior symptomatology and the poor efficiency of actual therapeutics are the main causes. Drug resistance in pancreatic cancer is largely caused by an active stroma contributing to tumor progression [3]. Therefore, developing new therapeutic strategies has become an urgent need.

Modulation of redox homeostasis in cancer cells has emerged as a new opportunity for tumor intervention. Induction of reactive oxygen species (ROS) by these compounds may affect all the redox dependent pathways in the cell, which can be detrimental to cells. Antioxidant enzymes are often induced to eliminate elevated ROS production. Due to metabolic transformation, cancer cells have an increased and maximized antioxidant capacity in order to evade the ROS-induced cell death. For instance, the expression of mutant oncogenic KrasG12D is commonly present in pancreatic ductal adenocarcinoma (PDAC), resulting in an elevated basal state of the transcription factor, nuclear factor E2-related factor 2 (NRF2) to mount an antioxidant response [4,5]. Therefore, even a slight additional

ROS induction, using redox modulators, would lead to the killing of cancer cells [6,7], and provides an interesting therapeutic approach that has been established as a means of successful anti-cancer therapy [8–11].

Redox modulating selenium (Se) compounds have gained substantial attention in the last decade as potential cancer therapeutic agents [12]. Several Se compounds have shown high selectivity and sensitivity in malignant cells [13]. Depending on the compound use, they have been reported to induce different types of cell death, including apoptosis, autophagy, necrosis, or necroptosis.

Importantly, along with their active metabolites that execute their biological activity, the dosage and chemical form of Se compounds highly determine their efficacy [12]. Methylselenol (CH_3SeH) is considered a key metabolite in the anticancer activity of Se compounds. However, the in situ production or alternatively, the use of precursors, is required due to the high reactivity and volatility of this molecule.

In search of novel CH_3SeH precursors, we previously synthesized a series of methylselenoesters that were active ($GI_{50} < 10$ μM at 72 h) against a panel of cancer cell lines [14]. Herein, we studied the mechanism of action of the two lead compounds with the additional synthesis of new analogs (ethyl, pentyl, and benzyl derivatives) (Figure 1). This study uncovers a novel cell death mechanism for these Se-compounds as entosis inducers. Entosis was first described under anchorage-independent conditions and the loss of β1-integrin (CD29) signaling [15]. However, it has also been described in adherent cells [16–18] and recently, aberrant mitosis [16] and glucose deprivation [19] have been identified as other possible triggers.

During entosis, the stiffer cell (hereafter target cell) actively participates in its own internalization, via adherent junctions and the actin cytoskeleton that play a pivotal role in this process. Ultimately, the target cell is killed through lysosomal enzyme-mediated degradation, using the autophagy machinery, but independent of autophagosome formation [20]. The death subroutine might swift to apoptosis in the absence of autophagy-dependent nutrient recycling, or eventually, the internalized cell might divide or be released [21].

Methylseleninic acid (MSA) and the novel CH_3SeH precursors induce cell detachment through downregulation of cell division control protein 42 homolog (CDC42) and its downstream effector CD29 [22]. Cell-cell adhesion molecules such as N-cadherin were upregulated after treatment and facilitated cell clustering, which finally ended with cell-in-cell invasion and the degradation of the inner cell. To our knowledge, this is the first time that Se compounds have been reported to induce this type of cell death.

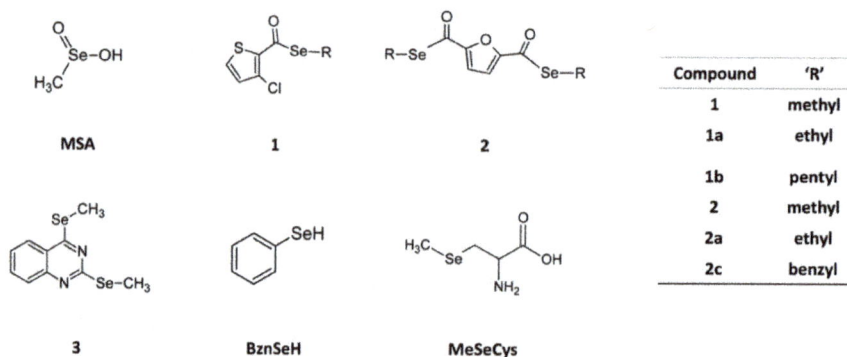

Compound	'R'
1	methyl
1a	ethyl
1b	pentyl
2	methyl
2a	ethyl
2c	benzyl

Figure 1. Chemical structures of the compounds. MSA, and compounds **1** and **2** were the primary focus of this study, whereas remaining compounds were used for comparative analysis in some experiments. MSA: methylseleninic acid; R: substituent; BznSeH: benzeneselenol; MeSeCys: methylselenocysteine.

2. Results

2.1. MSA, and Compounds **1** and **2** Reduce Panc-1 Cell Viability Both in 2D and 3D Cultures

Initial characterization of the compounds was performed through viability assays in 2D and 3D cultures of Panc-1 cells, given that 3D cultures have been demonstrated to mimic tumor behavior more efficiently than traditional monolayer (2D) cultures. Panc-1 cells were treated with increasing concentrations of MSA, and compounds **1** or **2** for 72 h. Cell viability was then determined. All three compounds were cytotoxic, with compound 2 being the most potent compound in 2D cultures. The compounds had IC50 values in the low micromolar range in 2D cultures (2.28, 3.31, and 1.43 µM for MSA, and compounds **1** and **2**, respectively). However, cells grown as spheroids (3D) were consistent with previously reported data [23], and more resistant and higher doses of the compounds were required to reduce cell proliferation and induce cell death (Figure 2A,B).

To further study the induced cell death in 3D cultures, spheroids were stained with Hoechst and propidium iodide (PI) after 72 h treatment. While Hoechst stains the nucleus of all cells, PI only penetrates and stains damaged membranes of dying cells. As shown in Figure 2C, the three compounds were not only able to induce cell death, but the cell death was observed in the core of the spheroid, suggesting that these compounds were able to reach to the core of the sphere.

The selenoester entity could be easily hydrolyzed by a nucleophile such as water, rendering the corresponding carboxylic acids and releasing CH_3SeH, which is believed to be a key molecule in Se activity (Figure 2D). To exclude the possibility that the toxicity was from the linked moieties, the analog carboxylic acids of compounds **1** (**1′**) and **2** (**2′**) were selectively tested as a proof-of-concept. As seen in Figure 2E, they did not induce any cell death compared to the Se-containing molecules.

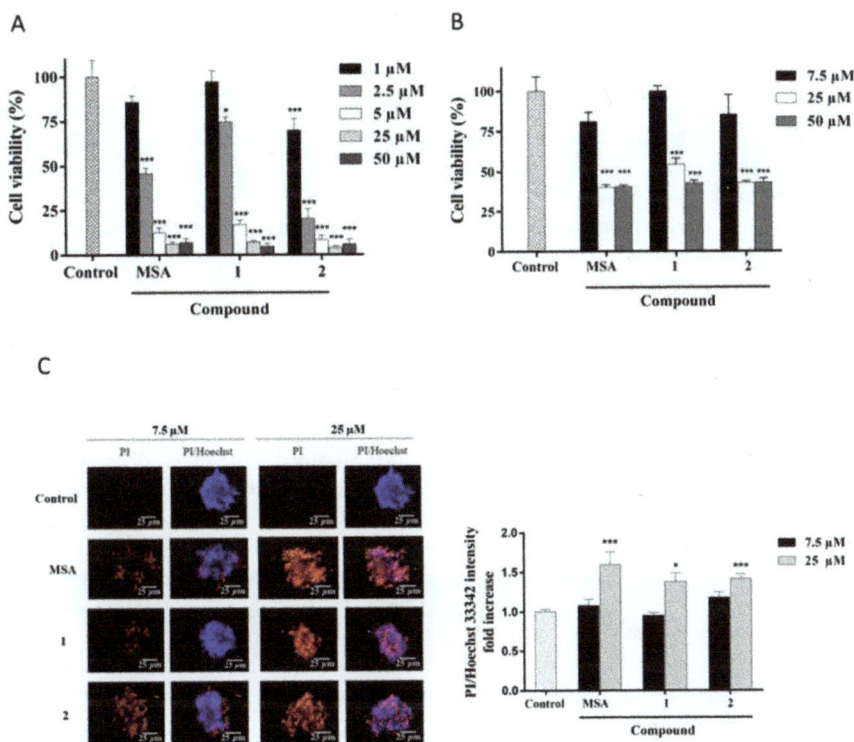

Figure 2. *Cont.*

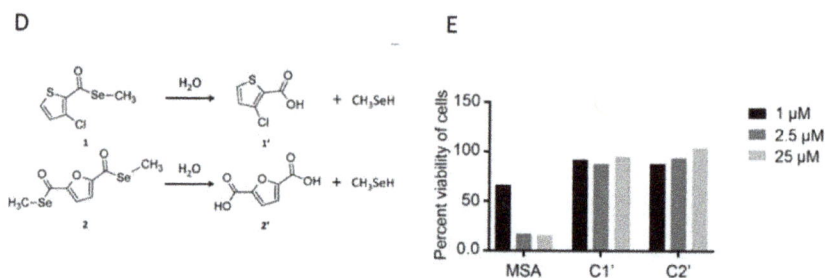

Figure 2. Compounds **1** and **2** and MSA decrease cell viability in 2D and 3D Panc-1 cultures. (**A**) Panc-1 cells (2D cultures) were treated with different concentrations of the compounds for 72 h followed by the determination of cell viability by the MTT (3-(4,5-Dimethylthiazol-2-yl)-2,5-Diphenyltetrazolium Bromide) assay. Results represent mean ± SEM of at least three independent experiments performed in quadruplicate. (**B**) Panc-1 spheroids (3D cultures) were treated with different concentrations of the compounds for 72 h, after which cell viability was determined using the acid phosphatase (APH) assay. Results represent mean ± SEM of at least three independent experiments performed in quadruplicate. (**C**) Representative confocal images of Panc-1 spheroids stained with Hoechst 33342 and PI after 72 h treatment with 7.5 µM and 25 µM of respective compounds. 10× objective magnification images were acquired from the Operetta® High-Content Imaging System and processed by Colombus™ analysis software. The adjacent graph represents a quantitative analysis of PI/Hoechst fluorescence. Results represent mean ± SEM (*n* = 4). (**D**) Potential hydrolysis reaction of compounds **1** and **2**. (**E**) 2D cell viability after treatment with the corresponding carboxylic acid for 72 h. Statistical significance compared to control: * $p < 0.05$, *** $p < 0.001$.

2.2. MSA, and Compounds 1 and 2 Induce Cell Detachment and Compromise Reattachment Abilities by Promoting an Aberrant Adhesive Repertoire

In order to study the early effects of this particular cell death, a concentration of 5 µM of respective compounds was chosen for further experiments in 2D cultures. Post 6 h treatment of Panc-1 cells, morphological changes like rounding of the cells and cellular detachment from culture flasks were observed. At 24 h, almost all the cells were detached, had acquired a refringent morphology, and were grouped in a grape-like manner (Figure 3A). Trypan blue exclusion, however, indicated that the floating cells were still alive at that particular time point (Figure 3B). To examine if the aberrant cellular detachment was irreversible, an adhesion assay was performed wherein the floating cells were washed to remove traces of the compounds and reseeded in fresh medium. The cells were then allowed to reattach to culture flasks for 3 h. Nevertheless, their reattachment abilities after treatment were observed to be compromised, with a clear loss of ability to re-adhere, especially in the case of compound **2** (Figure 3C).

As a next step, the effect of respective compounds on different cellular adhesion markers was analyzed. Post 24 h treatment, the expression of CD29, known to mediate adhesion to the extracellular matrix [24], was significantly reduced, as observed after flow cytometry analysis (Figure 3D), explaining the loss of cellular adhesion caused by these compounds. Moreover, the expression of N-cadherin, a cell-cell adhesion marker [25], showed a considerable increase after treatment with respective compounds, which explains the observed grape-like cellular clumping after detachment (Figure 3D).

In order to assess the fate of the detached cells, i.e., if they were able to recover or eventually go into the cell death mode, a clonogenic assay was performed. As illustrated in Figure 3E, post 24 h treatment with respective compounds, the cells displayed a significant decrease in colony formation compared to the control.

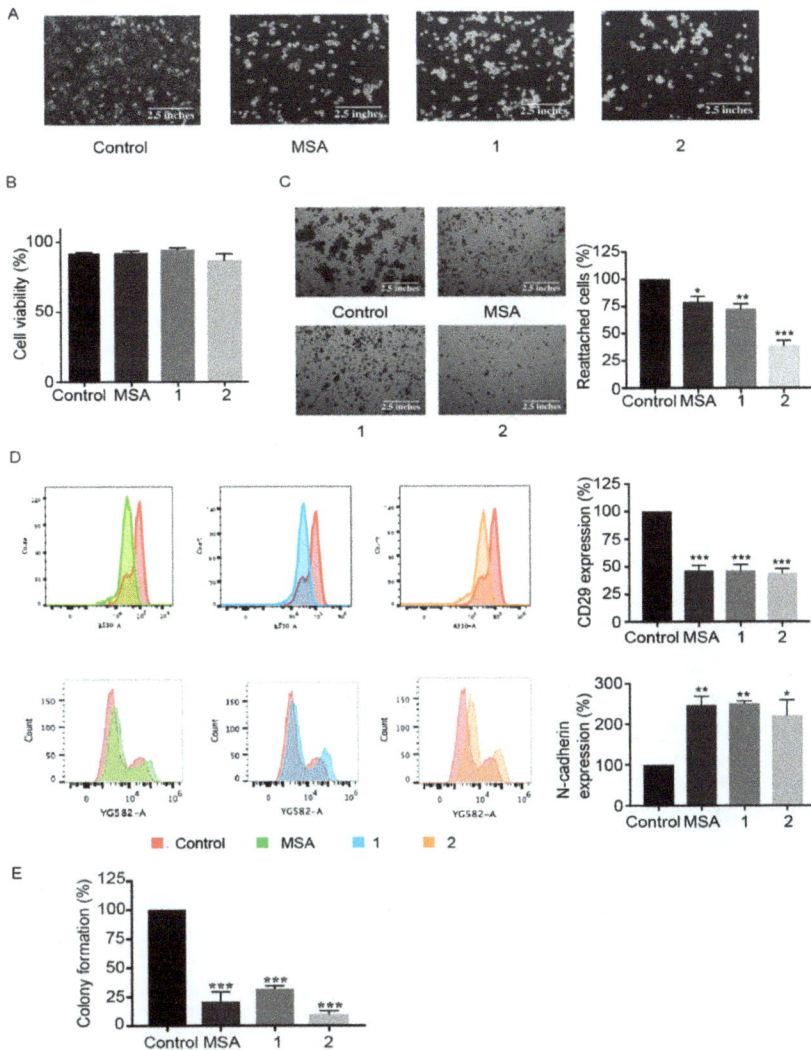

Figure 3. Compounds **1** and **2** and MSA induce loss of cellular adhesion prior to cell death and impair the colony forming ability. Panc-1 cells were treated with 5 μM of MSA, or compounds **1** or **2** for 24 h. (**A**) Representative phase-contrast images of treatment-induced cell detachment with respective compounds. (**B**) The viability of the floating cells was assessed using a trypan blue exclusion assay. (**C**) Adhesion assay. After 24 h of treatment with respective compounds, the non-adherent but viable cells, were collected and an adhesion assay was performed for 3 h, following which the adherent cells were stained with Coomassie Brilliant Blue. Representative phase-contrast microscopic images and graphical representation of percentage of non-adherent cells reattaching the tissue culture treated plates. Error bars indicate mean ± SEM of three biological replicates. (**D**) The expression levels of adhesion proteins, CD29, and N-Cadherin, post 24 h treatment with respective compounds as analyzed by flow cytometry and its graphical representation. Error bars indicate mean ± SEM of three biological replicates. (**E**) Clonogenic assay. Post 24 h treatment, with respective compounds, the non-adherent cells were collected and re-seeded in 24 well plates to check for the colony forming ability of these cells. Reduced colony forming ability indicates cell death. Error bars indicate mean ± SEM of three biological replicates. Statistical significance as compared to control * $p < 0.05$, ** $p < 0.01$, *** $p < 0.001$.

2.3. MSA, Compounds 1 and 2 Induce Entosis.

Two well described cell death pathways that have been reported to be initiated by the loss of cell adhesion are anoikis and entosis. Whereas anoikis is triggered exclusively upon adhesion loss and is coursed through caspase activation, entosis is characterized by active cell invasion, leading to endophagocytosis and the formation of cell-in-cell structures, and has been described both in suspension and adherent cells.

To distinguish the programmed cell death mode, we analyzed the expression of total poly (ADP-ribose) polymerase 1 (PARP) and cleaved PARP, wherein MSA and compounds **1** and **2** slightly increased the 89 kD cleaved fragment at 72 h (Supplementary Figure S1A). PARP has been reported to be cleaved by caspases, cathepsins, and calpains [26–29]. In order to rule out the possibility of apoptosis, the expression of caspase 9 and cleaved caspase 9 (upstream marker for apoptosis) was analyzed after 48 h treatment with these compounds. We observed no expression of cleaved caspase 9, suggesting that no activation of the caspase cascade was induced (Supplementary Figure S1A). Additionally, cells were treated with the broad pancaspase inhibitor z-VAD-fmk, along with respective compounds. Treatment with z-VAD-fmk did not prevent the cellular detachment, as well as cell death, induced by these compounds, as observed by brightfield microscopy and the trypan blue exclusion assay, respectively (Supplementary Figure S1B,C), suggesting a caspase-independent mechanism.

Furthermore, the expression of cathepsins, a structurally and catalytically distinguished class of proteases, was checked. A context-depending role has been described for cathepsins, with either tumor-promoting or suppressing activities. They have not only been reported to function as apoptotic mediators, but also to be related to entosis [15] and cell cannibalism [30].

Both cathepsin B (CatB) and cathepsin D (CatD) have been reported to play an important role in entosis [15,20,31]. Increased expression of CatB was observed, indicating that lysosomal degradation is implied in cell death induced by these compounds. Unexpectedly, CatD levels were downregulated (Figure 4A). Another family with a prominent role in entosis is the Rho family of GTPases, master regulators of the actin cytoskeleton. Therefore, the protein levels of CDC42 and RhoA were determined. Whereas CDC42 levels were decreased, RhoA levels remained unchanged (Figure 4A). To further confirm entosis, cell fate was tracked once detached. Cells were labeled with green or red fluorescent dyes, seeded, and treated with the compounds. Visualization by confocal microscopy revealed cell-in-cell internalization (Figure 4B). A time-lapse experiment also recorded live confirmed morphological changes during the formation of cell-in-cell structures and the ultimate degradation of the target cell (Supplementary videos 1–3).

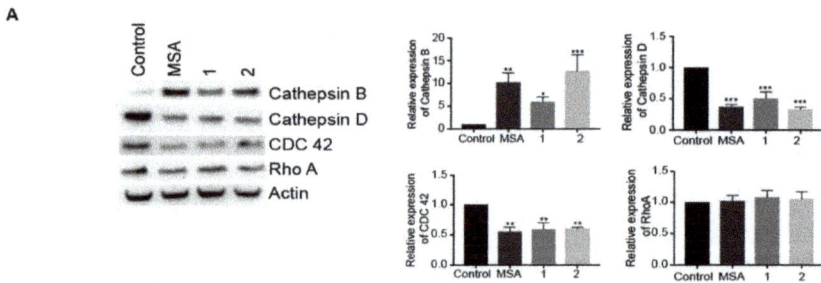

Figure 4. *Cont.*

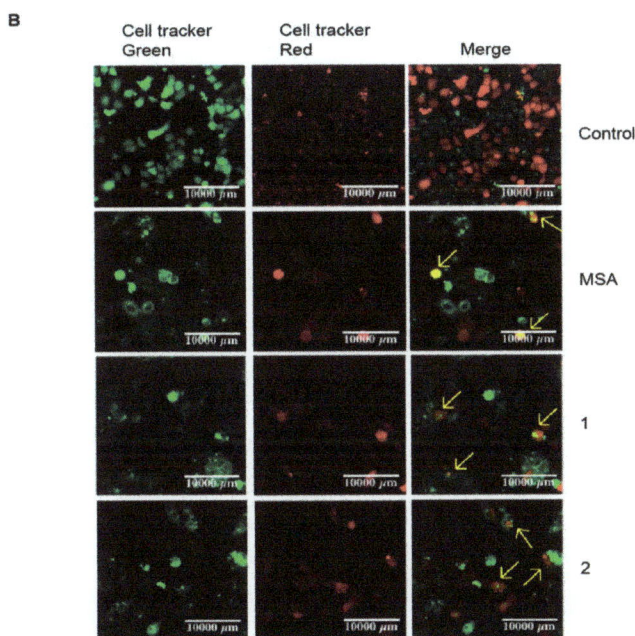

Figure 4. MSA, and compounds **1** and compound **2** induce entosis in Panc-1 cells. (**A**) Western blot analyses of Cathepsin B, Cathepsin D, CDC42, and Rho A upon treatment with MSA or compound **1** or compound **2** for 24 h. Beta actin was used as a loading control. The corresponding graphs display a quantitative analysis of western blots performed using the ImageJ program and GraphPad Prism software. Error bars indicate mean ± SEM of three biological replicates. (**B**) Panc-1 cells stained with CellTracker Red or Green and further mixed (1:1) were treated with MSA or compound **1** or compound **2** for 18 h, followed by live imaging using a confocal microscope, Zeiss LSM800. The arrows indicate cell-in-cell formations i.e., cells undergoing entosis. Statistical significance as compared to control * *p* < 0.05, ** *p* < 0.01, *** *p* < 0.001.

2.4. Cell Detachment Is Not Restricted to Selenomethylated Compounds and Does Not Correlate with the Cytotoxic Potential of the Compound

In order to distinguish the type of Se compound that could cause this phenomenon, other commercial Se derivatives together with other newly synthesized analogs of compounds **1** and **2** were analyzed. To evaluate if this effect was exclusive to methylated forms of Se or unrestricted to other alkyl or aromatic derivatives, the ethyl derivative for compounds **1** and **2** (**1a** and **2a**, respectively), the pentyl derivative for compound **1** (**1b**), and the benzyl analog for compound **2** (**2c**), were synthesized to cover different alkyl lengths and additional substituents. Methylselenocysteine (MeSeCys) was also selected as another CH_3SeH precursor and benzeneselenol (BznSeH) as an additional aromatic selenol for a comparative analysis. In addition, compound **3** was chosen, a previously synthesized selenide in our laboratory, as a proof-of-concept compound without a labile bond between the core of the molecule and the methylseleno residue [32], and therefore less prone to release it (Figure 1).

First, cell proliferation and cell death were evaluated. As illustrated in Figure 5A, all the compounds were able to reduce cell proliferation. However, a longer chain or the substitution with a benzyl residue impaired the cytostatic activity of the compounds. In general trends, and considering the 72 h time point, the potency to reduce proliferation decreased according to the following order: methyl > ethyl > pentyl or benzyl. Compound **2a** stopped proliferation at 24 h, while the methylated analog (**2**) achieved a reduction at 72 h. BznSeH and compound **3** had the highest cytostatic potential, with both of them inhibiting proliferation at 24 h of treatment.

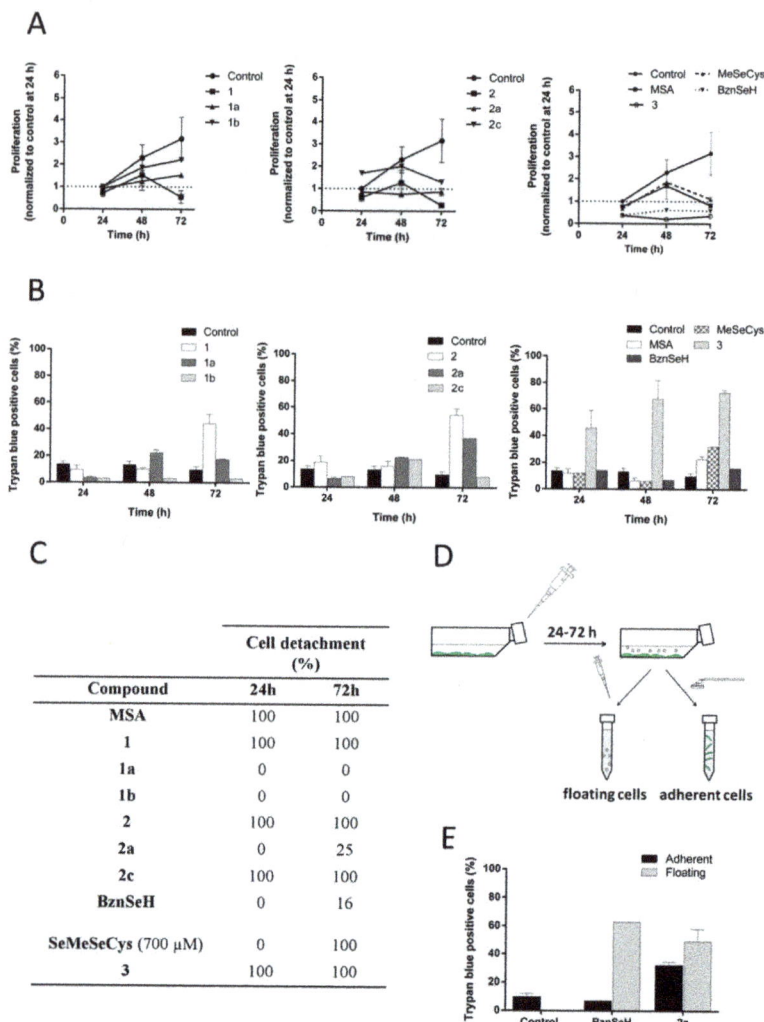

Figure 5. Evaluation of the antiproliferative, cytotoxic, and de-adhesive properties of other Se analogs. Cells were seeded and incubated for 24 h before starting treatments with the compounds. Cell proliferation (**A**) and cell death based on Trypan blue exclusion (**B**) were analyzed after treatment with a 5 μM dose. For MeSeCys, a 700 μM dose was used. For BznSeH and compound **2a**, inducing a floating and attached population, proliferation, and cell death were calculated without taking into account the two populations in this case. (**C**) Cell detachment quantification after treatment. (**D**) Procedure scheme to evaluate the attached and floating population. Floating cells were collected with a pipette and remaining cells were considered as attached and slightly scrapped. (**E**) Cell death comparison in the floating and adherent populations induced by compound **2a** and BznSeH.

The methyl derivatives were more cytotoxic at 72 h than analogs with a longer alkyl chain or the benzyl moiety (Figure 5B). Almost all the compounds showed similar activity at 24 and 48 h, with the exception of compound **1b**, which was not cytotoxic. BznSeH, which was highly cytostatic, did not induce noteworthy cell death, with only 15% of dead cells at 72 h. MeSeCys, which has been reported to have a similar activity to MSA in vivo, required a considerably higher concentration to achieve similar cell death induction in vitro, due to the need of metabolic processing to release CH_3SeH, consistent with previous reports [33]. Compound **3**, on the other hand, was the most potent compound, inducing cell death at 24 h treatment.

In addition to cell proliferation and cell death, the ability of the compounds to induce cell detachment, and ultimately the same cell death mechanism as the methyl analogs, was analyzed. Most of the compounds completely detached the cells or completely remained ineffective, as illustrated in Figure 5C. However, some compounds induced two populations, and in this case, the procedure schematized in Figure 5D was followed. Compounds **1a** and **1b** were unable to detach cells. whereas compounds **2c** and **3** had detached all the cells at 24 h. MeSeCys detached all the cells at 72 h, and a concomitant increase in cell death was observed at that time point. However, cell detachment potential was not correlated with cell death induction in the case of compound **2c**, which was almost innocuous. BznSeH and compound **2a** induced mixed populations, with attached and floating cell fractions. (Figure 5D,E). Nevertheless, they had considerably less detachment potential than the methylated analogs, with only 16 and 24% of detached cells at 72 h, respectively.

3. Discussion

In this study, we demonstrate that MSA and two novel methylselenoesters induce entosis after provoking cell detachment in Panc-1 cells, revealing a new and unexplored cell death mechanism for Se compounds.

Compounds **1** and **2** and MSA reduced cell proliferation in both 2D and 3D cultures. Treatment with the compounds led to a unique phenotype, characterized by changes in morphology and cell detachment from the culture plate prior to cell death. Detached cells were alive at 24 h, but their reattachment capability and the colony forming ability had been dramatically compromised. We dismissed the possibility that the compounds were promoting anchorage-independent survival, and instead induced cellular death, as confirmed by the MTT assay and the expression of cleaved PARP in 2D cultures and PI staining in 3D spheroids.

Cell adhesion is gaining more attention due to its implication in cancer metastasis and progression, in addition to drug resistance. Importantly, these results are in accordance with recent investigations revealing that MSA targeted adhesion molecules in a leukemic cancer cell line, whereas inorganic selenite affected other gene sets, indicating an interesting type-dependent effect of Se compounds [34]. To further confirm the compound-induced adhesion disturbance, levels of different adhesion molecules were screened 24 h after treatment. We found that the expression of CD29 was significantly reduced. This integrin has been linked to gemcitabine resistance and a poor outcome in pancreatic cancer [35]. Moreover, its knockdown has been reported to inhibit cell adhesion, migration, proliferation, and metastasis of pancreatic cancer, unveiling CD29 as a potential therapeutic target [36].

The loss of CD29 signaling and consequent detachment from culture plate trigger entosis [16]. Entosis is primarily the engulfment of one live cell into another live cell. In our study, the detached cells post 24 h treatment were observed to be viable. Also, the formation of adherent junctions has been shown to be crucial for entosis initiation [15]. This kind of cell-cell contacts are mediated though cadherins, which are calcium-dependent molecules that play central roles in cancer progression. We found increased N-cadherin levels, which could explain cell clumping after detachment and the ultimate invasion of one cell into another. Although E-cadherin usually forms adherens junctions in epithelial cells, the coexpression of E- and N-cadherin has been reported in adherens junctions of endoderm-derived epithelial tissues and tumors, such as pancreatic ducts [37]. In addition, Panc-1 cells express very low basal levels of E-cadherin and, according to Cano et al. [38], it cannot be discarded that pancreatic homotypic cell-in-cell formation might rely on N-cadherin-mediated cell contacts.

Although N-cadherin is usually linked to a more aggressive phenotype, it has been reported as a tumor suppressor in some types of cancers [39,40].

In addition, an upstream regulator of CD29 [22] and member of the Rho family of GTPases, CDC42, was also observed to be downregulated after treatment. CDC42 is overexpressed by 21% in pancreatic cancer [41] and the depletion of CDC42 enhances mitotic deadhesion and depends on Rho A activation in human bronchial epithelial cells [16]. Although it plays a crucial role in adherent entosis, it was reported to have no effect on suspension cells [16]. Consequently, treatment with the Se compounds affects CDC42 expression and mediates cell detachment through CD29 regulation.

Entotic cells mainly die through lysosomal-dependent pathways, although a swift to apoptosis can occur. In a floating population, different types of cell death have been reported to coexist [42].

Herein, we found that cell death induced by these Se compounds was caspase-independent, with a slight increase in PARP cleavage. We found increased levels of CatB in cell-in-cell structures undergoing entotic death, in concordance with previous reports [15]. By contrast, CatD, an interplayer between autophagy and apoptosis, was clearly downregulated. CatD can function as an anti-apoptotic mediator by increasing autophagy, revealing its two-faceted role [43]. In addition, CatD enhances anchorage-independent cell proliferation [44], and it is therefore quite interesting that it becomes down-regulated by compounds inducing cell detachment. Although cathepsins can mediate apoptosis, high levels of cathepsins have also been related to cancer progression. Pancreatic cancer patients, for instance, display a higher CatD concentration than healthy controls [45], and besides, elevated levels have been shown to promote cell dissemination in pancreatic cancer in vivo [46].

Cell detachment could be caused by CH_3SeH, one of the key metabolites in Se cytotoxicity, which has been reported to cause cell detachment in different cancer cell lines [47,48], along with a decrease in CD29 expression [48]. MSA is a penultimate precursor and compounds **1** and **2** bear this moiety. However, we ruled out that this effect was exclusive to the methylated form of Se, given that other compounds were able to induce the same phenotype. Lengthening the alkyl chain or the substitution over an arylselenol in general dramatically decreased the percentage of detached cells. However, the substitution of methyl for benzyl (compound **2c**) induced similar deadhesive effects. Intriguingly, despite induced cell detachment by this compound, it did not lead to cell death. The decreased cytotoxic effects are consistent with previous reports, showing the impaired cytotoxic activity of selenobenzyl derivatives with respect to their corresponding methylated analogs [49]. Hence, it is clear that detachment per se does not trigger death signaling, and it will be interesting to investigate the additional signaling pathways that the methyl and benzylseleno moieties are differentially able to activate, in order to avoid anchorange-independent cell growth.

In summary, we report a novel mechanism of action for MSA and two methylselenoesters: the induction of cell detachment through CDC42 and CD29 down-regulation leading to cell-in-cell formation (entosis) and death of the inner cell. However, these compounds need to be further evaluated in in vivo studies to gain an in-depth insight into their administration, hepatic metabolism for bioavailability and absorption, distribution, metabolism, and excretion properties. Additionally, the therapeutic potential of these compounds would be governed by the balance between their toxicity and efficacy profiles. Therefore, further research to fully dissect the relationship between structure, detachment abilities, and cell death induction of organic Se derivatives is required in order to understand the complex Se biochemistry.

4. Materials and Methods

4.1. Cell Culture

Panc-1 cells were obtained from the American Type Culture Collection (ATCC) and cultured in DMEM:F12 (Gibco™, ThermoFisher Scientific, Paisley, Scotland), 10% FBS (HyClone™, GE Healthcare Life Sciences, Logan, UT, USA), and 1% glutamine (Gibco) at 37 °C under 5% CO_2. The 3D spheroids were cultured following the protocol described by Longati et al. [23]. Briefly, phenol red-free

DMEM:F12 (Gibco™, ThermoFisher Scientific, Paisley, Scotland), 10% FBS (HyClone™, GE Healthcare Life Sciences, Logan, UT, USA), and 0.24% methylcellulose were used. On day 0, 2500 Panc-1 cells in 50 µL volume were seeded in a low adherent 96-well round bottom microplate (Falcon™, ThermoFisher Scientific, Stockholm, Sweden). On day 4, treatments were added, diluted in 50 µL of medium.

4.2. 2D Viability Assay

Cell viability after treatment was assessed by the MTT (3-(4,5-dimethylthiazol-2-yl)-2,5-diphenyltetrazolium bromide) (Sigma Aldrich®, Stockholm, Sweden) assay. 6000 Panc-1 cells were seeded in 96-well plates. Cells were treated with increasing concentrations of the compounds. Dilutions of the compounds in cell medium were freshly prepared from a 0.01 M stock in DMSO. After 72 h treatment, 50 µL of MTT solution in PBS (2 mg/mL) was added and cells were incubated at 37 °C under 5% CO_2 for 4 h. Medium was removed and 150 µL of DMSO was added to dissolve the formazan crystals. Absorbance was read at 590 nm in a VersaMax microplate reader (Molecular Devices, San Jose, CA, USA). Viability is expressed as the percentage of untreated cells.

4.3. 3D Viability Assay

3D viability after 72 h treatment was analyzed with the acid phosphatase assay, following a previously described protocol [23]. Briefly, 70 µL of medium was carefully removed and 60 µL of PBS along with 100 µL APH buffer (1.5 M sodium acetate pH = 5.2, 0.1% TritonX-100) containing a final concentration of freshly prepared 2 mg/mL p-nitrophenyl phosphate were added. Cells were incubated for 5 h at 37 °C under 5% CO_2 and then 10 µL of NaOH 1M was added to stop the reaction. Absorbance was read at 405 nm in a VersaMax microplate reader (Molecular Devices, San Jose, CA, USA). Viability is expressed as the percentage of untreated cells.

4.4. Fluorescent Staining

Spheroid formation was developed in a Gravity Trap™ ULA plate (InSphero Europe GmbH, Waldshut, Germany), following the manufacturer's protocols. Briefly, on day 0, the plate was pre-wetted with 40 µL of medium before seeding 2000 Panc-1 cells in 75 µL phenol red-free DMEM:F12, 10% FBS, and 1% glutamine. The plate was centrifuged for 2 min at 250× *g*. On day 4, cells were treated, adding 25 µL of the corresponding compound in medium. Dilutions were freshly prepared from a 0.1 M DMSO stock. On day 7, cells were stained with 1 µM Hoechst 33342 (Molecular Probes®, Life Technologies™, Eugene, OR, USA) for 2 h, and 2 µM PI (Molecular Probes®, Life Technologies™, Eugene, OR, USA) for 1 h at 37 °C under 5% CO_2. Spheroids were then carefully washed once with PBS and fixed with paraformaldehyde (4%) at RT. Imaging was performed on the Operetta® High-content Imaging System (PerkinElmer, San Jose, CA, USA) (confocal mode, 10× objective magnification, 0.3 objective NA, 35 µm focus height) and processed by the Colombus™ (PerkinElmer, San Jose, CA, USA) analysis software.

4.5. Western Blotting

Protein lysate containing 20 µg of proteins was separated on a Bolt 4–12% Bis-Tris Gel (Novex™, ThermoFisher Scientific, Goteborg, Sweden) and transferred to a nitrocellulose membrane using the iBlot Gel Transfer Device (ThermoFisher Scientific, Goteborg, Sweden). Incubation with primary antibody (Cathepsin B (D1C7Y), Cell Signaling, Leiden, The Netherlands, Catalog no. 31718; Cathepsin D, BD Biosciences, San Jose, CA, USA, Catalog no. 610800; CDC42, Abcam, Cambridge, UK, Catalog no. ab155940; RhoA (67B89), Cell Signaling, Leiden, The Netherlands, Catalog no. 2117; PARP, Cell Signaling, Leiden, The Netherlands, Catalog no. 9542; Caspase 9, Bioss, Nordic BioSite, Stockholm, Sweden, Catalog no. bs-0049R; beta actin, Sigma-Aldrich, Stockholm, Sweden, Catalog no. A5441) diluted in TBST containing 3.5% bovine serum albumin (BSA) was done overnight at 4 °C. Secondary antibodies, goat anti-rabbit IgG HRP (Southern Biotech, Stockholm, Sweden Catalog no. 4030-05), or goat anti mouse IgG HRP (Southern Biotech, Stockholm, Sweden Catalog no. 1030-05) were incubated for 1 h.

Membranes were developed using the AmershamTM ECLTM Start Western Blotting Detection Reagent (GE Healthcare Life Sciences, Logan, UT, USA) and bands were visualized using the Bio-Rad Quantity One imaging system (Bio-Rad, Stockholm, Sweden). Images were quantified using ImageJ software.

4.6. Adhesion Assay

0.5×10^6 cells were seeded in 25 cm^2 flasks and incubated at 37 °C under 5% CO$_2$ 24 h. Media was changed and cells were treated with 5 μM of compounds, after which floating cells were collected, centrifuged, and seeded at a density of 40,000 cells in 400 μL of fresh medium/well in a 24-well plate. The control cells were scrapped and seeded at the same density in 24-well plates. The cells were allowed to adhere to the surface of the plates. Cells were incubated at 37 °C, 5% CO$_2$ for 3 h, when 95% of the control cells adhered to the plate, after which they were fixed using 4% paraformaldehyde (PFA). The cells were further stained with 200 μL Coomassie blue staining solution (0.2% Coomassie Blue Brilliant R-250, 10% Acetic Acid and 40% Methanol) for 1 h at room temperature. The cells were then washed with PBS and further incubated for 1 h with 0.5 mL elution buffer (0.1 N NaOH and 50% Methanol). Furthermore, 0.5 mL of developing solution containing 10% Trichloroacetic acid (TCA) was added into the wells. Following this, 200 μL of the mix was transferred to a 96-well plate and further absorbance was read at 595 nm using the plate reader Infinite® M200 Pro, Tecan, Mannedorf, Switzerland.

4.7. Flow Cytometry

0.5×10^6 cells were seeded in 25 cm^2 flasks and allowed to attach for 24 h. After that, medium was replaced and cells were treated with the corresponding compounds or vehicle (DMSO) for 24 h. Cells were collected, washed with PBS-staining buffer (1% BSA, 0.01% NaN$_3$, 1% FBS), and stained for 30 min at 4 °C and darkness in 50 μL PBS-staining buffer with the corresponding antibody: CD29/integrin 1-β (FITC conjugate, clone MEM-101A, Life Technologies, Eugene, OR, USA), CD325/N-cadherin (PE conjugate, clone 8C11, Life Technologies, Eugene, OR, USA). Cells were washed once with PBS-staining buffer and resuspended in fixation buffer (PBS, 1% paraformaldehyde, 2% FBS) until being read in a BD FACSCalibur™ (BD Biosciences, San Jose, CA, USA).

4.8. Clonogenic Assay

0.5×10^6 cells were seeded in 25 cm^2 flasks and incubated at 37 °C under 5% CO$_2$ for 24 h. Media was changed and cells were treated with 5 μM of compounds for 24 h, after which floating cells were collected, centrifuged, and seeded at a density of 1000 cells in total volume of 2 mL/well in a six-well plate. The control cells were checked for colony formation for five days. A group of 50 cells were considered as one colony. The plates were later stained with crystal violet according to Franken et al. [50].

4.9. Chemical Synthesis

The NMR spectra (^{1}H and ^{13}C) were recorded on a Bruker 400 Ultrashield™ spectrometer (Rheinstetten, Germany) and are provided in the supplementary material. The samples were solved in CDCl$_3$ and TMS was used as an internal standard. IR spectra were obtained on a Thermo Nicolet FT-IR Nexus spectrophotometer (Thermo Nicolet, Madison, WI, USA) using KBr pellets for solids or NaCl plates for oil compounds. The HRMS spectra were recorded on a Thermo Scientific Q Exactive Focus mass spectrometer (Thermo Scientific™, Waltham, MA, USA) by direct infusion. For TLC assays, Alugram SIL G7UV254 sheets (Macherey-Nagel; Düren, Germany) were used. Column chromatography was performed with silica gel 60 (E. Merck KGaA, Darmstadt, Germany). Chemicals were purchased from E. Merck KGaA (Darmstadt, Germany), Panreac Química S.A. (Montcada i Reixac, Barcelona, Spain), Sigma-Aldrich Quimica, S.A. (Alcobendas, Madrid, Spain), and Acros Organics (Janssen Pharmaceuticalaan, Geel, Belgium).

4.9.1. Procedure for Compounds **1** and **2**

Compounds **1** and **2** were synthesized as described in our previous work [14], under the references **5** and **15**, respectively.

Procedure for Compounds **1a**, **2a**, **1b** and **2c**

The chemical synthesis was carried out following an already described procedure [51,52] with some modifications. Briefly, the corresponding carboxylic acid was chlorinated by a reaction with $SOCl_2$. Se powder reacted with $NaBH_4$ (1:2) in water or ethanol (1:1) and N_2 atmosphere to form NaHSe. The corresponding acyl chloride dissolved in *N,N*-dimethylformamide (2 mL) or chloroform (2 mL) was then added and the reaction was stirred at room temperature until the reaction took place (20 min–3.5 h). The reaction was followed by IR or TLC. The mixture was filtered and the intermediate was further alkylated with the corresponding halide until discoloration of the mixture. The product was extracted with methylene chloride and dried over Na_2SO_4. The solvent was eliminated under rotatory evaporation and the residue was purified through column chromatography.

Ethyl 3-Chlorothiophen-2-Carboselenoate (**1a**)

From 3-chlorothiophen-2-carboxylic acid (1.5 mmol), Se powder (1.5 mmol), $NaBH_4$ (3 mmol), and ethyl iodide (1.5 mmol). A yellow oil was obtained, which was further purified through column chromatography using methylene chloride as the eluent. Yield: 11%. ^{1}H NMR (400 MHz, $CDCl_3$): δ 1.52 (t, 3H, –CH_3, $J_{CH3-CH2}$ = 7.5 Hz), 3.11 (q, 2H, –CH_2–), 7.06 (d, 1H, H_4 J_{4-5} = 5.3 Hz), 7.54 ppm (d, 1H, H_5). ^{13}C NMR (100 MHz, $CDCl_3$): δ 15.7 (–CH_3), 20.5 (–CH_2–), 128.0 (C_4), 130.6 (C_2), 131.0 (C_5), 137.7 (C_3), 184.3 ppm (–C=O). IR (KBr): ν 3105 (w, C–H_{arom}), 2962–2867 (s, C–H_{aliph}), 1649 cm^{-1} (s, –C=O). HRMS calculated for $C_7H_8ClOSSe$ (M + H): 254.91441, found: 254.91418.

Diethyl 2,5-Furandicarboselenoate (**2a**)

From 2,5-furandicarboxylic acid (1.74 mmol), Se powder (3.48 mmol), $NaBH_4$ (7.1 mmol), and ethyl iodide (3.48 mmol). Conditions: 45 min reaction with NaHSe and 2 h reaction with ethyl iodide. A yellow solid was obtained, which was purified through column chromatography using ethyl acetate/hexane (1:10) as the eluent. Yield: 10%; m.p.: 35–36 °C. ^{1}H NMR (400 MHz, $CDCl_3$): δ 1.5 (t, 6H, 2–CH_3, $J_{CH2-CH3}$ = 7.5 Hz), 3.11 (q, 4H, –CH_2–), 7.17 ppm (s, 2H, H_3 + H_4). ^{13}C NMR (100 MHz, $CDCl_3$): δ 15.8 (–CH_3), 19.2 (–CH_2–), 114.7 (C_3 + C_4), 153.5 (C_2 + C_5), 183.5 ppm (–C=O). IR (KBr): ν 3143 (w, C–H_{arom}), 2961–2860 (s, C–H_{aliph}), 1649 cm^{-1} (s, –C=O). HRMS calculated for $C_{10}H_{13}O_3Se_2$ (M + H): 340.91896; found: 340.91891.

Pentyl 3-Chlorothiophen-2-Carboselenoate (**1b**)

From 3-chlorothiophen-2-carboxylic acid (2 mmol), Se powder (2 mmol), $NABH_4$ (2.15 mmol), and pentyl iodide (2.15 mmol). Under N_2 atmosphere, absolute ethanol (10 mL) was added to a mixture of $NaBH_4$ and selenium cooled by an ice bath, with magnetic stirring. After the formation of NaHSe was achieved, the ice bath was removed and the following reactions were carried out at room temperature. Before adding an excess of pentyl iodide, the mixture was filtered. Conditions: 20 min reaction with NaHSe and 20 min reaction with pentyl iodide. The solvent was eliminated under rotatory evaporation. The product was purified through column chromatography using a gradient elution of ethyl acetate: hexane. An orange oil was obtained. Yield: 53%. ^{1}H NMR (400 MHz, $CDCl_3$): δ 0.83 (t, 3H, –CH_3, $J_{CH3-CH2}$ = 7.1 Hz), 1.27–1.35 (m, 4H, γ + δCH_2), 1.66–1.77 (m, 2H, ßCH_2), 3.02 (t, 2H, αCH_2, $J_{CH2-CH2}$ = 7.4 Hz,), 6.96 (d, 1H, H_4, J_{4-5} = J_{5-4} = 5.3 Hz), 7.44 ppm (d, 1H, H_5). ^{13}C NMR (100 MHz, $CDCl_3$): δ 12.94 (–CH_3), 21.19 (δCH_2), 25.56 (αCH_2), 28.89 (ßCH_2), 31.18 (γCH_2), 126.82 (C_4), 129.50 (C_2), 129.76 (C_5), 136.56 (C_3), 183.18 ppm (–CO). IR (KBr): ν 2922 –2852 (s, C–H_{alif}), 1669 cm^{-1} (s, –C=O).

Dibenzyl 2,5-Furandicarboselenoate (2c)

From 2,5-furandicarboxylic acid (1.74 mmol), Se powder (3.48 mmol), NABH$_4$ (7.1 mmol), and benzyl bromide (3.48 mmol). Conditions: 1.5 h reaction with NaHSe and 3.5 h reaction with benzyl bromide. The product was extracted with methylene chloride, further washed with water, dried over Na$_2$SO$_4$. The solvent was eliminated under rotatory evaporation. A yellow oil was obtained, which was precipitated and washed with diethyl ether. A yellow solid was obtained. Yield: 25%; m.p.: 114–115 °C. ^{1}H NMR (400 MHz, CDCl$_3$): δ 4.34 (s, 4H, 2–CH$_2$–), 7.2 (s, 2H, H$_3$ + H$_4$), 7.22–7.35 ppm (m, 10 H, H$_{arom}$). ^{13}C NMR (100 MHz, CDCl$_3$): δ 28.5 (–CH$_2$–), 115.0 (C$_3$ + C$_4$), 127.3 (C$_{4'}$), 128.7 (C$_{2'}$ + C$_{6'}$), 129.1(C$_{3'}$ + C$_{5'}$), 138.21 (C$_{1'}$), 153.2 (C$_2$ + C$_5$), 182.8 (–C=O). IR (KBr): ν 3123–3088 (s, C–H$_{arom}$), 1677 cm^{-1} (s, –C=O). HRMS C$_{20}$H$_{16}$O$_3$Se$_2$Na (M + Na$^+$): calculated 486.9322; found 486.9430.

4.9.2. Procedure for Compound 3

Compound **3** was synthesized in a previous work [32], under the reference **3c**.

4.10. Timelapse

0.5×10^6 cells were seeded in 25 cm^2 flasks and incubated at 37 °C under 5% CO$_2$ 24 h. Media was changed and cells were treated with 5 μM of compounds for 24 h, after which floating cells were collected, centrifuged, and seeded at a density of 50,000 cells in total volume of 100 μL/well in a 96-well plate. Post 48 h of treatment, the cells were imaged live for another 24 h in Operetta and images were captured every 5 min.

4.11. Confocal

Monolayer cultures were stained with CellTracker Red or Green (Invitrogen) for 1 h at 37 °C in the absence of serum. After this, 0.4×10^5 cells stained with each of the cell trackers were mixed (1:1) and seeded onto 25 cm^2 flasks for 24 h, followed by treatment with 5 μM of respective compounds for 18 h. Cells were imaged live using an LSM 800 confocal microscope (Zeiss, Oberkochen, Germany).

4.12. Statistical Analysis

One-way ANOVA followed by Dunnet's test was performed using GraphPad 6.01 (GraphPad Software, San Diego, CA, USA). (* $p < 0.05$, ** $p < 0.01$, *** $p < 0.001$).

Supplementary Materials: Supplementary materials can be found at http://www.mdpi.com/1422-0067/19/10/2849/s1.

Author Contributions: Conceptualization, A.P.F., C.S., P.K., and N.D.-A; Methodology, A.P.F., P.K., N.D.-A., and J.A.P.; Validation, A.P.F., P.K., and N.D.-A.; Formal Analysis, P.K and N.D.-A.; Investigation, P.K. and N.D.-A.; Resources, A.P.F.; C.S., and J.A.P; Writing-Original Draft Preparation, N.D.-A. and P.K.; Writing-Review & Editing, A.P.F.; P.K., N.D.-A., and C.S.; Visualization, P.K. and N.D-A; Supervision, A.P.F.; Project Administration, A.P.F.; Funding Acquisition, A.P.F.

Funding: This article has been financially supported by The Swedish Cancer Society (Cancerfonden). The research leading to these results has also received funding from "la Caixa" Banking Foundation through a grant to N.D.-A, who additionally received a mobility scholarship from Asociación de Amigos de la Universidad de Navarra.

Acknowledgments: We thank Pablo Garnica for his help with compound **1b**.

Conflicts of Interest: The authors declare no conflict of interest.

Abbreviations

BznSeH	Benzeneselenol
CatB	Cathepsin B
CatD	Cathepsin D
CD29	β1-integrin
CDC42	Cell division control protein 42 homolog
PARP	DNA damage-responsive enzymes poly(ADP-ribose) polymerase

MSA	Methylseleninic acid
MeSeCys	Methylselenocysteine
CH$_3$SeH	Methylselenol
Nrf2	nuclear factor erythroid 2 related factor 2
PDAC	Pancreatic ductal adenocarcinoma
PI	Propidium iodide
ROS	Reactive oxygen species
Se	Selenium

References

1. Siegel, R.; Ma, J.; Zou, Z.; Jemal, A. Cancer statistics, 2014. *CA Cancer J. Clin.* **2014**, *64*, 9–29. [CrossRef] [PubMed]

2. Rahib, L.; Smith, B.D.; Aizenberg, R.; Rosenzweig, A.B.; Fleshman, J.M.; Matrisian, L.M. Projecting cancer incidence and deaths to 2030: The unexpected burden of thyroid, liver, and pancreas cancers in the united states. *Cancer Res.* **2014**, *74*, 2913–2921. [CrossRef] [PubMed]

3. Bijlsma, M.F.; van Laarhoven, H.W.M. The conflicting roles of tumor stroma in pancreatic cancer and their contribution to the failure of clinical trials: A systematic review and critical appraisal. *Cancer Metastasis Rev.* **2015**, *34*, 97–114. [CrossRef] [PubMed]

4. Al Saati, T.; Clerc, P.; Hanoun, N.; Peuget, S.; Lulka, H.; Gigoux, V.; Capilla, F.; Béluchon, B.; Couvelard, A.; Selves, J.; et al. Oxidative Stress Induced by Inactivation of TP53INP1 Cooperates with KrasG12D to Initiate and Promote Pancreatic Carcinogenesis in the Murine Pancreas. *Am. J. Pathol.* **2013**, *182*, 1996–2004. [CrossRef] [PubMed]

5. Kong, B.; Qia, C.; Erkan, M.; Kleeff, J.; Michalski, C.W. Overview on how oncogenic Kras promotes pancreatic carcinogenesis by inducing low intracellular ROS levels. *Front. Physiol.* **2013**. [CrossRef] [PubMed]

6. Diehn, M.; Cho, R.W.; Lobo, N.A.; Kalisky, T.; Dorie, M.J.; Kulp, A.N.; Qian, D.; Lam, J.S.; Ailles, L.E.; Wong, M.; et al. Association of reactive oxygen species levels and radioresistance in cancer stem cells. *Nature* **2009**, *458*, 780–783. [CrossRef] [PubMed]

7. Gorrini, C.; Harris, I.S.; Mak, T.W. Modulation of oxidative stress as an anticancer strategy. *Nat. Rev. Drug Discov.* **2013**, *12*, 931–947. [CrossRef] [PubMed]

8. Cairns, R.A.; Harris, I.; Mccracken, S.; Mak, T.W. Cancer cell metabolism. *Cold Spring Harb. Symp. Quant. Biol.* **2011**. [CrossRef] [PubMed]

9. Cairns, R.; Harris, I.; Mak, T. Regulation of cancer cell metabolism. *Nat. Rev. Cancer* **2011**, *11*, 85–95. [CrossRef] [PubMed]

10. Sosa, V.; Moline, T.; Somoza, R.; Paciucci, R.; Kondoh, H.; Me, L.L. Oxidative stress and cancer: An overview. *Ageing Res. Rev.* **2013**, *12*, 376–390. [CrossRef] [PubMed]

11. Chaiswing, L.; St Clair, W.H.; St Clair, D.K. Redox Paradox: A novel approach to therapeutics-resistant cancer. *Antioxid. Redox Signal.* **2018**. [CrossRef] [PubMed]

12. Fernandes, A.P.; Gandin, V. Selenium compounds as therapeutic agents in cancer. *Biochim. Biophys. Acta* **2015**, *1850*, 1642–1660. [CrossRef] [PubMed]

13. Gandin, V.; Khalkar, P.; Braude, J.; Fernandes, A.P. Organic selenium compounds as potential chemotherapeutic agents for improved cancer treatment. *Free Radic. Biol. Med.* **2018**. [CrossRef] [PubMed]

14. Díaz-Argelich, N.; Encío, I.; Plano, D.; Fernandes, A.P.; Palop, J.A.; Sanmartín, C. Novel Methylselenoesters as Antiproliferative Agents. *Molecules* **2017**, *22*, 1288. [CrossRef] [PubMed]

15. Overholtzer, M.; Mailleux, A.A.; Mouneimne, G.; Normand, G.; Schnitt, S.J.; King, R.W.; Cibas, E.S.; Brugge, J.S. A Nonapoptotic Cell Death Process, Entosis, that Occurs by Cell-in-Cell Invasion. *Cell* **2007**, *131*, 966–979. [CrossRef] [PubMed]

16. Durgan, J.; Tseng, Y.Y.; Hamann, J.C.; Domart, M.C.; Collinson, L.; Hall, A.; Overholtzer, M.; Florey, O. Mitosis can drive cell cannibalism through entosis. *eLife* **2017**, *6*, 1–26. [CrossRef] [PubMed]

17. Garanina, A.S.; Kisurina-Evgenieva, O.P.; Erokhina, M.V.; Smirnova, E.A.; Factor, V.M.; Onishchenko, G.E. Consecutive entosis stages in human substrate-dependent cultured cells. *Sci. Rep.* **2017**, *7*, 1–12. [CrossRef] [PubMed]

18. Wan, Q.; Liu, J.; Zheng, Z.; Zhu, H.; Chu, X.; Dong, Z.; Huang, S.; Du, Q. Regulation of myosin activation during cell-cell contact formation by Par3-Lgl antagonism: Entosis without matrix detachment. *Mol. Biol. Cell* **2012**, *23*, 2076–2091. [CrossRef] [PubMed]

19. Hamann, J.C.; Surcel, A.; Chen, R.; Teragawa, C.; Albeck, J.G.; Robinson, D.N.; Overholtzer, M. Entosis Is Induced by Glucose Starvation. *Cell Rep.* **2017**, *20*, 201–210. [CrossRef] [PubMed]

20. Florey, O.; Kim, S.E.; Sandoval, C.P.; Haynes, C.M.; Overholtzer, M. Autophagy machinery mediates macroendocytic processing and entotic cell death by targeting single membranes. *Nat. Cell Biol.* **2011**, *13*, 1335–1343. [CrossRef] [PubMed]

21. Krishna, S.; Overholtzer, M. Mechanisms and consequences of entosis. *Cell. Mol. Life Sci.* **2016**, *73*, 2379–2386. [CrossRef] [PubMed]

22. Reymond, N.; Im, J.H.; Garg, R.; Vega, F.M.; Borda d'Agua, B.; Riou, P.; Cox, S.; Valderrama, F.; Muschel, R.J.; Ridley, A.J. Cdc42 promotes transendothelial migration of cancer cells through β1 integrin. *J. Cell Biol.* **2012**, *199*, 653–668. [CrossRef] [PubMed]

23. Longati, P.; Jia, X.; Eimer, J.; Wagman, A.; Witt, M.-R.; Rehnmark, S.; Verbeke, C.; Toftgård, R.; Löhr, M.; Heuchel, R.L. 3D pancreatic carcinoma spheroids induce a matrix-rich, chemoresistant phenotype offering a better model for drug testing. *BMC Cancer* **2013**, *13*, 95. [CrossRef] [PubMed]

24. Seguin, L.; Desgrosellier, J.S.; Weis, S.M.; Cheresh, D.A. Integrins and cancer: Regulators of cancer stemness, metastasis, and drug resistance. *Trends Cell Biol.* **2015**, *25*, 234–240. [CrossRef] [PubMed]

25. Van Roy, F. Beyond E-cadherin: Roles of other cadherin superfamily members in cancer. *Nat. Rev. Cancer* **2014**, *14*, 121–134. [CrossRef] [PubMed]

26. McGinnis, K.M.; Gnegy, M.E.; Park, Y.H.; Mukerjee, N.; Wang, K.K.W. Procaspase-3 and Poly(ADP)ribose Polymerase (PARP) Are Calpain Substrates. *Biochem. Biophys. Res. Commun.* **1999**, *263*, 94–99. [CrossRef] [PubMed]

27. Gobeil, S.; Boucher, C.C.; Nadeau, D.; Poirier, G.G. Characterization of the necrotic cleavage of poly(ADP-ribose) polymerase (PARP-1): Implication of lysosomal proteases. *Cell Death Differ.* **2001**, *8*, 588–594. [CrossRef] [PubMed]

28. Shalini, S.; Dorstyn, L.; Dawar, S.; Kumar, S. Old, new and emerging functions of caspases. *Cell Death Differ.* **2015**, *22*, 526–539. [CrossRef] [PubMed]

29. Chaitanya, G.; Alexander, J.S.; Babu, P. PARP-1 cleavage fragments: Signatures of cell-death proteases in neurodegeneration. *Cell Commun. Signal.* **2010**, *8*, 31. [CrossRef] [PubMed]

30. Lugini, L.; Matarrese, P.; Tinari, A.; Lozupone, F.; Federici, C.; Iessi, E.; Gentile, M.; Luciani, F.; Parmiani, G.; Rivoltini, L.; et al. C annibalism of live lymphocytes by human metastatic but not primary melanoma cells. *Cancer Res.* **2006**, *66*, 3629–3638. [CrossRef] [PubMed]

31. Khalkhali-Ellis, Z.; Goossens, W.; Margaryan, N.V.; Hendrix, M.J.C. Cleavage of histone 3 by cathepsin D in the involuting mammary gland. *PLoS ONE* **2014**, *9*, e103230. [CrossRef] [PubMed]

32. Moreno, E.; Plano, D.; Lamberto, I.; Font, M.; Encío, I.; Palop, J.A.; Sanmartín, C. Sulfur and selenium derivatives of quinazoline and pyrido[2,3-d]pyrimidine: Synthesis and study of their potential cytotoxic activity in vitro. *Eur. J. Med. Chem.* **2012**, *47*, 283–298. [CrossRef] [PubMed]

33. Ip, C.; Thompson, H.J.; Zhu, Z.; Ganther, H.E. In vitro and in vivo studies of methylseleninic acid: Evidence that a monomethylated selenium metabolite is critical for cancer chemoprevention. *Cancer Res.* **2000**, *60*, 2882–2886. [CrossRef] [PubMed]

34. Khalkar, P.; Ali, H.A.; Codó, P.; Argelich, N.D.; Martikainen, A.; Arzenani, M.K.; Lehmann, S.; Walfridsson, J.; Ungerstedt, J.; Fernandes, A.P. Selenite and methylseleninic acid epigenetically affects distinct gene sets in myeloid leukemia: A genome wide epigenetic analysis. *Free Radic. Biol. Med.* **2018**, *117*, 247–257. [CrossRef] [PubMed]

35. Yang, D.; Shi, J.; Fu, H.; Wei, Z.; Xu, J.; Hu, Z.; Zhang, Y.; Yan, R.; Cai, Q. Integrin B1 modulates tumour resistance to gemcitabine and serves as an independent prognostic factor in pancreatic adenocarcinomas. *Tumor Biol.* **2016**, *37*, 12315–12327. [CrossRef] [PubMed]

36. Grzesiak, J.J.; Cao, H.S.T.; Burton, D.W.; Kaushal, S.; Vargas, F.; Clopton, P.; Snyder, C.S.; Deftos, L.J.; Hoffman, R.M.; Bouvet, M. Knockdown of the β1 integrin subunit reduces primary tumor growth and inhibits pancreatic cancer metastasis. *Int. J. Cancer* **2011**, *129*, 2905–2915. [CrossRef] [PubMed]

37. Straub, B.K.; Rickelt, S.; Zimbelmann, R.; Grund, C.; Kuhn, C.; Iken, M.; Ott, M.; Schirmacher, P.; Franke, W.W. E-N-cadherin heterodimers define novel adherens junctions connecting endoderm-derived cells. *J. Cell Biol.* **2011**, *195*, 873–887. [CrossRef] [PubMed]

38. Cano, C.E.; Sandí, M.J.; Hamidi, T.; Calvo, E.L.; Turrini, O.; Bartholin, L.; Loncle, C.; Secq, V.; Garcia, S.; Lomberk, G.; et al. Homotypic cell cannibalism, a cell-death process regulated by the nuclear protein 1, opposes to metastasis in pancreatic cancer. *EMBO Mol. Med.* **2012**, *4*, 964–979. [CrossRef] [PubMed]

39. Su, Y.; Li, J.; Shi, C.; Hruban, R.H.; Radice, G.L. N-cadherin functions as a growth suppressor in a model of K-ras-induced PanIN. *Oncogene* **2016**, *35*, 3335–3341. [CrossRef] [PubMed]

40. Camand, E.; Peglion, F.; Osmani, N.; Sanson, M.; Etienne-Manneville, S. N-cadherin expression level modulates integrin-mediated polarity and strongly impacts on the speed and directionality of glial cell migration. *J. Cell Sci.* **2012**, *125*, 844–857. [CrossRef] [PubMed]

41. Maldonado, M.D.M.; Dharmawardhane, S. Targeting Rac and Cdc42 GTPases in Cancer. *Cancer Res.* **2018**, *78*, 3101–3111. [CrossRef] [PubMed]

42. Ishikawa, F.; Ushida, K.; Mori, K.; Shibanuma, M. Loss of anchorage primarily induces non-apoptotic cell death in a human mammary epithelial cell line under atypical focal adhesion kinase signaling. *Cell Death Dis.* **2015**, *6*, e1619. [CrossRef] [PubMed]

43. Hah, Y.S.; Noh, H.S.; Ha, J.H.; Ahn, J.S.; Hahm, J.R.; Cho, H.Y.; Kim, D.R. Cathepsin D inhibits oxidative stress-induced cell death via activation of autophagy in cancer cells. *Cancer Lett.* **2012**, *323*, 208–214. [CrossRef] [PubMed]

44. Glondu, M.; Liaudet-Coopman, E.; Derocq, D.; Platet, N.; Rochefort, H.; Garcia, M. Down-regulation of cathepsin-D expression by antisense gene transfer inhibits tumor growth and experimental lung metastasis of human breast cancer cells. *Oncogene* **2002**, *21*, 5127–5134. [CrossRef] [PubMed]

45. Park, H.-D.; Kang, E.-S.; Kim, J.-W.; Lee, K.-T.; Lee, K.H.; Park, Y.S.; Park, J.-O.; Lee, J.; Heo, J.S.; Choi, S.H.; et al. Serum CA19-9, cathepsin D, and matrix metalloproteinase-7 as a diagnostic panel for pancreatic ductal adenocarcinoma. *Proteomics* **2012**, *12*, 3590–3597. [CrossRef] [PubMed]

46. Dumartin, L.; Whiteman, H.J.; Weeks, M.E.; Hariharan, D.; Dmitrovic, B.; Iacobuzio-Donahue, C.A.; Brentnall, T.A.; Bronner, M.P.; Feakins, R.M.; Timms, J.F.; et al. AGR2 is a novel surface antigen that promotes the dissemination of pancreatic cancer cells through regulation of cathepsins B and D. *Cancer Res.* **2011**, *71*, 7091–7102. [CrossRef] [PubMed]

47. Jiang, C.; Wang, Z.; Ganther, H.; Lu, J. Caspases as key executors of methyl selenium-induced apoptosis (anoikis) of DU-145 prostate cancer cells. *Cancer Res.* **2001**, *61*, 3062–3070. [PubMed]

48. Kim, A.; Oh, J.H.; Park, J.M.; Chung, A.S. Methylselenol generated from selenomethionine by methioninase downregulates integrin expression and induces caspase-mediated apoptosis of B16F10 melanoma cells. *J. Cell. Physiol.* **2007**, *212*, 386–400. [CrossRef] [PubMed]

49. Ibáñez, E.; Plano, D.; Font, M.; Calvo, A.; Prior, C.; Palop, J.A.; Sanmartín, C. Synthesis and antiproliferative activity of novel symmetrical alkylthio- and alkylseleno-imidocarbamates. *Eur. J. Med. Chem.* **2011**, *46*, 265–274. [CrossRef] [PubMed]

50. Franken, N.A.P.; Rodermond, H.M.; Stap, J.; Haveman, J.; van Bree, C. Clonogenic assay of cells in vitro. *Nat. Protoc.* **2006**, *1*, 2315–2319. [CrossRef] [PubMed]

51. Domínguez-Álvarez, E.; Plano, D.; Font, M.; Calvo, A.; Prior, C.; Jacob, C.; Palop, J.A.; Sanmartín, C. Synthesis and antiproliferative activity of novel selenoester derivatives. *Eur. J. Med. Chem.* **2014**, *73*, 153–166. [CrossRef] [PubMed]

52. Klayman, D.L.; Griffin, T.S. Reaction of Selenium with Sodium Borohydride in Protic Solvents. A Facile Method for the Introduction of Selenium into Organic Molecules. *J. Am. Chem. Soc.* **1973**, *2*, 197–199. [CrossRef]

International Journal of
Molecular Sciences

MDPI

Review

Role of Hypoxic Stress in Regulating Tumor Immunogenicity, Resistance and Plasticity

Stéphane Terry [1,*], Rania Faouzi Zaarour [2], Goutham Hassan Venkatesh [2], Amirtharaj Francis [2], Walid El-Sayed [2], Stéphanie Buart [1], Pamela Bravo [1], Jérome Thiery [1,*] and Salem Chouaib [1,2,*]

[1] INSERM UMR 1186, Integrative Tumor Immunology and Genetic Oncology, Gustave Roussy, EPHE, Fac. de médecine—Univ. Paris-Sud, University Paris-Saclay, Villejuif F-94805, France; stephanie.buart@gustaveroussy.fr (S.B.); pamelisa.b@gmail.com (P.B.)
[2] Thumbay Research Institute of Precision Medicine, Gulf Medical University, Ajman 4184, United Arab Emirates; dr.rania@gmu.ac.ae (R.F.Z.); dr.goutham@gmu.ac.ae (G.H.V.); francis@gmu.ac.ae (A.F.); walidshaaban@gmu.ac.ae (W.E.-S.)
* Correspondence: stephane.terry@gustaveroussy.fr (S.T.); jerome.thiery@gustaveroussy.fr (J.T.); salem.chouaib@gustaveroussy.fr or salem.chouaib@gmu.ac.ae (S.C.); Tel.: +971-509-804-120 (S.C.)

Received: 7 September 2018; Accepted: 4 October 2018; Published: 6 October 2018

Abstract: Hypoxia, or gradients of hypoxia, occurs in most growing solid tumors and may result in pleotropic effects contributing significantly to tumor aggressiveness and therapy resistance. Indeed, the generated hypoxic stress has a strong impact on tumor cell biology. For example, it may contribute to increasing tumor heterogeneity, help cells gain new functional properties and/or select certain cell subpopulations, facilitating the emergence of therapeutic resistant cancer clones, including cancer stem cells coincident with tumor relapse and progression. It controls tumor immunogenicity, immune plasticity, and promotes the differentiation and expansion of immune-suppressive stromal cells. In this context, manipulation of the hypoxic microenvironment may be considered for preventing or reverting the malignant transformation. Here, we review the current knowledge on how hypoxic stress in tumor microenvironments impacts on tumor heterogeneity, plasticity and resistance, with a special interest in the impact on immune resistance and tumor immunogenicity.

Keywords: hypoxia; tumor microenvironment; tumor heterogeneity; cancer; cancer stem cells; EMT; cell plasticity; DNA damage and repair; immune evasion; HIF

1. Introduction

The tumor microenvironment (TME) is a complex system that consists of the extracellular matrix (ECM) and numerous cell types including fibroblasts, adipose cells, immune cells, endothelial cells as well as components of the blood and lymphatic vascular networks and the nervous system. TME plays an important role in tumor development and progression [1–3]. It involves soluble factors and metabolic changes. Among these metabolic changes, hypoxia plays a pivotal role in shaping the TME [4,5]. In such a system, hypoxia appears as an essential metabolic element to control cellular plasticity and tumor heterogeneity [6,7]. It is well established that hypoxic stress is a feature of most solid tumors and is associated with poor clinical outcomes in various cancer types [2,8–11]. Hypoxia arises due to a combination of excessive oxygen consumption by growing tumor cells and the disorganized tumor-associated vasculature [3]. Considerable evidence now suggests that hypoxia plays an important role in tumor progression, affecting both metastatic spread and selection of cells with more aggressive phenotypes [7,12,13]). This is at least partly explained by the fact that hypoxia can promote cancer cell stemness, invasion or metastatic capacities via the activation of hypoxic cascades and hypoxia-inducible factors (HIFs). To date, the mechanisms at play are still far from being understood. The adaptive responses to hypoxia are regulated by HIFs. The master regulator of the hypoxic response is the hypoxia-inducible factor 1 (HIF-1). In mammalian cells, the response to hypoxia depends in large part on the activation of HIF-1, a heterodimeric transcription factor consisting of a hypoxia-inducible HIF-1α subunit and a constitutively expressed HIF-1β subunit [14]. HIF-1 transactivates target genes containing *cis* acting hypoxia response elements that contain the HIF-1-binding site sequence. HIF-1α protein levels are tightly regulated by the cellular pO2. Under hypoxic stress, hypoxia-dependent stabilization of HIF dimers allows for the induction of numerous genes regulating various biological processes and functions in cells, including angiogenesis, cell survival, proliferation, pH regulation, and metabolism [4].

2. Hypoxia Induced Tumor Plasticity and Heterogeneity

Tumors contain distinct cell types that collectively create microenvironmental conditions controlling the tumor growth and its evolution. Insufficient concentration of oxygen in the growing tumor generates hypoxic stress, which can lead to metabolic, epigenetics and phenotypic reprogramming of the cells coincident with fluctuations in the composition of the microenvironment [15,16], while potentially affecting the functions, the phenotype and/or the number of microenvironmental cell components [5,6]. As a corollary, hypoxia should be considered as a driver of cell plasticity, since it can promote the capacity of a cell to shift from its original cellular state to a distinct cellular state. One interesting unanswered question is the impact of hypoxic stress on tumor heterogeneity. It is well established that tumors exhibit substantial heterogeneity with potential consequences on their evolution in time and response to treatments [17–20]. So far, the extent of this heterogeneity has been only partially explored, especially in relation to the diverse mutational landscapes found in tumors [17]. Clearly, more work is now needed to explore and define the phenotypic heterogeneity of the various cell types. The advent of single-cell approaches offers a unique opportunity to gain insights into tumor heterogeneity [21–24]. Recently, using breast tumors, Azizi and colleagues nicely showed that environmental factors, including hypoxia present in the tumor, but marginal in the normal tissue, were linked to the increased diversity of immune phenotypic states of T cells, myeloid cells and Natural killer (NK) cells [23]. Tumor-resident T cells appeared to be particularly responsive to such regulation, as shown by the increased number of gene signatures activated in highly hypoxic tumors. The findings also suggest that various degrees of hypoxia, inflammation, and nutrient supply, or a combination of these factors in the local microenvironment could lead to a spectrum of phenotypic states while promoting the enrichment of certain subpopulations such as the Treg subset. The work of Palazon et al. recently revealed the essential role of HIF-1α in regulating the effector state of CD8+ T cells [25]. Hypoxia stimulated the production of the cytolytic molecule granzyme B in a HIF-1α- but not HIF-2α-dependent fashion. Importantly, hypoxia through HIF-1α also increased the expression of activation-related costimulatory molecules CD137, OX40, and GITR, and checkpoint receptors PD-1, TIM3, and LAG3. This may have important implications for tumor immunology. Further experimental data from these investigators already denote the importance of the HIF1/VEGF-A axis to promote vascularization and T cell infiltration.

Aside from its impact on stromal components, the cell plasticity of cancer cells represents a major source of phenotypic heterogeneity in the tumor. Here again, HIFs, angiogenesis and inflammatory factors such as VEGF, or TGF-β (induced and activated under hypoxic conditions), might exert important regulatory functions. A prime example of this notion comes from the numerous studies demonstrating that all these factors can stimulate epithelial-mesenchymal transition (EMT) and/or support a mesenchymal state [13,26,27]. It is also well established that certain cancer cells have the capacity to transit between epithelial and mesenchymal phenotypes, or states, via epithelial-mesenchymal transition (EMT), or the reverse process, mesenchymal-epithelial transition (MET) [26]. In such a scenario, cancer cell plasticity is tightly regulated by signals perceived from the TME and anatomic sites. Notably, hypoxic stress might enable other types of phenotypic changes. For instance, HIF-1α and hypoxia could contribute to the neuroendocrine transformation of prostate tumors and adenocarcinoma cells through cooperation with the transcription FoxA2, reduced Notch-mediated signaling, and induction of neuronal and neuroendocrine gene programs in the cells [28–30]. Despite substantial evidence for a role of hypoxia in triggering EMT programs, the exact mechanisms at play remain relatively unclear. Both promoting and suppressing roles of hypoxia have been described in human and in mouse laboratory models [31–35]. In fact, our knowledge of what really occurs in patient tumors is still fragmentary. In this regard, the study of Puram et al. is particularly valuable [36]. These investigators profiled transcriptomes of ~6000 single cells from 18 head and neck squamous cell carcinomas. This included the analysis of 2216 malignant cells allowing the study of intra-tumoral phenotypic diversity of the cells. They found that malignant cells varied within and between tumors in their expression of signatures related to cell cycle, stress, hypoxia, epithelial differentiation, and partial EMT. One notable aspect of the findings was the strong correlation found between hypoxia and EMT signatures in the individual tumors. Similarly, we recently explored the relationship between hypoxia status and EMT-TF expression levels by analyzing lung adenocarcinomas included in the TCGA-LUAD project [37]. In this large cohort, hypoxia signatures, as well as HIF1A mRNA expression, were significantly and positively correlated with EMT-TF expression levels. In an attempt to better model the impact of hypoxia in non-small cell lung cancer (NSCLC), we exploited the primary NSCLC IGR-Heu cells and observed that EMT-related phenotypic changes were particularly exacerbated when hypoxic stress was maintained for a prolonged period. Moreover, under these experimental conditions, the shift towards a mesenchymal phenotype was only observed in a fraction of stressed cells. While some cells undergo EMT, others do not shift towards the EMT spectrum. Therefore, despite long-term exposure to hypoxic stress, a high proportion of clones retained epithelial features contributing to expand the phenotypic diversity in the cancer cell population (Figure 1) [37]. It is also interesting to keep in mind that in vivo, cancer cells may be exposed to chronic or intermittent hypoxic stresses, and depending on their location, to various hypoxia levels [38]. The propensity of hypoxic stress to generate cancer cell heterogeneity was further illustrated by the recent study of Lehmann and colleagues [39]. In their attempt to dissect how plasticity of tumor cell migration and EMT is involved in the early metastatic steps, they identified the hypoxia/HIF-1 axis as an inducer of amoeboid detachment and the production of heterogeneous cell subsets whose phenotype and migration were dependent or independent of Twist-mediated EMT. Taken together, these reports underscore the importance of hypoxic stress in mediating tumor plasticity and heterogeneity.

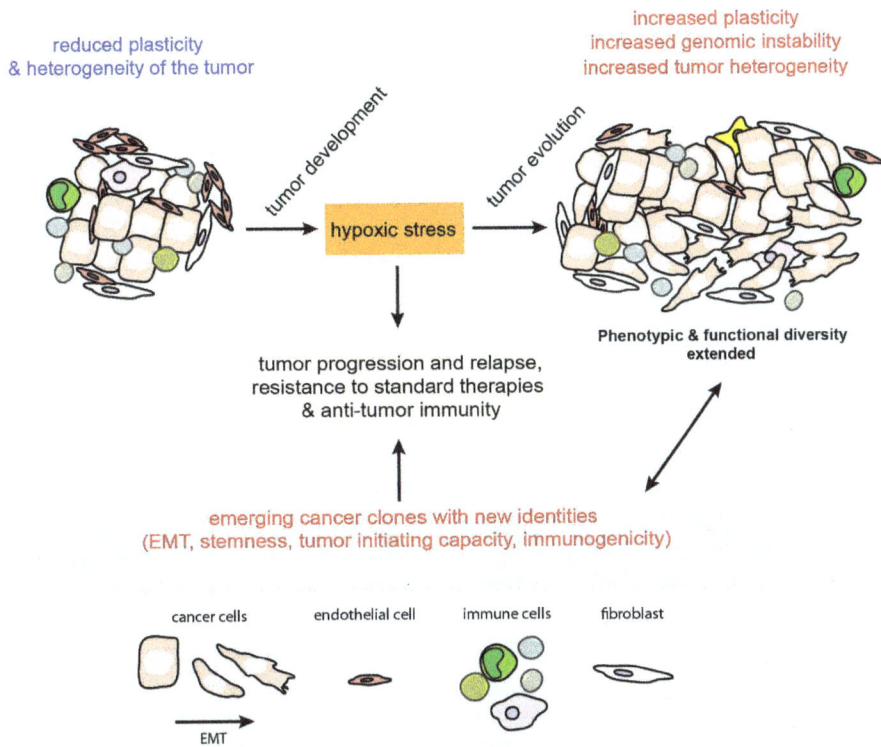

Figure 1. Tumors contain distinct cell types that collectively enable tumor growth and progression. Hypoxic stress can contribute by increasing cell plasticity, genomic instability and phenotypic heterogeneity of certain carcinoma cells, leading to intra-tumor heterogeneity and the emergence of cancer clones resistant to therapies and anti-tumor immunity.

3. Impact of Plasticity and Heterogeneity on Tumor Immune Escape

Evidence is accumulating that tumor plasticity and heterogeneity might be key determinants in the emergence of therapy resistant cancer clones (Figure 1) [19,40]. Considering the relationship between tumors and the immune system, it becomes quite clear that EMT or dedifferentiation can turn even highly immunogenic cancer clones into poorly immunogenic cancer variants resistant to T cell immune attacks through various mechanisms accompanying their phenotypic reprogramming [41–43]. This includes defects in the antigen-presentation machinery involving major histocompatibility complex (MHC) class I molecules, defects in immune recognition following loss of adhesion molecules, gain or loss of immune-modulatory factors and secretion of immunosuppressive substances, or gain of anti-apoptotic properties by the cancer cells against cytotoxic immune effectors. Thus, the acquisition of a more mesenchymal phenotype by cancer cells has been associated with deficiencies in the MHC I antigen presentation pathway [44–47], downregulation of E-Cadherin [37], which could be critical for the recognition of cancer cells by tumor infiltrating lymphocytes (TILs) expressing [48,49], hyperactivity of TGF-beta signaling [45], or increased expression of programmed death-ligand 1 (PD-L1) [46,50,51]. Such immune resistant phenotypes are not only relevant for resistance to T-cell-mediated killing. Numerous reports showed evidence of a link between acquisition of mesenchymal features by cancer cells and their relative protection from NK-cell-mediated lysis [37,52,53], or phagocytosis through direct or indirect mechanisms [54]. Ricciardi and colleagues observed that exposing carcinoma cells to inflammatory cytokines not only promotes EMT in these cells but also confers a number of immunomodulatory properties, including interference with proliferation, differentiation and apoptosis of NK, T and B cell populations [55]. On the other hand, immune cells such as macrophages and NK cells can also mediate EMT of cancer cells, and presumably, could influence immune resistant states of carcinoma cells [56–58].

A study by Huergo-Zapico et al. recently showed that NK-cells could mediate EMT programs in melanoma cells, simultaneously potentiating the immune resistance capacity of the latter. On the contrary, data from at least two studies using various model systems have raised the possibility that EMT induction could increase cancer cell susceptibility to NK cells through up-regulation of NKG2D ligands or cell adhesion molecule 1 (CADM1) [59,60]. This further highlights the contextual nature of the events. For a better understanding of these discrepancies, we suggest that special attention should be paid to the dynamic and the continuum of EMT states, as well as on the timing and the nature of the EMT inducers used to manipulate laboratory models. Considering the role of hypoxic stress in our recent study, we demonstrated that a prolonged hypoxic stress (1% O_2) promotes EMT in the NSCLC IGR-Heu population in a manner that depends on the hypoxia effector HIF-1-α [37]. As mentioned above, while some cells experienced profound phenotypic changes toward mesenchymal states, others do not, thus generating cancer cell heterogeneity in the cancer cell population. This was reflected by the presence of a mixture of cells moving along the EMT spectrum with more epithelial or mesenchymal phenotypes. Among the cancer subclones emerging from this hypoxic stress, those with a more mesenchymal phenotype had an increased propensity to resist attacks by cytotoxic lymphocytes as compared to the more epithelial counterpart. This was illustrated by their reduced susceptibility to both cytotoxic T cells (CTL) and NK cell-mediated lysis [37]. In another study, hypoxia-induced EMT of hepatocellular carcinoma cells promoted an immunosuppressive TME by stimulating expression of indoleamine 2,3-dioxygenase (IDO) in monocyte-derived macrophages [61].

Work by Zhang indicates that HIF-1α can stimulate CD47 expression, an important factor for maintaining plasticity of the cells, which also enables breast cancer cells to avoid phagocytosis by macrophages [62]. CD47 hampers the "eat me signal" on cancer cells by interacting with SIRP on macrophages impairing phagocytosis. More recently, Noman and colleagues identified CD47 as a direct target of SNAI1 and ZEB1 [54]. They observed that the CD47 blockade sensitized cancer cells to phagocytosis, particularly in breast cancer cells with Mesenchymal features. In Triple-negative breast cancers (TNBCs), a heterogeneous group of breast tumors that can present many of the salient features found during EMT, the recent report by Samanta et al. revealed that several immuno-modulatory molecules including CD47, PD-L1 and CD73 are direct HIF target genes in TNBC cells [63]. Thus, CD47 expression could reduce killing by macrophages, whereas CD73 and PD-L1 mediate independent mechanisms to inhibit the T-cell effector functions. The coordinate transcriptional induction of these factors was especially observed in cells exposed to certain chemotherapeutic agents such as carboplatin, doxorubicin, gemcitabine, or paclitaxel. Taken together, this data gives great insight into how plasticity of the cancer cells can be linked to a multi-resistant phenotype involving resistance to chemotherapy and immune resistance. The high amount of TGF-β (another HIF target gene) produced by certain carcinoma cells, or the stromal compartment, could also be crucial in dampening the immune response in tumors [20,64–67]. Moreover, interactions between the different contingents should be highlighted. For instance, carcinoma cells with a mesenchymal or a partial EMT features could cooperate and interact with cancer associated-fibroblasts to regulate their phenotype, and presumably immune suppression [36]. Substantial evidence also indicates the role of HIF-mediated immune plasticity in shaping anti-tumor immunity [5,6,68]. As mentioned above, HIF1 could be a major regulator of effector CD8+ T cell functions [25]. An interesting study by Hatfield and colleagues reported that hypoxia reversal via supplemental oxygenation had significant anti-tumor effects in mouse models, resulting in long-term survival of the mice [69]. Importantly, the observed effects were mainly attributed to the presence of T and natural killer cells. Investigators further showed an association with increased intratumoral infiltration, reduced immunosuppression by regulatory T cells and inhibition of tumor-reactive CD8 T cells concomitant with increased pro-inflammatory cytokines and decreased immunosuppressive substances including TGF-β. It is known that dendritic cell differentiation and maturation is impaired under hypoxia, with negative effects on their T-cell activating functions [70]. The work of Facciabene and colleagues invoked the role hypoxia in the recruitment of Tregs through inducing expression of chemokine CC-chemokine ligand 28 (CCL28), which in turn, promotes angiogenesis and tumor tolerance [71]. Further research also indicates the direct role of HIF-1α in regulating the functionality and plasticity of T-regs [72,73]. Myeloid-derived suppressor cells (MDSCs) and tumor-associated macrophages (TAMs) are also known to contribute to tumor-mediated immune escape [74]. Eubank and colleagues showed evidence for a role of HIF-1 and HIF-2 in the promotion of macrophage angiogenic property [75]. HIF-1α could also regulate their inhibitory functions on T cells [4,5]. Interestingly, the study of Corzo et al. showed that hypoxia via HIF-1α could somehow extend the suppressive function of tumor MDSCs while redirecting their differentiation toward macrophages in the TME [76]. Finally, we showed that hypoxia could regulate the tumor MDSC functions by direct transcriptional induction of the programmed death-ligand 1 (PD-L1) in these cells, resulting in increased MDSC-mediated T cell tolerance [77].

4. Mechanisms of Hypoxia-Induced Cancer Stem Cells

CSCs are a subpopulation of cancer cells that have the ability to self-renew, to divide, to give rise to another malignant stem cell and to drive tumor growth and heterogeneity [78,79]. Hypoxia is a significant culprit for the development of tumor cell resistance to therapy, which is in part due to the generation of cancer stem cells (CSCs) [80–82]. Both HIF-1α and HIF-2α have been found to contribute to the mechanisms involved in mediating stemness [80,82–84]. Despite numerous studies in cancer model systems, the molecular mechanisms underlying the CSC generation, downstream of HIFs have not yet been completely elucidated. So far, they have been explored in various cancer models. HIF proteins can directly or indirectly regulate the expression of genes involved in the initiation and maintenance of stem cells such as (OCT4, SOX2, KLF4, MYC, NANOG, CRIPTO, Wnt or NOTCH) [80,85–88]. In addition to their essential functions during embryonic development, these genes could exert diverse functions in cancer. In certain human tumors, they might represent valuable tools to predict recurrence and tumor plasticity, although such prognostic value is far from being established [79,89–94].

In response to hypoxia, HIF-2α was shown to upregulate OCT4 and SOX2 expression resulting in an increase in the migratory capacity of glioma cells [95,96]. In the study of Tang et al., increased levels of HIF-1α in colorectal cancer cells was associated with increased chemoresistance through the GLI2 transcription factor, which coincides with an increase in cancer stem cells [97]. Similarly, HIF-1α and HIF-2α have been shown to increase the expression of the stem cell marker CD133 in glioblastoma cells concurrent with increased chemoresistance [98]. In breast cancer cells, HIF-1α and HIF-2α increased NANOG mRNA by stimulating expression of AlkB homolog 5 (ALKBH5), an m(6)A demethylase able to demethylate NANOG mRNA [99]. HIF-1 was required for the activation of the p38 MAP kinase pathway and inhibition of ERK signaling resulting in stabilization of NANOG, KLF4, and enrichment of breast cancer stem cells [100]. Clearly, understanding how these different signaling mechanisms interact to drive tumor progression and therapy resistance under variable oxygenation conditions will be critical to the efforts to develop more effective cancer therapies.

5. EMT at the Crossroad of Stemness

EMT has been proposed to drive invasion, resistance to therapy, and spreading of cancer to distant sites [13,26,27]. Cells that are committed to EMT also exhibit numerous attributes that are known to be characteristics of stemness [101]. Although cancer stem cells account for only a small part of the tumor bulk, they are assumed to be the main players involved in therapeutic resistance, cancer relapse, and distant metastasis. Hypoxia and HIF proteins likely contribute to the molecular link between EMT and stemness (Figure 1). Indeed, HIFs are not only involved in the regulation of stem cell factors, in response to hypoxic stress HIF1 protein activates the expression of EMT-transcription factors TWIST1 or ZEB1, which ultimately promotes EMT [31,102]. HIF1 can also help cells transition to a more mesenchymal phenotype by regulating the lysyl oxidases LOX and LOXL2, leading to repression of E-cadherin [34,103]. Other studies have reported that Notch signaling and the EMT-TF SNAIL could be involved in this network as well [35]. It is important to note that cancer cells undergoing EMT in response to hypoxia will not only gain mesenchymal properties, but also may acquire stem cell-like features [104]. Signaling pathways leading to EMT involves TGF-β, STAT3, miR-210 among others (Notch, Nanog) [26,104–106]. TGF-β expression is regulated by HIF-1, and in turn, TGF-β plays an important role in stabilizing HIF-1 [37,107]. TGF-β has been described as having a dual function both in suppressing as well as promoting cancer stem cell populations [108]. The effect of TGF-β has also been correlated with the stage of the cancer; at early stages TGF-β has anti-growth effects, whereas at late stages, it promotes the development of aggressive growth [109]. Interestingly, in breast cancer, stem cell-like cells obtained after TGF-β exposure showed resistance to radiation therapy [110]. Likewise, renal cell carcinoma cells having acquired a stem-like phenotype after TGF-β-induced EMT showed an increase in chemoresistance [111]. In gynecologic cancer patients, the use of chemotherapy can induce TGF-β signaling resulting in reduced chemosensitivity [112]. In primary lung cancer cells, TGF-β exposure led to an increase in cancer stem cell population through repression of miRNA138 [113] while in colon cancer, TGF-β seems to play a key role in angiogenesis, tumor growth and metastasis [114]. On the other hand, in the context of hepatocellular carcinoma, TGF-β resulted in a decrease in cell survival of stem-like side populations [115]. Collectively, these studies highlight the importance of HIFs and TGF-β in the regulation of EMT, and provide support for the development of strategies exploiting these pathways to overcome therapy resistance.

It should be noted that STAT3 also plays an important role in the regulation of cancer stem cells and therapy resistance [116,117]. STAT3 has been demonstrated to be a potent stabilizer of HIF-1 in multiple cancer cell models [118–121]. Moreover, at the molecular level, STAT3 signaling is complex and cooperates with several other pathways implicated in cancer growth. This has recently been reviewed by Galoczova et al. [116]. There is currently a need and ample room to better explore STAT3 implications in EMT, cancer stem cells and tumor resistance to therapy. Of particular interest is the development of strategies for STAT3 inhibition, which has been shown to induce apoptotic cell death of STAT3 dependent cancer cells [122]. MicroRNAs also deserve particular attention. They are small non-coding RNAs that function in post-transcriptional regulation of gene expression and in mRNA silencing. Recent studies unraveled the role of hypoxia in the regulation of microRNA machinery components Drosha and Dicer in cancer cells with important consequences for miRNA biogenesis and tumor progression [123,124]. In particular, this work points to the role of hypoxia in promoting EMT and stem cell phenotypes through mechanisms involving oxygen-dependent H3K27me3 demethylases KDM6A/B or HIF-1 target ETS1/ELK1, which ultimately may lead to derepression of certain EMT-TFs such as the miR-200 target ZEB1. On the other hand, miR-210 is highly induced in response to hypoxic stress and it regulates HIF expression [125,126]. miR-210 is known to have important functions during cancer progression, with both promoting and suppressive roles [127]. Inhibition of miR-210 through small molecules results in inhibition of tumorigenesis in a mouse model for triple negative breast cancer [128]. In ovarian cancer cells, it was found to be a promoter of EMT by causing a decrease in E-cadherin, and increase in vimentin [129]. In lung cancer cells, miR-210 was found to regulate the susceptibility of cancer cells to lysis by cytotoxic T cells [126].

6. Hypoxia, DNA Repair and Genomic Instability

Hypoxic regions are heterogeneous within tumors and the hypoxic phase can vary with time and intensity (acute and/or chronic). Accumulating evidence demonstrates that this component of the TME can be associated with an increase in genomic instability of tumor cells, covering a wide range of alterations, from point mutations to chromosomal instability. The magnitude of genetic aberrations such as increases in gene mutations and gene amplifications due to variation in severity of hypoxia can be 5–1000-fold [130]. Indeed, several studies have suggested that hypoxia can induce DNA damage, alter cell cycle checkpoints and/or the sensing and repair of DNA damage, and consequently favor genetic instability (Figure 1) [131–133]. In this regard, several teams have documented an increase in the rate of DNA mutations in cells exposed to in vitro or in vivo hypoxic conditions, mostly using reporter assays [134–136]. The origins of these hypoxia-induced DNA mutations are probably multiple, emerging from hypoxia-mediated oncogene amplification, induction of DNA damages or DNA replication stress, deregulation of DNA damage checkpoint signaling, interference with DNA repair or escape from cell death [132]. Importantly, cycles of hypoxia and re-oxygenation are common phenomena seen in solid tumors and characterized by an increase in the intracellular free radical species [133,137,138], which are also strongly associated with accumulation of DNA damage [133]. However, in the absence of re-oxygenation (chronic hypoxia), hypoxia-induced genetic instability mostly arises from the influence of hypoxic conditions on DNA-damage repair pathways or the induction of a replicative stress, without detectable induced DNA damage [131]. It will be important to further investigate this intriguing possibility, especially in vivo.

In the case of DNA damage, the G1/S and the G2/M checkpoints kinases ataxia telangiectasia mutated [139], ATM-Rad3-related (ATR) and CHK2/CHK1, respectively, transmit signals to the effector molecules such as p53, p21 and CDC25 to prevent cell cycle progression or to initiate programmed cell death. Interestingly, emerging evidence suggests that different severities and durations of hypoxia may have different effects on cell cycle checkpoint controls. For example, oxygen levels such as 0.2% can bypass ATM or ATR and cell cycle checkpoint signaling allowing the propagation of tumor cells with potentially altered DNA that can contribute to genomic instability [140]. Furthermore, it has been proposed that hypoxia can exert selective pressure that leads to expansion of tumor cells with reduced apoptotic capacity due to, for example, TP53 mutations [141], which is considered as the guardian of genome integrity. As mentioned above, DNA repair pathways, especially homologous recombination (HR), mismatch repair (MMR), non-homologous end joining (NHEJ) and base-excision repair (BER) have also been shown to be compromised under hypoxic conditions [131,132]. For example, it was demonstrated that hypoxia can decrease the expression of the HR-related protein RAD51 in a HIF-1α independent manner [142]. Similarly, reduced expression of the NHEJ-related proteins DNA-PKcs, Ku70, Ku80 and DNA-ligase IV has been observed in hypoxic conditions [143]. Hypoxia has been also shown to transcriptionally downregulate the MMR genes MLH1, MSH2 and MSH6 [144] and a hypoxia driven microsatellite instability (MSI) has been proposed [132]. This hypoxic modulation of DNA repair pathways is thus thought to be of major importance in the genomic instability induced by chronic hypoxic conditions.

The induction of DNA damage, the alteration of DNA-damage cell cycle checkpoints and a functional decrease in the DNA repair pathways under hypoxic conditions probably contribute to "mutator" phenotypes in hypoxic cells and to genomic instability, which might have important effects on the anti-tumor immune response and tumor immunogenicity. For example, recent studies have provided new insights into how specific genomic alterations deriving from genome instability can impact on immune evasion of antitumor immunity [43]. Moreover, recent findings demonstrate the role of double strand break repair pathway in up-regulation of PD-L1 expression by cancer cells [145]. However, the influence of hypoxia-induced DNA-damages in PD-L1 expression is currently unknown. Importantly, a potential mutational burden in hypoxic cells could also be linked to their immunogenicity. Indeed, during the past few years, several groups have identified cancer rejection antigens formed by peptides that are entirely absent from normal human tissues, so-called "neo-antigens". Such neo-antigens are solely created by tumor-specific DNA alterations/mutations that result in the formation of novel protein sequences. As compared with non-mutated self-antigens, neo-antigens are thought to be of particular relevance to tumor control, as the quality of the T cell pool that is available for these antigens is not affected by central tolerance [146]. As a result, neo-antigens appear to represent ideal targets for T cell-based cancer immunotherapy [147]. In this regard, tumors harboring deleterious mutations in the DNA repair pathways were found to carry a high number of candidate neo-antigens, which is associated with a clinical benefit from immune checkpoint inhibitor therapy (anti-PD1) (Le, 2015 #2213), indicating that a high burden of tumor neo-antigens correlates with a durable response to anti-immune checkpoint-based immunotherapy. Two recent studies revealed that mutations and/or loss of the DNA repair mechanism leads to increased mutational load, thus resulting in enhanced neo-antigen generation in cancer cells [148,149]). Nevertheless, the hypothesis that hypoxia-induced DNA damages/genomic instability can lead to a high mutational burden and high numbers of neo-antigens, increasing the potential immunogenicity of hypoxic cells, has never been validated.

7. Therapeutic Targeting of Hypoxia in Cancer

In view of the importance of the link between cancer stem cells, cell plasticity and therapeutic implications in cancer development, targeting the hypoxic niches may offer a great advantage in anti-cancer therapy. This is because targeting hypoxic niches results in eliminating diverse cell populations including cancer stem cells, and preventing the commitment of certain highly plastic cells to an EMT program [27,150].

In support of this idea, it was shown that oxygen administration to patients does transiently relieve tumor hypoxia, and as a result, improve therapy [151]. As such, detection of hypoxic areas in vivo is an essential first step. Recent development of two-photon molecular probes in detecting hypoxia in tissue in vivo and in vitro recently demonstrated some efficacy in detecting hypoxia in deep tumor tissue [152]. However, its effective use in the human situation needs to be established. Another important issue to be addressed is for the drugs to be targeted to the hypoxic zones. Several approaches for targeting hypoxic tumor cells are being explored including hypoxia-activated prodrugs, gene therapy, recombinant anaerobic bacteria and specific targeting of HIFs, or targeting pathways important in hypoxic cells such as the mTOR and UPR pathways [27,153,154]. Hypoxia activated prodrugs are drugs that are converted to their active state under a hypoxic environment. These have been developed and used in combination with chemotherapy or targeted therapy [155]. Recombinant anaerobic bacteria have been considered as gene delivery vehicles to cancer cell sites and spare normal tissue. The *Clostridium* strain that expresses prodrug-converting enzymes has been used, allowing for high therapeutic doses in the tumor [156]. A combination of hypoxia-activated drugs with nanotechnology can be used to enhance tumor specific delivery of anti-cancer agents to the hypoxic tumor zone.

HIF1, being presumably the most powerful factor in the hypoxic response represents an ideal target for therapy. In this regard, the development of selective HIF-1α antagonists remains an important clinical challenge [157,158]. Nonetheless, molecule inhibitory drugs reducing HIF-1α levels, or targeting HIF1 stability/activity may provide interesting benefits in anti-cancer therapy. Nanoparticle formulations containing amino bisphosphonate zoledronic have been successfully used in combination with doxorubicin to sensitize cancer cells to multidrug resistance through inhibiting HIF-1 [159,160]. Inhibitors affecting HIF protein translation include cardiac glycosides, PX-478 or topoisomerase I inhibitors [161–167]. Translation of HIF1 mRNA is known to be controlled by the PI3K/AKT/mTOR pathway. Inhibition of this pathway could decrease HIF expression and the resultant tumorigenesis [168–170]. As an alternative, targeting pathways downstream of HIF signaling includes the use of anti-VEGF therapy (monoclonal antibodies targeting VEGF (bevacizumab) or small molecule inhibitors targeting the VEGF receptor), which has been used for anti-angiogenic/vascular normalization effects in certain medical indications including ovarian, renal, lung or colorectal cancers, in combination with chemotherapy [171]. Finally, recent studies give promise to our ultimate ability to design specific inhibitors of HIFs. A new class of HIF antagonists are currently being tested and have already proven to selectively target HIF-2α with relatively low toxicity compared to current anti-angiogenic drugs [158,172–174].

8. Conclusions

Expansion of resistant cancer cell clones during cancer treatment is a major issue for cancer therapy. It reflects a clonal evolution resulting from genomic instability, cellular plasticity and activation of stemness pathways, as well as complex regulatory networks orchestrated by the TME. Tumor microenvironmental hypoxia is a relevant example that demonstrates how microenvironmental parameters can interfere and neutralize immune cell functions. Converging evidence now suggests its potential value as a prognostic factor as well as a predictive factor owing to its multiple contributions to chemoresistance, radio resistance, angiogenesis, resistance to cell death, altered metabolism, genomic instability, cell plasticity and various immune-related aspects. There is a clear rationale to develop efficient ways to target microenvironmental hypoxia to prevent tumor evolution and the emergence of therapy resistance. However, information on the mechanisms at play is still fragmentary and may vary in a contextual manner. Despite insightful experimental studies using in vitro or in vivo models, the challenge remains for scientists and clinicians alike to gain a better understanding of how human tumors respond to hypoxia. It will also be critical to develop specific agents for targeting hypoxia and associated pathways. This has the potential to provide innovative cancer therapies that can enhance antitumor immunity and overcome the barriers of treatment resistance, tumor tolerance and escape from immune surveillance. In the era of cancer immunotherapy, current strategies such as immune checkpoint blockade have focused on attempting to target immune cells directly to boost the immune system of the host. Is it possible to use therapeutic targets derived from the hypoxic TME and associated pathways as new therapeutic solutions for immunotherapy of cancer? This question merits further investigation. An important challenge will be to determine the best combination strategies as well as the optimal timing and sequence of these combinations.

We are still at the beginning of an exciting period of discovery, and integrating the manipulation of hypoxic stress in cancer immunotherapy may lead to more durable and effective cancer immunotherapy approaches in the future.

Author Contributions: Writing-Original Draft Preparation, S.T., R.F.Z., G.H.V., A.F., W.E.-S., S.B., P.B., J.T., S.C.; Writing-Review & Editing, S.T., R.F.Z., G.H.V., J.T., S.C.; Visualization, S.T.; Supervision, S.T., S.C.

Funding: This work was supported by la Ligue Contre le Cancer (EL2015.LNCC/SaC), Institut National du Cancer (PLBIO15-266), and the SIRIC-SOCRATE program.

Conflicts of Interest: The authors declare no conflict of interest.

References

1. Balkwill, F.R.; Capasso, M.; Hagemann, T. The tumor microenvironment at a glance. *J. Cell. Sci.* **2012**, *125*, 5591–5596. [CrossRef] [PubMed]
2. Semenza, G.L. Defining the role of hypoxia-inducible factor 1 in cancer biology and therapeutics. *Oncogene* **2010**, *29*, 625–634. [CrossRef] [PubMed]
3. Semenza, G.L. Hypoxia-inducible factors in physiology and medicine. *Cell* **2012**, *148*, 399–408. [CrossRef] [PubMed]
4. Majmundar, A.J.; Wong, W.J.; Simon, M.C. Hypoxia-inducible factors and the response to hypoxic stress. *Mol. Cell* **2010**, *40*, 294–309. [CrossRef] [PubMed]
5. Noman, M.Z.; Hasmim, M.; Messai, Y.; Terry, S.; Kieda, C.; Janji, B.; Chouaib, S. Hypoxia: A key player in antitumor immune response. A review in the theme: cellular responses to hypoxia. *Am. J. Physiol. Cell Physiol.* **2015**, *309*, C569–C579. [CrossRef] [PubMed]
6. Terry, S.; Buart, S.; Chouaib, S. Hypoxic stress-induced tumor and immune plasticity, suppression, and impact on tumor heterogeneity. *Front. Immunol.* **2017**, *8*, 1625. [CrossRef] [PubMed]
7. Taddei, M.L.; Giannoni, E.; Comito, G.; Chiarugi, P. Microenvironment and tumor cell plasticity: An easy way out. *Cancer Lett.* **2013**, *341*, 80–96. [CrossRef] [PubMed]

8. Mouriaux, F.; Sanschagrin, F.; Diorio, C.; Landreville, S.; Comoz, F.; Petit, E.; Bernaudin, M.; Rousseau, A.P.; Bergeron, D.; Morcos, M. Increased HIF-1α expression correlates with cell proliferation and vascular markers CD31 and VEGF-A in uveal melanoma. *Invest. Ophthalmol. Vis. Sci.* **2014**, *55*, 1277–1283. [CrossRef] [PubMed]

9. Chen, X.; Iliopoulos, D.; Zhang, Q.; Tang, Q.; Greenblatt, M.B.; Hatziapostolou, M.; Lim, E.; Tam, W.L.; Ni, M.; Chen, Y.; et al. XBP1 promotes triple-negative breast cancer by controlling the HIF1α pathway. *Nature* **2014**, *508*, 103–107. [CrossRef] [PubMed]

10. Zhou, J.; Huang, S.; Wang, L.; Yuan, X.; Dong, Q.; Zhang, D.; Wang, X. Clinical and prognostic significance of HIF-1α overexpression in oral squamous cell carcinoma: A meta-analysis. *World J. Surg. Oncol.* **2017**, *15*, 104. [CrossRef] [PubMed]

11. Buart, S.; Terry, S.; Noman, M.Z.; Lanoy, E.; Boutros, C.; Fogel, P.; Dessen, P.; Meurice, G.; Gaston-Mathe, Y.; Vielh, P.; et al. Transcriptional response to hypoxic stress in melanoma and prognostic potential of GBE1 and BNIP3. *Oncotarget* **2017**, *8*, 108786–108801. [CrossRef] [PubMed]

12. Rankin, E.B.; Giaccia, A.J. Hypoxic control of metastasis. *Science* **2016**, *352*, 175–180. [CrossRef] [PubMed]

13. Tsai, Y.P.; Wu, K.J. Hypoxia-regulated target genes implicated in tumor metastasis. *J. Biomed. Sci.* **2012**, *19*, 102. [CrossRef] [PubMed]

14. Wang, G.L.; Semenza, G.L. Purification and characterization of hypoxia-inducible factor 1. *J. Biol. Chem.* **1995**, *270*, 1230–1237. [CrossRef] [PubMed]

15. Xie, H.; Simon, M.C. Oxygen availability and metabolic reprogramming in cancer. *J. Biol. Chem.* **2017**, *292*, 16825–16832. [CrossRef] [PubMed]

16. Tsai, Y.P.; Wu, K.J. Epigenetic regulation of hypoxia-responsive gene expression: Focusing on chromatin and DNA modifications. *Int. J. Cancer* **2014**, *134*, 249–256. [CrossRef] [PubMed]

17. McGranahan, N.; Swanton, C. Clonal heterogeneity and tumor evolution: past, present, and the future. *Cell* **2017**, *168*, 613–628. [CrossRef] [PubMed]

18. Junttila, M.R.; de Sauvage, F.J. Influence of tumour micro-environment heterogeneity on therapeutic response. *Nature* **2013**, *501*, 346–354. [CrossRef] [PubMed]

19. Holzel, M.; Bovier, A.; Tuting, T. Plasticity of tumour and immune cells: A source of heterogeneity and a cause for therapy resistance? *Nat. Rev. Cancer* **2013**, *13*, 365–376. [CrossRef] [PubMed]

20. Fridman, W.H.; Zitvogel, L.; Sautes-Fridman, C.; Kroemer, G. The immune contexture in cancer prognosis and treatment. *Nat. Rev. Clin. Oncol.* **2017**, *14*, 717–734. [CrossRef] [PubMed]

21. Tirosh, I.; Izar, B.; Prakadan, S.M.; Wadsworth, M.H., 2nd; Treacy, D.; Trombetta, J.J.; Rotem, A.; Rodman, C.; Lian, C.; Murphy, G.; Fallahi-Sichani, M.; et al. Dissecting the multicellular ecosystem of metastatic melanoma by single-cell RNA-seq. *Science* **2016**, *352*, 189–196. [CrossRef] [PubMed]

22. Costa, A.; Kieffer, Y.; Scholer-Dahirel, A.; Pelon, F.; Bourachot, B.; Cardon, M.; Sirven, P.; Magagna, I.; Fuhrmann, L.; Bernard, C.; et al. Heterogeneity and immunosuppressive environment in human breast cancer. *Cancer Cell* **2018**, *33*, 463–479. [CrossRef] [PubMed]

23. Azizi, E.; Carr, A.J.; Plitas, G.; Cornish, A.E.; Konopacki, C.; Prabhakaran, S.; Nainys, J.; Wu, K.; Kiseliovas, V.; Setty, M.; et al. Single-cell map of diverse immune phenotypes in the breast tumor microenvironment. *Cell* **2018**, *174*, 1293–1308. [CrossRef] [PubMed]

24. Lavin, Y.; Kobayashi, S.; Leader, A.; Amir, E.D.; Elefant, N.; Bigenwald, C.; Remark, R.; Sweeney, R.; Becker, C.D.; Levine, J.H.; et al. Innate immune landscape in early lung adenocarcinoma by paired single-cell Analyses. *Cell* **2017**, *169*, 750–765. [CrossRef] [PubMed]

25. Palazon, A.; Tyrakis, P.A.; Macias, D.; Velica, P.; Rundqvist, H.; Fitzpatrick, S.; Vojnovic, N.; Phan, A.T.; Loman, N.; Hedenfalk, I.; et al. An HIF-1α/VEGF-A Axis in cytotoxic t cells regulates tumor progression. *Cancer Cell* **2017**, *32*, 669–683. [CrossRef] [PubMed]

26. Nieto, M.A.; Huang, R.Y.; Jackson, R.A.; Thiery, J.P. EMT: 2016. *Cell* **2016**, *166*, 21–45. [CrossRef] [PubMed]

27. Paolicchi, E.; Gemignani, F.; Krstic-Demonacos, M.; Dedhar, S.; Mutti, L.; Landi, S. Targeting hypoxic response for cancer therapy. *Oncotarget* **2016**, *7*, 13464–13478. [CrossRef] [PubMed]

28. Qi, J.; Nakayama, K.; Cardiff, R.D.; Borowsky, A.D.; Kaul, K.; Williams, R.; Krajewski, S.; Mercola, D.; Carpenter, P.M.; Bowtell, D.; et al. Siah2-dependent concerted activity of HIF and FoxA2 regulates formation of neuroendocrine phenotype and neuroendocrine prostate tumors. *Cancer Cell* **2010**, *18*, 23–38. [CrossRef] [PubMed]

29. Danza, G.; Di Serio, C.; Rosati, F.; Lonetto, G.; Sturli, N.; Kacer, D.; Pennella, A.; Ventimiglia, G.; Barucci, R.; Piscazzi, A.; et al. Notch signaling modulates hypoxia-induced neuroendocrine differentiation of human prostate cancer cells. *Mol. Cancer Res.* **2012**, *10*, 230–238. [CrossRef] [PubMed]

30. Lin, T.P.; Chang, Y.T.; Lee, S.Y.; Campbell, M.; Wang, T.C.; Shen, S.H.; Chung, H.J.; Chang, Y.H.; Chiu, A.W.; Pan, C.C.; et al. REST reduction is essential for hypoxia-induced neuroendocrine differentiation of prostate cancer cells by activating autophagy signaling. *Oncotarget* **2016**, *7*, 26137–26151. [CrossRef] [PubMed]

31. Yang, M.-H.; Wu, M.-Z.; Chiou, S.-H.; Chen, P.-M.; Chang, S.-Y.; Liu, C.-J.; Teng, S.-C.; Wu, K.-J. Direct regulation of TWIST by HIF-1α promotes metastasis. *Nature Cell Biol.* **2008**, *10*, 295–305. [CrossRef] [PubMed]

32. Scortegagna, M.; Martin, R.J.; Kladney, R.D.; Neumann, R.G.; Arbeit, J.M. Hypoxia-inducible factor-1α suppresses squamous carcinogenic progression and epithelial-mesenchymal transition. *Cancer Res.* **2009**, *69*, 2638–2646. [CrossRef] [PubMed]

33. Luo, D.; Wang, J.; Li, J.; Post, M. Mouse Snail Is a Target Gene for HIF. *Mol. Cancer Res.* **2011**, *9*, 234–245. [CrossRef] [PubMed]

34. Schietke, R.; Warnecke, C.; Wacker, I.; Schodel, J.; Mole, D.R.; Campean, V.; Amann, K.; Goppelt-Struebe, M.; Behrens, J.; Eckardt, K.U.; et al. The lysyl oxidases LOX and LOXL2 are necessary and sufficient to repress E-cadherin in hypoxia: Insights into cellular transformation processes mediated by HIF-1. *J. Biol. Chem.* **2010**, *285*, 6658–6669. [CrossRef] [PubMed]

35. Sahlgren, C.; Gustafsson, M.V.; Jin, S.; Poellinger, L.; Lendahl, U. Notch signaling mediates hypoxia-induced tumor cell migration and invasion. *Proc. Natl. Acad. Sci. USA* **2008**, *105*, 6392–6397. [CrossRef] [PubMed]

36. Puram, S.V.; Tirosh, I.; Parikh, A.S.; Patel, A.P.; Yizhak, K.; Gillespie, S.; Rodman, C.; Luo, C.L.; Mroz, E.A.; Emerick, K.S.; et al. Single-cell transcriptomic analysis of primary and metastatic tumor ecosystems in head and neck cancer. *Cell* **2017**, *171*, 1611–1624. [CrossRef] [PubMed]

37. Terry, S.; Buart, S.; Tan, T.Z.; Gros, G.; Noman, M.Z.; Lorens, J.B.; Mami-Chouaib, F.; Thiery, J.P.; Chouaib, S. Acquisition of tumor cell phenotypic diversity along the EMT spectrum under hypoxic pressure: Consequences on susceptibility to cell-mediated cytotoxicity. *OncoImmunology* **2017**, *6*, e1271858. [CrossRef] [PubMed]

38. Harrison, L.; Blackwell, K. Hypoxia and anemia: Factors in decreased sensitivity to radiation therapy and chemotherapy? *Oncologist* **2004**, *9*, 31–40. [CrossRef] [PubMed]

39. Lehmann, S.; Te Boekhorst, V.; Odenthal, J.; Bianchi, R.; van Helvert, S.; Ikenberg, K.; Ilina, O.; Stoma, S.; Xandry, J.; Jiang, L.; et al. Hypoxia induces a HIF-1-dependent transition from collective-to-amoeboid dissemination in epithelial cancer cells. *Curr. Biol.* **2017**, *27*, 392–400. [CrossRef] [PubMed]

40. Keener, A.B. Shapeshifters in cancer: How some tumor cells change phenotype to evade therapy. *Nat. Med.* **2016**, *22*, 1194–1196. [CrossRef] [PubMed]

41. Landsberg, J.; Kohlmeyer, J.; Renn, M.; Bald, T.; Rogava, M.; Cron, M.; Fatho, M.; Lennerz, V.; Wolfel, T.; Holzel, M.; et al. Melanomas resist T-cell therapy through inflammation-induced reversible dedifferentiation. *Nature* **2012**, *490*, 412–416. [CrossRef] [PubMed]

42. Terry, S.; Savagner, P.; Ortiz-Cuaran, S.; Mahjoubi, L.; Saintigny, P.; Thiery, J.P.; Chouaib, S. New insights into the role of EMT in tumor immune escape. *Mol. Oncol.* **2017**, *11*, 824–846. [CrossRef] [PubMed]

43. Spranger, S.; Gajewski, T.F. Impact of oncogenic pathways on evasion of antitumour immune responses. *Nat. Rev. Cancer* **2018**, *18*, 139–147. [CrossRef] [PubMed]

44. Akalay, I.; Janji, B.; Hasmim, M.; Noman, M.Z.; Andre, F.; De Cremoux, P.; Bertheau, P.; Badoual, C.; Vielh, P.; Larsen, A.K.; et al. Epithelial-to-mesenchymal transition and autophagy induction in breast carcinoma promote escape from T-cell-mediated lysis. *Cancer Res.* **2013**, *73*, 2418–2427. [CrossRef] [PubMed]

45. Akalay, I.; Tan, T.Z.; Kumar, P.; Janji, B.; Mami-Chouaib, F.; Charpy, C.; Vielh, P.; Larsen, A.K.; Thiery, J.P.; Sabbah, M.; et al. Targeting WNT1-inducible signaling pathway protein 2 alters human breast cancer cell susceptibility to specific lysis through regulation of KLF-4 and miR-7 expression. *Oncogene* **2015**, *34*, 2261–2271. [CrossRef] [PubMed]

46. Dongre, A.; Rashidian, M.; Reinhardt, F.; Bagnato, A.; Keckesova, Z.; Ploegh, H.L.; Weinberg, R.A. Epithelial-to-Mesenchymal Transition Contributes to Immunosuppression in Breast Carcinomas. *Cancer Res.* **2017**, *77*, 3982–3989. [CrossRef] [PubMed]

47. Tripathi, S.C.; Peters, H.L.; Taguchi, A.; Katayama, H.; Wang, H.; Momin, A.; Jolly, M.K.; Celiktas, M.; Rodriguez-Canales, J.; Liu, H.; et al. Immunoproteasome deficiency is a feature of non-small cell lung cancer with a mesenchymal phenotype and is associated with a poor outcome. *Proc. Natl. Acad. Sci. USA* **2016**, *113*, E1555–E1564. [CrossRef] [PubMed]

48. Le Floc'h, A.; Jalil, A.; Vergnon, I.; Le Maux Chansac, B.; Lazar, V.; Bismuth, G.; Chouaib, S.; Mami-Chouaib, F. A E β 7 integrin interaction with E-cadherin promotes antitumor CTL activity by triggering lytic granule polarization and exocytosis. *J. Exp. Med.* **2007**, *204*, 559–570. [CrossRef] [PubMed]

49. Djenidi, F.; Adam, J.; Goubar, A.; Durgeau, A.; Meurice, G.; de Montpreville, V.; Validire, P.; Besse, B.; Mami-Chouaib, F. CD8+CD103+ tumor-infiltrating lymphocytes are tumor-specific tissue-resident memory T cells and a prognostic factor for survival in lung cancer patients. *J. Immunol.* **2015**, *194*, 3475–3486. [CrossRef] [PubMed]

50. Chen, L.; Gibbons, D.L.; Goswami, S.; Cortez, M.A.; Ahn, Y.H.; Byers, L.A.; Zhang, X.; Yi, X.; Dwyer, D.; Lin, W.; et al. Metastasis is regulated via microRNA-200/ZEB1 axis control of tumour cell PD-L1 expression and intratumoral immunosuppression. *Nat. Commun.* **2014**, *5*, 5241. [CrossRef] [PubMed]

51. Noman, M.Z.; Janji, B.; Abdou, A.; Hasmim, M.; Terry, S.; Tan, T.Z.; Mami-Chouaib, F.; Thiery, J.P.; Chouaib, S. The immune checkpoint ligand PD-L1 is upregulated in EMT-activated human breast cancer cells by a mechanism involving ZEB-1 and miR-200. *OncoImmunology* **2017**, *6*, e1263412. [CrossRef] [PubMed]

52. Hamilton, D.H.; Huang, B.; Fernando, R.I.; Tsang, K.Y.; Palena, C. WEE1 inhibition alleviates resistance to immune attack of tumor cells undergoing epithelial-mesenchymal transition. *Cancer Res.* **2014**, *74*, 2510–2519. [CrossRef] [PubMed]

53. Hamilton, D.H.; Griner, L.M.; Keller, J.M.; Hu, X.; Southall, N.; Marugan, J.; David, J.M.; Ferrer, M.; Palena, C. Targeting estrogen receptor signaling with fulvestrant enhances immune and chemotherapy-mediated cytotoxicity of human lung cancer. *Clin. Cancer Res.* **2016**, *22*, 6204–6216. [CrossRef] [PubMed]

54. Noman, M.Z.; Van Moer, K.; Marani, V.; Gemmill, R.M.; Tranchevent, L.C.; Azuaje, F.; Muller, A.; Chouaib, S.; Thiery, J.P.; Berchem, G.; et al. CD47 is a direct target of SNAI1 and ZEB1 and its blockade activates the phagocytosis of breast cancer cells undergoing EMT. *OncoImmunology* **2018**, *7*, e1345415. [CrossRef] [PubMed]

55. Ricciardi, M.; Zanotto, M.; Malpeli, G.; Bassi, G.; Perbellini, O.; Chilosi, M.; Bifari, F.; Krampera, M. Epithelial-to-mesenchymal transition (EMT) induced by inflammatory priming elicits mesenchymal stromal cell-like immune-modulatory properties in cancer cells. *Br. J. Cancer* **2015**, *112*, 1067–1075. [CrossRef] [PubMed]

56. Su, S.; Liu, Q.; Chen, J.; Chen, J.; Chen, F.; He, C.; Huang, D.; Wu, W.; Lin, L.; Huang, W.; et al. A positive feedback loop between mesenchymal-like cancer cells and macrophages is essential to breast cancer metastasis. *Cancer Cell* **2014**, *25*, 605–620. [CrossRef] [PubMed]

57. Liu, C.Y.; Xu, J.Y.; Shi, X.Y.; Huang, W.; Ruan, T.Y.; Xie, P.; Ding, J.L. M2-polarized tumor-associated macrophages promoted epithelial-mesenchymal transition in pancreatic cancer cells, partially through TLR4/IL-10 signaling pathway. *Lab. Investig.* **2013**, *93*, 844–854. [CrossRef] [PubMed]

58. Huergo-Zapico, L.; Parodi, M.; Cantoni, C.; Lavarello, C.; Fernandez-Martinez, J.L.; Petretto, A.; DeAndres-Galiana, E.J.; Balsamo, M.; Lopez-Soto, A.; Pietra, G.; et al. NK-cell editing mediates epithelial-to-mesenchymal transition via phenotypic and proteomic changes in melanoma cell lines. *Cancer Res.* **2018**, *78*, 3913–3925. [CrossRef] [PubMed]

59. Lopez-Soto, A.; Huergo-Zapico, L.; Galvan, J.A.; Rodrigo, L.; de Herreros, A.G.; Astudillo, A.; Gonzalez, S. Epithelial-mesenchymal transition induces an antitumor immune response mediated by NKG2D receptor. *J. Immunol.* **2013**, *190*, 4408–4419. [CrossRef] [PubMed]

60. Chockley, P.J.; Chen, J.; Chen, G.; Beer, D.G.; Standiford, T.J.; Keshamouni, V.G. Epithelial-mesenchymal transition leads to NK cell-mediated metastasis-specific immunosurveillance in lung cancer. *J. Clin. Investig.* **2018**, *128*, 1384–1396. [CrossRef] [PubMed]

61. Ye, L.Y.; Chen, W.; Bai, X.L.; Xu, X.Y.; Zhang, Q.; Xia, X.F.; Sun, X.; Li, G.G.; Hu, Q.D.; Fu, Q.H.; et al. Hypoxia-induced epithelial-to-mesenchymal transition in hepatocellular carcinoma induces an immunosuppressive tumor microenvironment to promote metastasis. *Cancer Res.* **2016**, *76*, 818–830. [CrossRef] [PubMed]

62. Zhang, H.; Lu, H.; Xiang, L.; Bullen, J.W.; Zhang, C.; Samanta, D.; Gilkes, D.M.; He, J.; Semenza, G.L. HIF-1 regulates CD47 expression in breast cancer cells to promote evasion of phagocytosis and maintenance of cancer stem cells. *Proc. Natl. Acad. Sci. USA* **2015**, *112*, E6215–E6223. [CrossRef] [PubMed]

63. Samanta, D.; Park, Y.; Ni, X.; Li, H.; Zahnow, C.A.; Gabrielson, E.; Pan, F.; Semenza, G.L. Chemotherapy induces enrichment of CD47(+)/CD73(+)/PDL1(+) immune evasive triple-negative breast cancer cells. *Proc. Natl. Acad. Sci. USA* **2018**, *115*, E1239–E1248. [CrossRef] [PubMed]

64. Thomas, D.A.; Massague, J. TGF-β directly targets cytotoxic T cell functions during tumor evasion of immune surveillance. *Cancer Cell* **2005**, *8*, 369–380. [CrossRef] [PubMed]

65. Hasmim, M.; Noman, M.Z.; Messai, Y.; Bordereaux, D.; Gros, G.; Baud, V.; Chouaib, S. Cutting edge: Hypoxia-induced Nanog favors the intratumoral infiltration of regulatory T cells and macrophages via direct regulation of TGF-β1. *J. Immunol.* **2013**, *191*, 5802–5806. [CrossRef] [PubMed]

66. Mariathasan, S.; Turley, S.J.; Nickles, D.; Castiglioni, A.; Yuen, K.; Wang, Y.; Kadel, E.E., III; Koeppen, H.; Astarita, J.L.; Cubas, R.; Jhunjhunwala, S.; et al. TGFβ attenuates tumour response to PD-L1 blockade by contributing to exclusion of T cells. *Nature* **2018**, *554*, 544–548. [CrossRef] [PubMed]

67. Tauriello, D.V.F.; Palomo-Ponce, S.; Stork, D.; Berenguer-Llergo, A.; Badia-Ramentol, J.; Iglesias, M.; Sevillano, M.; Ibiza, S.; Canellas, A.; Hernando-Momblona, X.; et al. TGFβ drives immune evasion in genetically reconstituted colon cancer metastasis. *Nature* **2018**, *554*, 538–543. [CrossRef] [PubMed]

68. Palazon, A.; Aragones, J.; Morales-Kastresana, A.; de Landazuri, M.O.; Melero, I. Molecular pathways: Hypoxia response in immune cells fighting or promoting cancer. *Clin. Cancer Res.* **2012**, *18*, 1207–1213. [CrossRef] [PubMed]

69. Hatfield, S.M.; Kjaergaard, J.; Lukashev, D.; Schreiber, T.H.; Belikoff, B.; Abbott, R.; Sethumadhavan, S.; Philbrook, P.; Ko, K.; Cannici, R.; et al. Immunological mechanisms of the antitumor effects of supplemental oxygenation. *Sci. Transl. Med.* **2015**, *7*, 277. [CrossRef] [PubMed]

70. Mancino, A.; Schioppa, T.; Larghi, P.; Pasqualini, F.; Nebuloni, M.; Chen, I.H.; Sozzani, S.; Austyn, J.M.; Mantovani, A.; Sica, A. Divergent effects of hypoxia on dendritic cell functions. *Blood* **2008**, *112*, 3723–3734. [CrossRef] [PubMed]

71. Facciabene, A.; Peng, X.; Hagemann, I.S.; Balint, K.; Barchetti, A.; Wang, L.P.; Gimotty, P.A.; Gilks, C.B.; Lal, P.; Zhang, L.; et al. Tumour hypoxia promotes tolerance and angiogenesis via CCL28 and T(reg) cells. *Nature* **2011**, *475*, 226–230. [CrossRef] [PubMed]

72. Lee, J.H.; Elly, C.; Park, Y.; Liu, Y.C. E3 Ubiquitin Ligase VHL Regulates Hypoxia-Inducible Factor-1α to Maintain Regulatory T Cell Stability and Suppressive Capacity. *Immunity* **2015**, *42*, 1062–1074. [CrossRef] [PubMed]

73. Clambey, E.T.; McNamee, E.N.; Westrich, J.A.; Glover, L.E.; Campbell, E.L.; Jedlicka, P.; de Zoeten, E.F.; Cambier, J.C.; Stenmark, K.R.; Colgan, S.P.; et al. Hypoxia-inducible factor-1 α-dependent induction of FoxP3 drives regulatory T-cell abundance and function during inflammatory hypoxia of the mucosa. *Proc. Natl. Acad. Sci. USA* **2012**, *109*, E2784–E2793. [CrossRef] [PubMed]

74. Antonios, J.P.; Soto, H.; Everson, R.G.; Moughon, D.; Orpilla, J.R.; Shin, N.P.; Sedighim, S.; Treger, J.; Odesa, S.; Tucker, A.; et al. Immunosuppressive tumor-infiltrating myeloid cells mediate adaptive immune resistance via a PD-1/PD-L1 mechanism in glioblastoma. *Neuro Oncol.* **2017**, *19*, 796–807. [CrossRef] [PubMed]

75. Eubank, T.D.; Roda, J.M.; Liu, H.; O'Neil, T.; Marsh, C.B. Opposing roles for HIF-1α and HIF-2α in the regulation of angiogenesis by mononuclear phagocytes. *Blood* **2011**, *117*, 323–332. [CrossRef] [PubMed]

76. Corzo, C.A.; Condamine, T.; Lu, L.; Cotter, M.J.; Youn, J.I.; Cheng, P.; Cho, H.I.; Celis, E.; Quiceno, D.G.; Padhya, T.; et al. HIF-1α regulates function and differentiation of myeloid-derived suppressor cells in the tumor microenvironment. *J. Exp. Med.* **2010**, *207*, 2439–2453. [CrossRef] [PubMed]

77. Noman, M.Z.; Desantis, G.; Janji, B.; Hasmim, M.; Karray, S.; Dessen, P.; Bronte, V.; Chouaib, S. PD-L1 is a novel direct target of HIF-1α, and its blockade under hypoxia enhanced MDSC-mediated T cell activation. *J. Exp. Med.* **2014**, *211*, 781–790. [CrossRef] [PubMed]

78. Reya, T.; Morrison, S.J.; Clarke, M.F.; Weissman, I.L. Stem cells, cancer, and cancer stem cells. *Nature* **2001**, *414*, 105–111. [CrossRef] [PubMed]

79. Tang, D.G. Understanding cancer stem cell heterogeneity and plasticity. *Cell Res.* **2012**, *22*, 457–472. [CrossRef] [PubMed]

80. Keith, B.; Simon, M.C. Hypoxia-inducible factors, stem cells, and cancer. *Cell* **2007**, *129*, 465–472. [CrossRef] [PubMed]

81. Carnero, A.; Lleonart, M. The hypoxic microenvironment: A determinant of cancer stem cell evolution. *Bioessays* **2016**, *38*, S65–S74. [CrossRef] [PubMed]

82. Ayob, A.Z.; Ramasamy, T.S. Cancer stem cells as key drivers of tumour progression. *J. Biomed. Sci.* **2018**, *25*, 20. [CrossRef] [PubMed]

83. Saito, S.; Lin, Y.C.; Tsai, M.H.; Lin, C.S.; Murayama, Y.; Sato, R.; Yokoyama, K.K. Emerging roles of hypoxia-inducible factors and reactive oxygen species in cancer and pluripotent stem cells. *Kaohsiung J. Med. Sci.* **2015**, *31*, 279–286. [CrossRef] [PubMed]

84. Peng, G.; Liu, Y. Hypoxia-inducible factors in cancer stem cells and inflammation. *Trends Pharmacol. Sci.* **2015**, *36*, 374–383. [CrossRef] [PubMed]

85. Covello, K.L.; Kehler, J.; Yu, H.; Gordan, J.D.; Arsham, A.M.; Hu, C.J.; Labosky, P.A.; Simon, M.C.; Keith, B. HIF-2α regulates Oct-4: Effects of hypoxia on stem cell function, embryonic development, and tumor growth. *Genes Dev.* **2006**, *20*, 557–570. [CrossRef] [PubMed]

86. Mazumdar, J.; Dondeti, V.; Simon, M.C. Hypoxia-inducible factors in stem cells and cancer. *J. Cell. Mol. Med.* **2009**, *13*, 4319–4328. [CrossRef] [PubMed]

87. Bianco, C.; Cotten, C.; Lonardo, E.; Strizzi, L.; Baraty, C.; Mancino, M.; Gonzales, M.; Watanabe, K.; Nagaoka, T.; Berry, C.; et al. Cripto-1 is required for hypoxia to induce cardiac differentiation of mouse embryonic stem cells. *Am. J. Pathol.* **2009**, *175*, 2146–2158. [CrossRef] [PubMed]

88. Westfall, S.D.; Sachdev, S.; Das, P.; Hearne, L.B.; Hannink, M.; Roberts, R.M.; Ezashi, T. Identification of oxygen-sensitive transcriptional programs in human embryonic stem cells. *Stem Cells Dev.* **2008**, *17*, 869–881. [CrossRef] [PubMed]

89. Wahab, S.M.R.; Islam, F.; Gopalan, V.; Lam, A.K. The Identifications and Clinical Implications of Cancer Stem Cells in Colorectal Cancer. *Clin. Colorectal Cancer* **2017**, *16*, 93–102. [CrossRef] [PubMed]

90. van Schaijik, B.; Davis, P.F.; Wickremesekera, A.C.; Tan, S.T.; Itinteang, T. Subcellular localisation of the stem cell markers OCT4, SOX2, NANOG, KLF4 and c-MYC in cancer: A review. *J. Clin. Pathol.* **2018**, *71*, 88–91. [CrossRef] [PubMed]

91. Yeung, T.M.; Gandhi, S.C.; Bodmer, W.F. Hypoxia and lineage specification of cell line-derived colorectal cancer stem cells. *Proc. Natl. Acad. Sci. USA* **2011**, *108*, 4382–4387. [CrossRef] [PubMed]

92. Mu, P.; Zhang, Z.; Benelli, M.; Karthaus, W.R.; Hoover, E.; Chen, C.C.; Wongvipat, J.; Ku, S.Y.; Gao, D.; Cao, Z.; et al. SOX2 promotes lineage plasticity and antiandrogen resistance in TP53- and RB1-deficient prostate cancer. *Science* **2017**, *355*, 84–88. [CrossRef] [PubMed]

93. Zayed, H.; Petersen, I. Stem cell transcription factor SOX2 in synovial sarcoma and other soft tissue tumors. *Pathol. Res. Pract.* **2018**, *214*, 1000–1007. [CrossRef] [PubMed]

94. Terry, S.; El-Sayed, I.Y.; Destouches, D.; Maille, P.; Nicolaiew, N.; Ploussard, G.; Semprez, F.; Pimpie, C.; Beltran, H.; Londono-Vallejo, A.; et al. CRIPTO overexpression promotes mesenchymal differentiation in prostate carcinoma cells through parallel regulation of AKT and FGFR activities. *Oncotarget* **2015**, *6*, 11994–12008. [CrossRef] [PubMed]

95. Koh, M.Y.; Lemos, R., Jr.; Liu, X.; Powis, G. The hypoxia-associated factor switches cells from HIF-1α- to HIF-2α-dependent signaling promoting stem cell characteristics, aggressive tumor growth and invasion. *Cancer Res.* **2011**, *71*, 4015–4027. [CrossRef] [PubMed]

96. Bhagat, M.; Palanichamy, J.K.; Ramalingam, P.; Mudassir, M.; Irshad, K.; Chosdol, K.; Sarkar, C.; Seth, P.; Goswami, S.; Sinha, S.; et al. HIF-2α mediates a marked increase in migration and stemness characteristics in a subset of glioma cells under hypoxia by activating an Oct-4/Sox-2-Mena (INV) axis. *Int. J. Biochem. Cell Biol.* **2016**, *74*, 60–71. [CrossRef] [PubMed]

97. Tang, Y.A.; Chen, Y.F.; Bao, Y.; Mahara, S.; Yatim, S.; Oguz, G.; Lee, P.L.; Feng, M.; Cai, Y.; Tan, E.Y.; et al. Hypoxic tumor microenvironment activates GLI2 via HIF-1α and TGF-β2 to promote chemoresistance in colorectal cancer. *Proc. Natl. Acad. Sci. USA* **2018**, *115*, E5990–E5999. [CrossRef] [PubMed]

98. Ahmed, E.M.; Bandopadhyay, G.; Coyle, B.; Grabowska, A. A HIF-independent, CD133-mediated mechanism of cisplatin resistance in glioblastoma cells. *Cell Oncol.* **2018**, *41*, 319–328. [CrossRef] [PubMed]

99. Zhang, C.; Samanta, D.; Lu, H.; Bullen, J.W.; Zhang, H.; Chen, I.; He, X.; Semenza, G.L. Hypoxia induces the breast cancer stem cell phenotype by HIF-dependent and ALKBH5-mediated m(6)A-demethylation of NANOG mRNA. *Proc. Natl. Acad. Sci. USA* **2016**, *113*, E2047–E2056. [CrossRef] [PubMed]

100. Lu, H.; Tran, L.; Park, Y.; Chen, I.; Lan, J.; Xie, Y.; Semenza, G.L. Reciprocal regulation of DUSP9 and DUSP16 expression by HIF1 controls ERK and p38 MAP kinase activity and mediates chemotherapy-induced breast cancer stem cell enrichment. *Cancer Res.* **2018**, *78*, 4191–4202. [CrossRef] [PubMed]

101. Scheel, C.; Weinberg, R.A. Cancer stem cells and epithelial-mesenchymal transition: Concepts and molecular links. *Semin Cancer Biol.* **2012**, *22*, 396–403. [CrossRef] [PubMed]

102. Zhang, W.; Shi, X.; Peng, Y.; Wu, M.; Zhang, P.; Xie, R.; Wu, Y.; Yan, Q.; Liu, S.; Wang, J. HIF-1α Promotes Epithelial-Mesenchymal Transition and Metastasis through Direct Regulation of ZEB1 in Colorectal Cancer. *PLoS ONE* **2015**, *10*, e0129603. [CrossRef] [PubMed]

103. Erler, J.T.; Bennewith, K.L.; Nicolau, M.; Dornhofer, N.; Kong, C.; Le, Q.T.; Chi, J.T.; Jeffrey, S.S.; Giaccia, A.J. Lysyl oxidase is essential for hypoxia-induced metastasis. *Nature* **2006**, *440*, 1222–1226. [CrossRef] [PubMed]

104. Kaur, G.; Sharma, P.; Dogra, N.; Singh, S. Eradicating cancer stem cells: Concepts, issues, and challenges. *Curr. Treat. Opt. Oncol.* **2018**, *19*, 20. [CrossRef] [PubMed]

105. Cai, Z.; Cao, Y.; Luo, Y.; Hu, H.; Ling, H. Signalling mechanism(s) of epithelial-mesenchymal transition and cancer stem cells in tumour therapeutic resistance. *Clin. Chim. Acta* **2018**, *483*, 156–163. [CrossRef] [PubMed]

106. Zavadil, J.; Cermak, L.; Soto-Nieves, N.; Bottinger, E.P. Integration of TGF-β/Smad and Jagged1/Notch signalling in epithelial-to-mesenchymal transition. *EMBO J.* **2004**, *23*, 1155–1165. [CrossRef] [PubMed]

107. McMahon, S.; Charbonneau, M.; Grandmont, S.; Richard, D.E.; Dubois, C.M. Transforming growth factor β1 induces hypoxia-inducible factor-1 stabilization through selective inhibition of PHD2 expression. *J. Biol. Chem.* **2006**, *281*, 24171–24181. [CrossRef] [PubMed]

108. Bellomo, C.; Caja, L.; Moustakas, A. Transforming growth factor β as regulator of cancer stemness and metastasis. *Br. J. Cancer* **2016**, *115*, 761–769. [CrossRef] [PubMed]

109. Tang, J.; Gifford, C.C.; Samarakoon, R.; Higgins, P.J. Deregulation of negative controls on TGF-β1 signaling in tumor progression. *Cancers* **2018**, *10*, 159. [CrossRef] [PubMed]

110. Konge, J.; Leteurtre, F.; Goislard, M.; Biard, D.; Morel-Altmeyer, S.; Vaurijoux, A.; Gruel, G.; Chevillard, S.; Lebeau, J. Breast cancer stem cell-like cells generated during TGFβ-induced EMT are radioresistant. *Oncotarget* **2018**, *9*, 23519–23531. [CrossRef] [PubMed]

111. Singla, M.; Kumar, A.; Bal, A.; Sarkar, S.; Bhattacharyya, S. Epithelial to mesenchymal transition induces stem cell like phenotype in renal cell carcinoma cells. *Cancer Cell Int.* **2018**, *18*, 57. [CrossRef] [PubMed]

112. Zhu, H.; Gu, X.; Xia, L.; Zhou, Y.; Bouamar, H.; Yang, J.; Ding, X.; Zwieb, C.; Zhang, J.; Hinck, A.P.; et al. A Novel TGFβ Trap Blocks Chemotherapeutics-Induced TGFβ1 Signaling and Enhances Their Anticancer Activity in Gynecologic Cancers. *Clin. Cancer Res.* **2018**, *24*, 2780–2793. [CrossRef] [PubMed]

113. Zhang, F.; Li, T.; Han, L.; Qin, P.; Wu, Z.; Xu, B.; Gao, Q.; Song, Y. TGFβ1-induced down-regulation of microRNA-138 contributes to epithelial-mesenchymal transition in primary lung cancer cells. *Biochem. Biophys. Res. Commun.* **2018**, *496*, 1169–1175. [CrossRef] [PubMed]

114. Chruscik, A.; Gopalan, V.; Lam, A.K. The clinical and biological roles of transforming growth factor β in colon cancer stem cells: A systematic review. *Eur. J. Cell Biol.* **2018**, *97*, 15–22. [CrossRef] [PubMed]

115. Kim, J.B.; Lee, S.; Kim, H.R.; Park, S.Y.; Lee, M.; Yoon, J.H.; Kim, Y.J. Transforming growth factor-β decreases side population cells in hepatocellular carcinoma in vitro. *Oncol Lett.* **2018**, *15*, 8723–8728. [CrossRef] [PubMed]

116. Galoczova, M.; Coates, P.; Vojtesek, B. STAT3, stem cells, cancer stem cells and p63. *Cell. Mol. Biol. Lett.* **2018**, *23*, 12. [CrossRef] [PubMed]

117. Groner, B.; von Manstein, V. Jak Stat signaling and cancer: Opportunities, benefits and side effects of targeted inhibition. *Mol. Cell. Endocrinol.* **2017**, *451*, 1–14. [CrossRef] [PubMed]

118. Jung, J.E.; Lee, H.G.; Cho, I.H.; Chung, D.H.; Yoon, S.H.; Yang, Y.M.; Lee, J.W.; Choi, S.; Park, J.W.; Ye, S.K.; et al. STAT3 is a potential modulator of HIF-1-mediated VEGF expression in human renal carcinoma cells. *FASEB J.* **2005**, *19*, 1296–1298. [CrossRef] [PubMed]

119. Xu, Q.; Briggs, J.; Park, S.; Niu, G.; Kortylewski, M.; Zhang, S.; Gritsko, T.; Turkson, J.; Kay, H.; Semenza, G.L.; et al. Targeting Stat3 blocks both HIF-1 and VEGF expression induced by multiple oncogenic growth signaling pathways. *Oncogene* **2005**, *24*, 5552–5560. [CrossRef] [PubMed]

120. Jung, J.E.; Kim, H.S.; Lee, C.S.; Shin, Y.J.; Kim, Y.N.; Kang, G.H.; Kim, T.Y.; Juhnn, Y.S.; Kim, S.J.; Park, J.W.; et al. STAT3 inhibits the degradation of HIF-1α by pVHL-mediated ubiquitination. *Exp. Mol. Med.* **2008**, *40*, 479–485. [CrossRef] [PubMed]

121. Pawlus, M.R.; Wang, L.; Hu, C.J. STAT3 and HIF1α cooperatively activate HIF1 target genes in MDA-MB-231 and RCC4 cells. *Oncogene* **2014**, *33*, 1670–1679. [CrossRef] [PubMed]

122. Wong, A.L.A.; Hirpara, J.L.; Pervaiz, S.; Eu, J.Q.; Sethi, G.; Goh, B.C. Do STAT3 inhibitors have potential in the future for cancer therapy? *Exp. Opin. Investig. Drugs* **2017**, *26*, 883–887. [CrossRef] [PubMed]

123. Rupaimoole, R.; Wu, S.Y.; Pradeep, S.; Ivan, C.; Pecot, C.V.; Gharpure, K.M.; Nagaraja, A.S.; Armaiz-Pena, G.N.; McGuire, M.; Zand, B.; et al. Hypoxia-mediated downregulation of miRNA biogenesis promotes tumour progression. *Nat. Commun.* **2014**, *5*, 5202. [CrossRef] [PubMed]

124. van den Beucken, T.; Koch, E.; Chu, K.; Rupaimoole, R.; Prickaerts, P.; Adriaens, M.; Voncken, J.W.; Harris, A.L.; Buffa, F.M.; Haider, S.; et al. Hypoxia promotes stem cell phenotypes and poor prognosis through epigenetic regulation of DICER. *Nat. Commun.* **2014**, *5*, 5203. [CrossRef] [PubMed]

125. Kelly, T.J.; Souza, A.L.; Clish, C.B.; Puigserver, P. A hypoxia-induced positive feedback loop promotes hypoxia-inducible factor 1α stability through miR-210 suppression of glycerol-3-phosphate dehydrogenase 1-like. *Mol. Cell. Biol* **2011**, *31*, 2696–2706. [CrossRef] [PubMed]

126. Noman, M.Z.; Buart, S.; Romero, P.; Ketari, S.; Janji, B.; Mari, B.; Mami-Chouaib, F.; Chouaib, S. Hypoxia-inducible miR-210 regulates the susceptibility of tumor cells to lysis by cytotoxic T cells. *Cancer Res.* **2012**, *72*, 4629–4641. [CrossRef] [PubMed]

127. Bavelloni, A.; Ramazzotti, G.; Poli, A.; Piazzi, M.; Focaccia, E.; Blalock, W.; Faenza, I. MiRNA-210: A Current Overview. *Anticancer Res.* **2017**, *37*, 6511–6521. [PubMed]

128. Costales, M.G.; Haga, C.L.; Velagapudi, S.P.; Childs-Disney, J.L.; Phinney, D.G.; Disney, M.D. Small molecule inhibition of microRNA-210 reprograms an oncogenic hypoxic circuit. *J. Am. Chem. Soc.* **2017**, *139*, 3446–3455. [CrossRef] [PubMed]

129. Ding, L.; Zhao, L.; Chen, W.; Liu, T.; Li, Z.; Li, X. miR-210, a modulator of hypoxia-induced epithelial-mesenchymal transition in ovarian cancer cell. *Int. J. Clin. Exp. Med.* **2015**, *8*, 2299–2307. [PubMed]

130. Bindra, R.S.; Glazer, P.M. Genetic instability and the tumor microenvironment: Towards the concept of microenvironment-induced mutagenesis. *Mutat. Res.* **2005**, *569*, 75–85. [CrossRef] [PubMed]

131. Scanlon, S.E.; Glazer, P.M. Multifaceted control of DNA repair pathways by the hypoxic tumor microenvironment. *DNA Repair* **2015**, *32*, 180–189. [CrossRef] [PubMed]

132. Luoto, K.R.; Kumareswaran, R.; Bristow, R.G. Tumor hypoxia as a driving force in genetic instability. *Genome Integr.* **2013**, *4*, 5. [CrossRef] [PubMed]

133. Bristow, R.G.; Hill, R.P. Hypoxia and metabolism. Hypoxia, DNA repair and genetic instability. *Nat. Rev. Cancer* **2008**, *8*, 180–192. [CrossRef] [PubMed]

134. Reynolds, T.Y.; Rockwell, S.; Glazer, P.M. Genetic instability induced by the tumor microenvironment. *Cancer Res.* **1996**, *56*, 5754–5757. [PubMed]

135. Li, C.Y.; Little, J.B.; Hu, K.; Zhang, W.; Zhang, L.; Dewhirst, M.W.; Huang, Q. Persistent genetic instability in cancer cells induced by non-DNA-damaging stress exposures. *Cancer Res.* **2001**, *61*, 428–432. [PubMed]

136. Keysar, S.B.; Trncic, N.; Larue, S.M.; Fox, M.H. Hypoxia/reoxygenation-induced mutations in mammalian cells detected by the flow cytometry mutation assay and characterized by mutant spectrum. *Radiat. Res.* **2010**, *173*, 21–26. [CrossRef] [PubMed]

137. Kondoh, M.; Ohga, N.; Akiyama, K.; Hida, Y.; Maishi, N.; Towfik, A.M.; Inoue, N.; Shindoh, M.; Hida, K. Hypoxia-induced reactive oxygen species cause chromosomal abnormalities in endothelial cells in the tumor microenvironment. *PLoS ONE* **2013**, *8*, e80349. [CrossRef] [PubMed]

138. Zhang, C.; Cao, S.; Toole, B.P.; Xu, Y. Cancer may be a pathway to cell survival under persistent hypoxia and elevated ROS: A model for solid-cancer initiation and early development. *Int. J. Cancer* **2015**, *136*, 2001–2011. [CrossRef] [PubMed]

139. Gentles, A.J.; Newman, A.M.; Liu, C.L.; Bratman, S.V.; Feng, W.; Kim, D.; Nair, V.S.; Xu, Y.; Khuong, A.; Hoang, C.D.; et al. The prognostic landscape of genes and infiltrating immune cells across human cancers. *Nat. Med.* **2015**, *21*, 938–945. [CrossRef] [PubMed]

140. Pires, I.M.; Bencokova, Z.; Milani, M.; Folkes, L.K.; Li, J.L.; Stratford, M.R.; Harris, A.L.; Hammond, E.M. Effects of acute versus chronic hypoxia on DNA damage responses and genomic instability. *Cancer Res.* **2010**, *70*, 925–935. [CrossRef] [PubMed]

141. Graeber, T.G.; Osmanian, C.; Jacks, T.; Housman, D.E.; Koch, C.J.; Lowe, S.W.; Giaccia, A.J. Hypoxia-mediated selection of cells with diminished apoptotic potential in solid tumours. *Nature* **1996**, *379*, 88–91. [CrossRef] [PubMed]

142. Bindra, R.S.; Schaffer, P.J.; Meng, A.; Woo, J.; Maseide, K.; Roth, M.E.; Lizardi, P.; Hedley, D.W.; Bristow, R.G.; Glazer, P.M. Down-regulation of Rad51 and decreased homologous recombination in hypoxic cancer cells. *Mol. Cell. Biol.* **2004**, *24*, 8504–8518. [CrossRef] [PubMed]

143. Tsuchimoto, T.; Sakata, K.; Someya, M.; Yamamoto, H.; Hirayama, R.; Matsumoto, Y.; Furusawa, Y.; Hareyama, M. Gene expression associated with DNA-dependent protein kinase activity under normoxia, hypoxia, and reoxygenation. *J. Radiat. Res.* **2011**, *52*, 464–471. [CrossRef] [PubMed]

144. Mihaylova, V.T.; Bindra, R.S.; Yuan, J.; Campisi, D.; Narayanan, L.; Jensen, R.; Giordano, F.; Johnson, R.S.; Rockwell, S.; Glazer, P.M. Decreased expression of the DNA mismatch repair gene Mlh1 under hypoxic stress in mammalian cells. *Mol.Cell. Biol* **2003**, *23*, 3265–3273. [CrossRef] [PubMed]

145. Sato, H.; Niimi, A.; Yasuhara, T.; Permata, T.B.M.; Hagiwara, Y.; Isono, M.; Nuryadi, E.; Sekine, R.; Oike, T.; Kakoti, S.; et al. DNA double-strand break repair pathway regulates PD-L1 expression in cancer cells. *Nat. Commun.* **2017**, *8*, 1751. [CrossRef] [PubMed]

146. Schumacher, T.N.; Schreiber, R.D. Neoantigens in cancer immunotherapy. *Science* **2015**, *348*, 69–74. [CrossRef] [PubMed]

147. Lu, Y.C.; Robbins, P.F. Cancer immunotherapy targeting neoantigens. *Semin. Immunol.* **2016**, *28*, 22–27. [CrossRef] [PubMed]

148. Germano, G.; Lamba, S.; Rospo, G.; Barault, L.; Magri, A.; Maione, F.; Russo, M.; Crisafulli, G.; Bartolini, A.; Lerda, G.; et al. Inactivation of DNA repair triggers neoantigen generation and impairs tumour growth. *Nature* **2017**, *552*, 116–120. [CrossRef] [PubMed]

149. Chae, Y.K.; Anker, J.F.; Bais, P.; Namburi, S.; Giles, F.J.; Chuang, J.H. Mutations in DNA repair genes are associated with increased neo-antigen load and activated T cell infiltration in lung adenocarcinoma. *Oncotarget* **2018**, *9*, 7949–7960. [CrossRef] [PubMed]

150. Pattabiraman, D.R.; Weinberg, R.A. Tackling the cancer stem cells—What challenges do they pose? *Nat. Rev. Drug Discov.* **2014**, *13*, 497–512. [CrossRef] [PubMed]

151. Jahanban-Esfahlan, R.; de la Guardia, M.; Ahmadi, D.; Yousefi, B. Modulating tumor hypoxia by nanomedicine for effective cancer therapy. *J. Cell. Physiol.* **2018**, *233*, 2019–2031. [CrossRef] [PubMed]

152. Liu, Y.; Liu, W.; Li, H.; Yan, W.; Yang, X.; Liu, D.; Wang, S.; Zhang, J. Two-photon fluorescent probe for detection of nitroreductase and hypoxia-specific microenvironment of cancer stem cell. *Anal. Chim. Acta* **2018**, *1024*, 177–186. [CrossRef] [PubMed]

153. Wigerup, C.; Pahlman, S.; Bexell, D. Therapeutic targeting of hypoxia and hypoxia-inducible factors in cancer. *Pharmacol. Ther.* **2016**, *164*, 152–169. [CrossRef] [PubMed]

154. Wilson, W.R.; Hay, M.P. Targeting hypoxia in cancer therapy. *Nat. Rev. Cancer* **2011**, *11*, 393–410. [CrossRef] [PubMed]

155. Baran, N.; Konopleva, M. Molecular pathways: Hypoxia-activated prodrugs in cancer therapy. *Clin. Cancer Res.* **2017**, *23*, 2382–2390. [CrossRef] [PubMed]

156. Cao, S.; Cripps, A.; Wei, M.Q. New strategies for cancer gene therapy: Progress and opportunities. *Clin. Exp. Pharmacol. Physiol.* **2010**, *37*, 108–114. [CrossRef] [PubMed]

157. Lee, K.; Zhang, H.; Qian, D.Z.; Rey, S.; Liu, J.O.; Semenza, G.L. Acriflavine inhibits HIF-1 dimerization, tumor growth, and vascularization. *Proc. Natl. Acad. Sci. USA* **2009**, *106*, 17910–17915. [CrossRef] [PubMed]

158. Martinez-Saez, O.; Gajate Borau, P.; Alonso-Gordoa, T.; Molina-Cerrillo, J.; Grande, E. Targeting HIF-2 α in clear cell renal cell carcinoma: A promising therapeutic strategy. *Crit. Rev. Oncol. Hematol.* **2017**, *111*, 117–123. [CrossRef] [PubMed]

159. Kopecka, J.; Porto, S.; Lusa, S.; Gazzano, E.; Salzano, G.; Giordano, A.; Desiderio, V.; Ghigo, D.; Caraglia, M.; De Rosa, G.; et al. Self-assembling nanoparticles encapsulating zoledronic acid revert multidrug resistance in cancer cells. *Oncotarget* **2015**, *6*, 31461–31478. [CrossRef] [PubMed]

160. Kopecka, J.; Porto, S.; Lusa, S.; Gazzano, E.; Salzano, G.; Pinzon-Daza, M.L.; Giordano, A.; Desiderio, V.; Ghigo, D.; De Rosa, G.; et al. Zoledronic acid-encapsulating self-assembling nanoparticles and doxorubicin: A combinatorial approach to overcome simultaneously chemoresistance and immunoresistance in breast tumors. *Oncotarget* **2016**, *7*, 20753–20772. [CrossRef] [PubMed]

161. Zhang, H.; Qian, D.Z.; Tan, Y.S.; Lee, K.; Gao, P.; Ren, Y.R.; Rey, S.; Hammers, H.; Chang, D.; Pili, R.; et al. Digoxin and other cardiac glycosides inhibit HIF-1α synthesis and block tumor growth. *Proc. Natl. Acad. Sci. USA* **2008**, *105*, 19579–19586. [CrossRef] [PubMed]

162. Schito, L.; Rey, S.; Tafani, M.; Zhang, H.; Wong, C.C.; Russo, A.; Russo, M.A.; Semenza, G.L. Hypoxia-inducible factor 1-dependent expression of platelet-derived growth factor B promotes lymphatic metastasis of hypoxic breast cancer cells. *Proc. Natl. Acad. Sci. USA* **2012**, *109*, E2707–E2716. [CrossRef] [PubMed]

163. Beppu, K.; Nakamura, K.; Linehan, W.M.; Rapisarda, A.; Thiele, C.J. Topotecan blocks hypoxia-inducible factor-1α and vascular endothelial growth factor expression induced by insulin-like growth factor-I in neuroblastoma cells. *Cancer Res.* **2005**, *65*, 4775–4781. [CrossRef] [PubMed]

164. Rapisarda, A.; Uranchimeg, B.; Sordet, O.; Pommier, Y.; Shoemaker, R.H.; Melillo, G. Topoisomerase I-mediated inhibition of hypoxia-inducible factor 1: Mechanism and therapeutic implications. *Cancer Res.* **2004**, *64*, 1475–1482. [CrossRef] [PubMed]

165. Rapisarda, A.; Uranchimeg, B.; Scudiero, D.A.; Selby, M.; Sausville, E.A.; Shoemaker, R.H.; Melillo, G. Identification of small molecule inhibitors of hypoxia-inducible factor 1 transcriptional activation pathway. *Cancer Res.* **2002**, *62*, 4316–4324. [PubMed]

166. Koh, M.Y.; Spivak-Kroizman, T.; Venturini, S.; Welsh, S.; Williams, R.R.; Kirkpatrick, D.L.; Powis, G. Molecular mechanisms for the activity of PX-478, an antitumor inhibitor of the hypoxia-inducible factor-1α. *Mol. Cancer Ther.* **2008**, *7*, 90–100. [CrossRef] [PubMed]

167. Welsh, S.; Williams, R.; Kirkpatrick, L.; Paine-Murrieta, G.; Powis, G. Antitumor activity and pharmacodynamic properties of PX-478, an inhibitor of hypoxia-inducible factor-1α. *Mol. Cancer Ther.* **2004**, *3*, 233–244. [PubMed]

168. Liang, S.; Medina, E.A.; Li, B.; Habib, S.L. Preclinical evidence of the enhanced effectiveness of combined rapamycin and AICAR in reducing kidney cancer. *Mol. Oncol.* **2018**. [CrossRef] [PubMed]

169. Peng, W.; Zhang, S.; Zhang, Z.; Xu, P.; Mao, D.; Huang, S.; Chen, B.; Zhang, C.; Zhang, S. Jianpi Jiedu decoction, a traditional Chinese medicine formula, inhibits tumorigenesis, metastasis, and angiogenesis through the mTOR/HIF-1α/VEGF pathway. *J. Ethnopharmacol.* **2018**, *224*, 140–148. [CrossRef] [PubMed]

170. Xing, Y.; Mi, C.; Wang, Z.; Zhang, Z.H.; Li, M.Y.; Zuo, H.X.; Wang, J.Y.; Jin, X.; Ma, J. Fraxinellone has anticancer activity in vivo by inhibiting programmed cell death-ligand 1 expression by reducing hypoxia-inducible factor-1α and STAT3. *Pharmacol. Res.* **2018**, *135*, 166–180. [CrossRef] [PubMed]

171. Feliz, L.R.; Tsimberidou, A.M. Anti-vascular endothelial growth factor therapy in the era of personalized medicine. *Cancer Chemother. Pharmacol.* **2013**, *72*, 1–12. [CrossRef] [PubMed]

172. Chen, W.; Hill, H.; Christie, A.; Kim, M.S.; Holloman, E.; Pavia-Jimenez, A.; Homayoun, F.; Ma, Y.; Patel, N.; Yell, P.; et al. Targeting renal cell carcinoma with a HIF-2 antagonist. *Nature* **2016**, *539*, 112–117. [CrossRef] [PubMed]

173. Cho, H.; Du, X.; Rizzi, J.P.; Liberzon, E.; Chakraborty, A.A.; Gao, W.; Carvo, I.; Signoretti, S.; Bruick, R.K.; Josey, J.A.; et al. On-target efficacy of a HIF-2α antagonist in preclinical kidney cancer models. *Nature* **2016**, *539*, 107–111. [CrossRef] [PubMed]

174. Cuvillier, O. The therapeutic potential of HIF-2 antagonism in renal cell carcinoma. *Transl. Androl. Urol.* **2017**, *6*, 131–133. [CrossRef] [PubMed]

International Journal of
Molecular Sciences

MDPI

Article

The Interaction of Selenium with Chemotherapy and Radiation on Normal and Malignant Human Mononuclear Blood Cells

Richard J. Lobb [1], Gregory M. Jacobson [2], Ray T. Cursons [2] and Michael B. Jameson [3,4,*]

[1] Tumour Microenvironment Laboratory, QIMR Berghofer Medical Research Institute,
Herston, QLD 4006, Australia; Richard.Lobb@qimrberghofer.edu.au
[2] Department of Biological Sciences, University of Waikato, Hamilton 3216, New Zealand;
greg.jacobson@waikato.ac.nz (G.M.J.); ray.cursons@waikato.ac.nz (R.T.C.)
[3] Oncology Department, Waikato Hospital, Hamilton 3204, New Zealand
[4] Waikato Clinical Campus, Faculty of Medical and Health Sciences, University of Auckland,
Hamilton 3204, New Zealand
* Correspondence: michael.jameson@waikatodhb.health.nz; Tel.: +64-7-839-8750

Received: 31 August 2018; Accepted: 11 October 2018; Published: 15 October 2018

Abstract: Selenium, a trace element with anticancer properties, can reduce harmful toxicities of chemotherapy and radiotherapy without compromising efficacy. However, the dose-response relationship in normal versus malignant human cells is unclear. We evaluated how methylseleninic acid (MSA) modulates the toxicity and efficacy of chemotherapy and radiation on malignant and non-malignant human mononuclear blood cells in vitro. We specifically investigated its effects on endoplasmic reticulum stress induction, intracellular glutathione concentration, DNA damage and viability of peripheral blood mononuclear cells and THP1 monocytic leukaemia cells in response to radiation, cytosine arabinoside or doxorubicin chemotherapy. MSA, at lower concentrations, induced protective responses in normal cells but cytotoxic effects in malignant cells, alone and in conjunction with chemotherapy or radiation. However, in normal cells higher concentrations of MSA were directly toxic and increased the cytotoxicity of radiation but not chemotherapy. In malignant cells higher MSA concentrations were generally more effective in combination with cancer treatments. Thus, optimal MSA concentrations differed between normal and malignant cells and treatments. This work supports clinical reports that selenium can significantly reduce dose-limiting toxicities of anticancer therapies and potentially improve efficacy of anticancer treatments. The optimal selenium compound and dose is not yet determined.

Keywords: selenium; glutathione; malignant; viability; DNA damage; ER stress

1. Introduction

Selenium (Se) is an essential trace element that is extensively studied in the prevention of numerous malignancies [1], although the majority of research on Se has focused on providing adequate nutritional intake in populations that have inherently low Se intake [2]. However, substantial preclinical data suggests that Se compounds, in supranutritional doses, have direct anticancer effects, mediated by various mechanisms including oxidative capability and modulation of immunological responses, angiogenesis, protein confirmation and DNA repair pathways [3,4]. These same mechanisms allow selenium compounds to act in synergy with cancer therapies and increase the efficacy of these treatments while reducing their normal tissue toxicities, as reviewed by Evans et al. [4]. Se compounds, when added to chemotherapy, resulted in improved tumour response rates and cures in human tumour xenograft animal models and reduced organ-specific toxicity [5–7]. Some aspects of these findings have been replicated in clinical trials, with various Se compounds ameliorating the toxicity of chemotherapy or radiotherapy [8–17], although the trials were not powered to evaluate overall treatment efficacy. These promising results argue for the initiation of larger clinical trials that can definitively assess the contributions of Se compounds to modulating both efficacy and toxicity of chemotherapy and radiation [4,18].

There is unquestionably a major unmet need in this regard. Despite many advances in supportive care, the toxicities of chemotherapy and radiotherapy still limit their efficacy, utility and acceptability to patients and clinicians, and result in poor quality of life for patients, treatment-related deaths and inadequate outcomes [19]. Apart from antiemetics and haemopoietic growth factors, few agents substantially prevent these toxicities, many are poorly-tolerated, and some reduce toxicities while compromising anticancer efficacy [20–22]. In contrast, Se compounds offer the potential, at optimum doses, of being well-tolerated agents that can improve both cancer outcomes and treatment toxicities.

In one study, Se-methyl-selenocysteine was more effective and dose-potent than seleno-l-methionine or sodium selenite in reducing cytotoxic chemotherapy-related mortality and augmenting its anticancer activity [6]. This may relate to the in vivo ability of Se-methyl-selenocysteine to directly generate methylselenol, a compound that is considered the active moiety for the observed effects of Se compounds in cancer cells [23–27]. In preclinical models Se-methyl-selenocysteine dosed at 0.2 mg/mouse/day optimises the mechanisms that mediate protection of normal tissues while enhancing tumour cytotoxicity [5,6,28]. In humans, however, this dose-response relationship has not been well-characterised, and thus the optimal type and dose of Se for use in clinical trials has not yet been determined [4]. Therefore, there is a need to provide a framework for characterising the divergent biological effects of Se in normal and malignant cells in humans, to inform future trials evaluating Se compounds in conjunction with anticancer treatments.

This investigation was undertaken to evaluate whether peripheral blood mononuclear cells (PBMCs) from healthy blood donors and a comparable malignant human cell line, THP1 monocytic leukaemia, could serve as an in vitro model to investigate the differential effects of Se on normal and malignant human mononuclear cells. Se has been previously shown to enhance apoptosis through the induction of endoplasmic reticulum (ER) stress in cancer cells [29], therefore we evaluated the induction of ER stress in both normal and malignant cells in response to Se treatment. Given that ER stress signalling can be induced in response to oxidative triggers, we also investigated the impact of Se on intracellular glutathione levels [30–32], a key component in maintaining redox homeostasis in the cell, and how this influences DNA damage and viability of normal and malignant cells to cytotoxic chemotherapy or radiation [33–35].

Instead of Se-methyl-selenocysteine, which does not generate methylselenol in vitro, we used methylseleninic acid (MSA), which directly provides methylselenol through non-enzymatic reduction, and enabled us to directly evaluate the impact of this active metabolite of Se compounds [25,27]. We used MSA at Se concentrations (2.5, 5 and 15 µM) that could be achieved in plasma in subsequent clinical trials, and were comparable to plasma levels in mice at effective doses [6]. MSA was used alone or in combination with cytotoxic chemotherapy drugs or gamma radiation to evaluate their interactions in normal and malignant cells.

We demonstrate that Se has divergent effects in normal and malignant human mononuclear cells, protecting normal cells from chemotherapy and radiation toxicity while enhancing their therapeutic effects against malignant cells. In this model we were also able to use analytical methods to demonstrate changes in biological pathways that mediate these effects of Se compounds, which could be incorporated into future clinical trials.

Int. J. Mol. Sci. **2018**, *19*, 3167

2. Results

2.1. Methylseleninic Acid (MSA) Induces Endoplasmic Reticulum (ER) Stress in Normal and Malignant Cells But Differentially Modulates Apoptosis

To investigate the induction of ER stress in normal and malignant cells we measured the cellular expression of 78 kDa glucose-regulated protein (GRP78) and phosphorylated eukaryotic initiation factor 2-alpha (phospho-EIF2α), and splicing of X-box binding protein 1 (XBP1), in response to exposure to increasing concentrations of MSA for 6 h. MSA induced ER stress in both normal and malignant cells, which was seen through an increase in the expression of GRP78, as well as an increase in the splicing of XBP1 (spliced: S-XBP1; unspliced: U-XBP1) and phosphorylation of EIF2α (Figure 1). Interestingly, when we assessed the effect of MSA on the apoptotic response induced by ER stress we found different patterns between normal and cancer cells (Figure 1). Caspase-8 was down-regulated by MSA in a concentration-dependent manner in normal PBMCs yet was upregulated in malignant THP1 cells at the same concentrations, with the maximal differential impact between normal and malignant cells at 5 μM MSA (Figure 1).

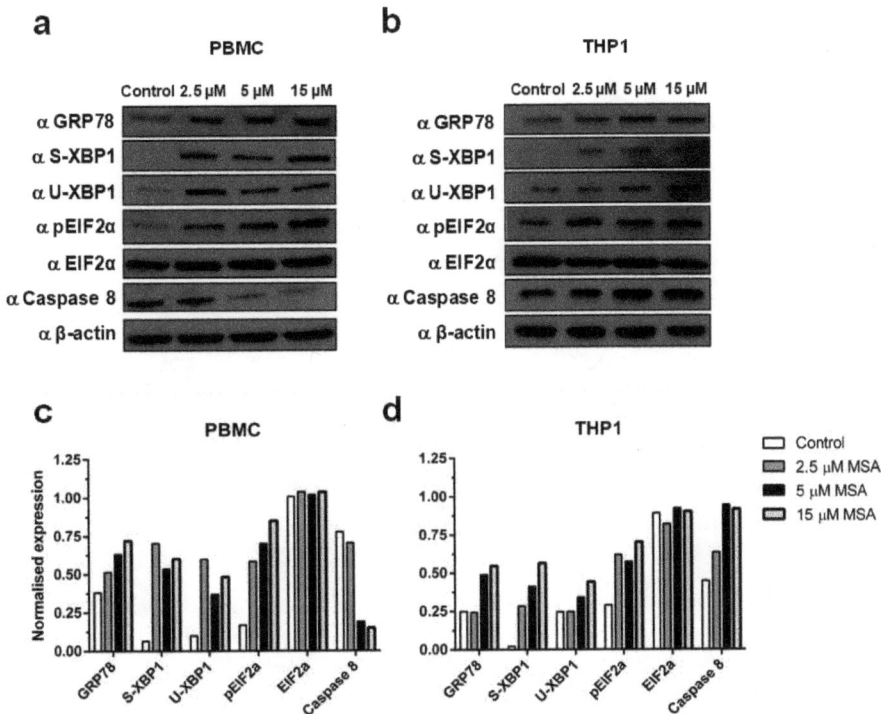

Figure 1. Selenium induces endoplasmic reticulum (ER) stress response in normal and malignant cells. (**a**) Concentration-dependent increase in ER stress proteins and decrease in caspase-8 in peripheral blood mononuclear cells (PBMCs) with 2.5, 5 and 15 μM methylseleninic acid (MSA) at 6 h; (**b**) Concentration-dependent increase in both ER stress proteins and caspase-8 in THP1 cells; (**c,d**) Quantification of protein expression in PBMC and THP1 cells.

2.2. MSA Has a Divergent Impact on Glutathione (GSH) Levels in Normal and Malignant Cells

To investigate the link between ER stress and generated oxidative stress we measured intracellular total GSH levels in normal and malignant cells. At 6 h we observed differential effects of MSA in normal and malignant cells (Figure 2a). MSA significantly increased total GSH levels in PBMC (Figure 2a) after 6 h in a concentration-dependent manner (a protective response). Conversely, THP1 cells had a baseline GSH level approximately 40-fold higher than PBMCs that was significantly reduced by MSA in a concentration-dependent manner after 6 h (Figure 2b).

We then tested the duration of the MSA-induced alteration on GSH levels in normal and malignant cells. The increase in GSH observed in PBMCs after 6 h of MSA treatment at 2.5 and 5 μM was maintained at 24 h but returned to baseline levels at 48 h (Figure 2c). However, at 15 μM MSA, the GSH concentration was less elevated at 24 h than at 6 h and also returned to baseline levels at 48 h. In THP1 cells, the depletion of GSH at 24 h was still significant but not concentration-dependent, whereas at 48 h the return of GSH levels towards baseline values was greater with 2.5 and 5 μM compared with 15 μM MSA (Figure 2d).

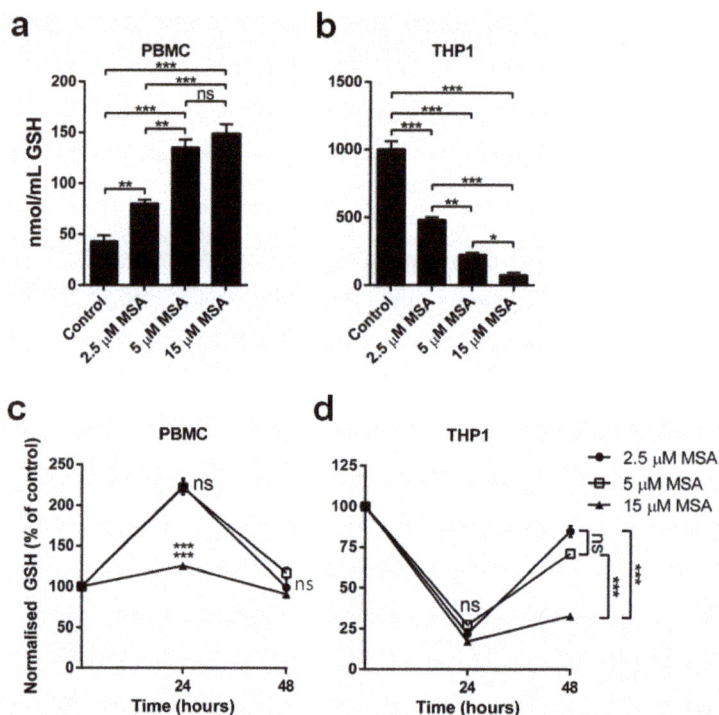

Figure 2. MSA has divergent impact on glutathione (GSH) levels in normal and malignant cells. (**a,b**) GSH quantification in PBMC and THP1 cells demonstrates that MSA significantly reduces GSH levels in THP1 cells and significantly increases GSH levels in PBMCs after 6 h ($n = 5$, ± SEM); (**c,d**) Timeline of GSH levels in PBMCs and THP1 cells after MSA treatments demonstrates GSH alterations are maintained for up to 24 h. $n = 3$, ± SEM, * $p < 0.05$, ** $p < 0.01$, *** $p < 0.001$, ns, not significant.

Next, we investigated if the MSA-induced GSH response in cells was maintained at 24 h after radiation and chemotherapy treatment. The GSH increase in normal PBMCs was maintained at 24 h when cells were also treated with 2 Gy radiation, cytosine arabinoside (AraC) or doxorubicin (Dox), though the maximum benefit was achieved with 2.5 μM MSA (Figure 3a–c). Furthermore, the depletion of GSH by MSA in malignant THP1 cells was still significantly reduced at 24 h after radiation and chemotherapy treatment, again without the advantage of higher MSA concentrations (Figure 3a–c).

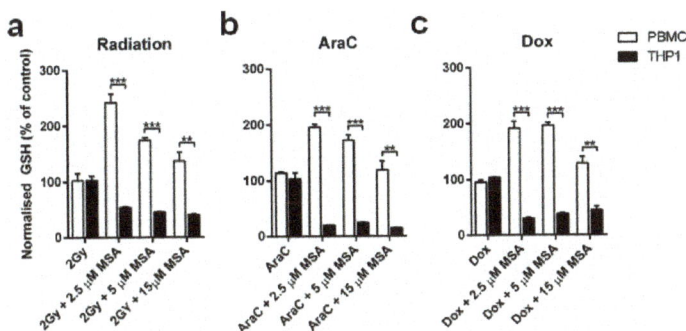

Figure 3. MSA-induced GSH alterations are maintained in the presence of therapeutic treatments. (a–c) GSH levels are significantly elevated in PBMCs at 24 h after radiation, AraC or Dox treatment, whereas GSH levels are significantly reduced in THP1 cells 24 h after treatment. $n = 3$, $\pm$ SEM, ** $p < 0.01$, *** $p < 0.001$.

2.3. MSA Reduces DNA Damage in Normal Cells While Increasing DNA Damage in Malignant Cells

Given the divergent effects of MSA on apoptosis induction and GSH expression in normal and malignant cells, we investigated if MSA would protect normal cells from DNA damage due to radiation or chemotherapy, while potentiating the DNA-damaging efficacy of these treatments in malignant cells. Using the comet assay (Figure 4a), this differential effect was pronounced with chemotherapy but not radiation. Treatment with MSA alone at the highest concentration, 15 µM, slightly increased DNA damage levels in normal cells but not in malignant cells, though the lower concentrations had no such effect (Figure 4b).

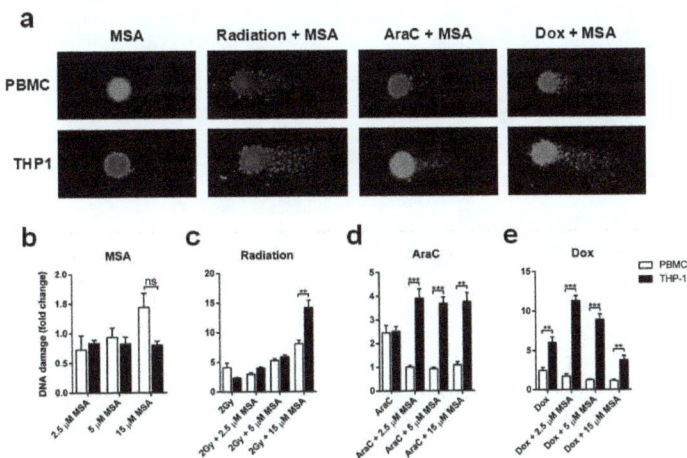

Figure 4. Selenium protects normal cells from DNA damage while enhancing DNA damage in malignant cells. (a) Representative image of comet assay (400× magnification) in PBMCs and THP1 cells treated with MSA alone or MSA in combination with radiation, AraC or Dox; (b) Quantification of DNA damage in PBMCs and THP1 cells: treatment with MSA 15 µM, but not lower concentrations, slightly increased DNA damage levels in normal but not in malignant cells; (c) DNA damage is increased in both PBMCs and THP1 cells exposed to 2 Gy radiation; (d,e) MSA is significantly protective against DNA damage in PBMCs while significantly increasing DNA damage in THP1 cells treated with AraC or Dox. $n = 3$, $\pm$ SEM, ** $p < 0.01$, *** $p < 0.001$, ns, not significant.

As expected, when PBMCs and THP1 cells were exposed to 2 Gy radiation DNA damage was elevated compared to controls (Figure 4c). While MSA at 2.5 µM reduced radiation-induced DNA damage in PBMC but not THP1 cells, higher concentrations of MSA progressively increased radiation-induced DNA damage in both normal and malignant cells (Figure 4c). DNA damage was significantly increased in PBMCs and THP1 cells when treated with AraC, however adding MSA protected the normal cells while increasing DNA damage in the malignant cells, without a clear concentration dependency (Figure 4d). Dox-induced DNA damage in PBMCs was not potentiated by MSA, while in contrast MSA increased Dox-induced DNA damage in THP1 cells (Figure 4e). However, this effect on THP1 cells was maximal at 2.5 µM MSA, and diminished at higher concentrations (Figure 4e).

2.4. MSA Treatment Protects Normal Cells While Potentiating Cell Death in Malignant Cells

We next investigated if the differences in DNA damage culminated in differences in cell viability. MSA alone significantly reduced the viability of THP1 cells with increasing MSA concentrations compared to PBMCs (Figure 5a). Although 2 Gy radiation alone did not affect viability of THP1 cells, adding MSA to radiation significantly reduced THP1 cell viability (Figure 5b). In agreement with the DNA damage induced by radiation, the addition of MSA to this treatment further reduced the viability of PBMCs (Figure 5b). However, when we assessed the combination of MSA with AraC or Dox we found significant differences (Figure 5c,d). Treatment with MSA at all concentrations provided significant protection of PBMCs while progressively increasing toxicity in THP1 cells in response to AraC or Dox treatment (Figure 5c,d).

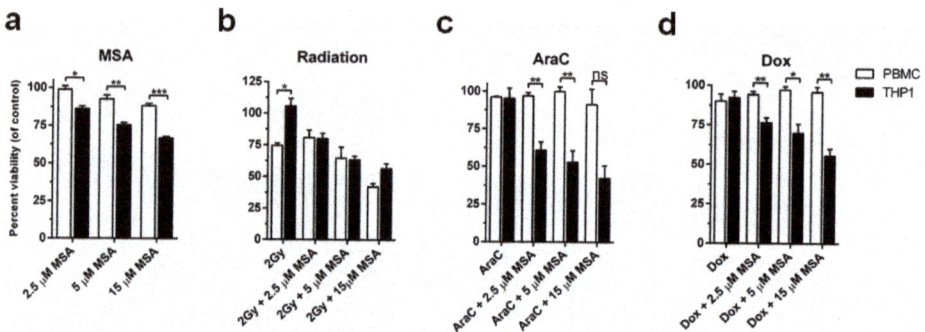

Figure 5. MSA protects normal cells and results in elevated cell killing of malignant cells after 48 h. (**a**) MSA treatment alone significantly reduces cell viability of malignant THP1 cells compared to normal PBMCs; (**b**) MSA does not significantly alter cell viability of PBMCs compared to THP1 cells in response to radiation; (**c,d**) MSA significantly protects normal PBMCs from cell death while enhancing the therapeutic activity of AraC or Dox in THP1 cells. $n = 3 \pm$ SEM, * $p < 0.05$, ** $p < 0.01$, *** $p < 0.001$, ns, not significant.

3. Discussion

The toxicity of anticancer therapies is a major ongoing clinical issue and developing agents that usefully modulate the toxicity and efficacy of chemotherapy and radiotherapy without compromising their efficacy is important. Preclinical work and some clinical trials suggest that Se compounds can achieve this, though the Se compounds and doses used have varied widely [4,6]. In the present study, we used an in vitro model of normal and malignant human mononuclear blood cells to investigate the dose-response relationship of Se in modulating the efficacy and toxicity of cancer treatments. We have shown important differences between normal and malignant cells in the dose-response relationship of Se to biological mechanisms that mediate cell survival and response to cancer treatments.

Se compounds have previously been shown to induce ER stress in a concentration- and time-dependent manner in prostate cancer cell lines, which leads to apoptosis in malignant cells [29]. In this study, we found that MSA induced apoptosis through caspase-8 expression in THP1 cells while reducing caspase-8 in PBMCs, in agreement with previous studies that have shown Se compounds induce apoptosis through caspase-8 activation [36]. Moreover, caspase-8-mediated apoptosis has been demonstrated to mediate the therapeutic synergy of Se compounds and chemotherapy treatment in various cancer settings [37,38]. The reduction in caspase-8 induced by MSA in PBMC in this study is consistent with the clinical data that Se compounds, at tested doses, are protective of normal tissues [10,16].

ER stress has been demonstrated to induce reactive oxygen species generation [39]. This results in the depletion of intracellular GSH, causing the cellular environment to become more oxidized, which is associated with increased apoptosis and necrosis [40–42]. GSH contributes to cellular resistance to anticancer treatments through covalent binding and inactivation of drugs [43–47]. Thus the 40-fold higher initial concentration of GSH present in THP-1 cells compared to PBMCs would protect the malignant cells against cytotoxic therapies, whereas the MSA-induced severe depletion of GSH in malignant cells shown in this study may contribute to the increased sensitivity to these treatments with MSA. These results are consistent with work showing that Se compounds inhibit the cisplatin-induced increase in GSH in ovarian cancer cells, thereby preventing chemoresistance [32]. These malignant cells may also have been sensitised to the effects of MSA due to their high concentrations of GSH, as GSH is a cofactor in the metabolic reduction of MSA to methylselenol [31]. This may not be relevant to Se compounds that generate methylselenol through other mechanisms.

Contrary to the effect of MSA seen in malignant cells, it induced a significant increase in GSH in normal cells, which is expected to protect them against cancer therapies. The simultaneous increase in GSH in normal cells and depletion of GSH in malignant cells may contribute to improving the therapeutic ratio of cancer treatment by reducing normal tissue toxicities while increasing the anticancer efficacy. This effect on GSH may mediate, at least in part, the observed ability of Se compounds to reduce the toxicity of chemotherapy and radiation in normal tissues [10,16].

Previous in vitro and in vivo studies have shown that DNA damage and inducible DNA damage is reduced with Se [33,34]. While we demonstrated that MSA reduced chemotherapy-induced DNA damage in normal cells, it was ineffective in protecting them against radiation-induced DNA damage and cytotoxicity. Of particular concern, at the highest concentration tested (15 μM), MSA significantly increased DNA damage from radiation in PBMC. In contrast, all concentrations of MSA increased the DNA damage and cytotoxicity of radiation and chemotherapy in the malignant cells.

This study supports previous work that demonstrated the potential therapeutic benefit of using Se in conjunction with cancer therapeutics, due to its differential effects on chemotherapy- or radiation-treated normal cells relative to malignant cells [48]. It is encouraging in this study that MSA generally protected normal cells while sensitising malignant cells to cytotoxic therapies, and that it informs about mechanisms that plausibly contribute to the reduction of clinically-significant toxicities seen in clinical trials with Se supplementation during cancer treatments [10–12,14].

A very important concern, however, has been raised by this study: in PBMCs the highest concentration of MSA proved toxic, and increased the cytotoxicity and DNA damage from radiation. This could increase the potential for second malignancies and other late complications of radiation, especially if using inorganic forms of Se that are associated with increased genotoxicity compared to several organic forms [49]. These outcomes have not been mentioned in clinical trials to date but the numbers evaluated have been small and follow-up is limited.

There is always a tension in cancer treatment between maximising efficacy while managing toxicities [19]. While Se has considerable and important potential to widen this usually narrow therapeutic window, data from this study strongly suggests that following the traditional cytotoxic therapy paradigm of using the maximum tolerated dose [50,51] may be inappropriate with Se compounds. However, our data also suggests that these interactions are treatment-specific, with greater vulnerability of normal cells when using the highest concentrations of MSA with radiation, but continued protection of normal cells from chemotherapy by MSA at all concentrations. Furthermore, there were marked differences in the concentration-dependence of the improved anticancer effects of chemotherapy or radiation with MSA on malignant THP1 cells. More modest concentrations of MSA proved equally effective to the highest one in terms of inducing ER stress and reducing GSH levels from radiation or chemotherapy, and at inducing DNA damage with chemotherapy drugs. The highest MSA concentration, however, was most effective at inducing DNA damage with radiation and at augmenting the cytotoxicity of radiation or chemotherapy.

4. Materials and Methods

4.1. Mononuclear Cell Isolation

PBMCs were isolated from buffy coats obtained from blood donations given by healthy individuals, and supplied by the New Zealand Blood Service at Waikato Hospital, Hamilton, New Zealand. Ethical approval for their use was granted by the Northern Y Health and Disability Ethics Committee (reference NTY/10/08/065/AM01, 16 August 2011). The mononuclear cell fraction was isolated via density gradient centrifugation using Histopaque® (St. Louis, MO, USA).

4.2. Cell Culture

PBMCs and THP1 cells were cultured in RPMI-1640 medium supplemented with 10% FBS, 1% penicillin (10,000 units/mL) and streptomycin (10,000 µg/mL) at 37 °C in 5% CO_2. Both cell lines were incubated either in the presence of MSA (2.5, 5 and 15 µM), cytosine arabinoside (AraC; 5 ng/mL), or doxorubicin (Dox; 20 nM) alone, as well as the combination of MSA and AraC or Dox. To assess the response to radiation, cells were irradiated with a total of 2 Gy with or without MSA. Cells were incubated with MSA for 6 h prior to treatment with chemotherapy or radiation.

4.3. Western Blot Analysis

Western blotting was carried out as previously described [52]. Briefly, total cell protein was isolated using radioimmunoprecipitation assay (RIPA) buffer (50 mM Tris pH 7.4, 150 mM NaCl, 1% Triton-X-100, 1% Na-deoxycholate, 0.1% sodium dodecyl sulfate (SDS), 1 mM ethylenediaminetetraacetic acid (EDTA), phosphatase inhibitors and protease cocktail inhibitors (Sigma Aldrich, St. Louis, MO, USA), and 1 mM phenylmethanesulfonylfluoride). Proteins were resolved by SDS-polyacriliamide gel (PAGE), transferred to nitrocellulose membranes, blocked in 5% non-fat powdered milk in tris-buffered saline-tween (TBS-T) and probed with antibodies. Protein bands were detected using a FUJIFILM Intelligent dark box II LAS-1000 system.

4.4. Measurement of GSH

Glutathione (GSH) assay kit was purchased from Sigma-Aldrich (St. Louis, MO, USA). The assay was carried out according to the manufacturer's directions. Total GSH was determined using a kinetic assay that measures the reduction of 5,5′-dithiobis-(2-nitrobenzoic) acid (DTNB) to 5-thio-2-nitrobenzoic acid (TNB) at 412 nm.

4.5. Comet Assay

DNA damage was assessed with the comet assay as previously described [53]. Preparation of slides was carried out by coating a pre-agarose-coated slide (1% normal melting point in PBS), with approximately 1000 cells in 0.5% low melting point agarose in PBS. Slides were lysed at 4 °C in a solution containing 1% Triton X-100, 2.5 M NaCl, 100 mM EDTA, and 10 mM Tris pH 10.0 for two hours. Slides were incubated for 20 min in an alkaline buffer (300 mM NaOH and 1 mM EDTA (pH > 13)) and electrophoresed for 20 min at 20 V and 300 mA at 4 °C in the same buffer. Slides were then neutralized and dried in 70% ethanol before being stained with SYBR Gold (Thermo Fisher Scientific, Waltham, MA, USA) and scored using the tail moment [53].

4.6. MTT Assay

Viability was measured with a tetrazolium salt as previously described [52]. The MTT (methyl-thiazol-tetrazolium) assay was used to assess the impact of treatments on cell viability in THP1 cells and PBMCs after 48 h. Cells were incubated with MTT for two hours, lysed in 20% SDS (w/v), 50% dimethylformamide (v/v) pH 4.7, and the absorbance was measured at 570 nm.

4.7. Statistical Analysis

GraphPad Prism version 6.0 (La Jolla, CA, USA) was used for all calculations. Multiple comparisons were controlled for using the Sidak-Bonferroni method. All experiments were performed at a minimum of 3 independent repeats. Differences with p-values less than 0.05 were considered significant.

5. Conclusions

Overall this study suggests that doses of Se compounds that achieve plasma Se concentrations in the range of 2.5–5 µM might achieve the optimal balance between enhancing efficacy and reducing the toxicity of radiation. It is possible that higher doses of Se might safely be used with some chemotherapy drugs. It is noteworthy that dosing to achieve plasma Se levels determined by this in vitro study would not apply to seleno-l-methionine, as it is non-specifically incorporated into the general protein pool, especially albumin, which gives disproportionately high plasma Se levels compared to dosing with equivalent elemental Se doses of sodium selenite or Se-methylselenocysteine [54,55].

The potential of Se to improve the efficacy and reduce toxicities of cancer treatments is important and deserves careful examination in clinical trials. However, when designing these trials, we need to be cognizant of the genotoxicity dose-dependence of the Se compounds to be used, with the potential for increased serious late toxicities of cancer treatments such as secondary malignancies, and evaluate this in our trials. Importantly, this study has demonstrated several laboratory methods that can be incorporated into clinical trials to enable investigators to evaluate the pharmacokinetic-pharmacodynamic relationship of the Se compounds being used in cancer patients. This will assist us in rationally determining the optimal dose and form of Se for use in combination with various cancer treatments in clinical trials; such trials are already underway [4,28].

Author Contributions: Study conception and design, R.J.L., R.T.C. and M.B.J.; methodology, R.J.L., G.M.J., R.T.C. and M.B.J.; acquisition of data: R.J.L. and G.M.J.; analysis and interpretation of data: R.J.L., G.M.J. and M.B.J.; writing—original draft preparation, R.J.L. and M.B.J.; writing—review and editing, R.J.L., G.M.J., R.T.C. and M.B.J.; funding acquisition, M.B.J.

Funding: This research was funded by Cycle for Life (Thames, New Zealand), Waikato Clinical School Summer Studentship and the Waikato/Bay of Plenty Division of the Cancer Society of New Zealand.

Acknowledgments: We are grateful to blood donors and the New Zealand Blood Service for provision of the buffy coats.

Conflicts of Interest: The authors declare no conflict of interest. The funders had no role in the design of the study; in the collection, analyses, or interpretation of data; in the writing of the manuscript, or in the decision to publish the results.

Abbreviations

AraC	Cytosine arabinoside
Dox	Doxorubicin
GSH	Glutathione
MSA	Methylseleninic acid
PBMC	Peripheral blood mononuclear cell

References

1. Rayman, M.P. Selenium in cancer prevention: A review of the evidence and mechanism of action. *Proc. Nutr. Soc.* **2005**, *64*, 527–542. [CrossRef] [PubMed]
2. Rayman, M.P. Selenium and human health. *Lancet* **2012**, *379*, 1256–1268. [CrossRef]
3. Lipinski, B. Sodium Selenite as an Anticancer Agent. *Anticancer Agents Med. Chem.* **2017**, *17*, 658–661. [CrossRef] [PubMed]
4. Evans, S.O.; Khairuddin, P.F.; Jameson, M.B. Optimising Selenium for Modulation of Cancer Treatments. *Anticancer Res.* **2017**, *37*, 6497–6509. [PubMed]
5. Cao, S.; Durrani, F.A.; Rustum, Y.M. Selective Modulation of the Therapeutic Efficacy of Anticancer Drugs by Selenium Containing Compounds against Human Tumor Xenografts. *Clin. Cancer Res.* **2004**, *10*, 2561–2569. [CrossRef] [PubMed]
6. Fakih, M.; Cao, S.; Durrani, F.A.; Rustum, Y.M. Selenium protects against toxicity induced by anticancer drugs and augments antitumor activity: A highly selective, new, and novel approach for the treatment of solid tumors. *Clin. Colorectal Cancer* **2005**, *5*, 132–135. [CrossRef] [PubMed]
7. Francescato, H.D.; Costa, R.S.; Camargo, S.M.R.; Zanetti, M.A.; Lavrador, M.A.; Bianchi, M.D. Effect of oral selenium administration on cisplatin-induced nephrotoxicity in rats. *Pharmacol. Res.* **2001**, *43*, 77–82. [CrossRef] [PubMed]
8. Song, M.; Kumaran, M.N.; Gounder, M.; Gibbon, D.G.; Nieves-Neira, W.; Vaidya, A.; Hellmann, M.; Kane, M.P.; Buckley, B.; Shih, W.; et al. Phase I trial of selenium plus chemotherapy in gynecologic cancers. *Gynecol. Oncol.* **2018**. [CrossRef] [PubMed]
9. Puspitasari, I.M.; Abdulah, R.; Yamazaki, C.; Kameo, S.; Nakano, T.; Koyama, H. Updates on clinical studies of selenium supplementation in radiotherapy. *Radiat. Oncol.* **2014**, *9*, 125. [CrossRef] [PubMed]
10. Sieja, K.; Talerczyk, M. Selenium as an element in the treatment of ovarian cancer in women receiving chemotherapy. *Gynecol. Oncol.* **2004**, *93*, 320–327. [CrossRef] [PubMed]
11. Asfour, I.A.; El Tehewi, M.M.; Ahmed, M.H.; Abdel-Sattar, M.A.; Moustafa, N.N.; Hegab, H.M.; Fathey, O.M. High-dose sodium selenite can induce apoptosis of lymphoma cells in adult patients with non-Hodgkin's lymphoma. *Biol. Trace Elem. Res.* **2009**, *127*, 200–210. [CrossRef] [PubMed]
12. Asfour, I.A.; Fayek, M.; Raouf, S.; Soliman, M.; Hegab, H.M.; El Desoky, H.; Saleh, R.; Moussa, M.A. The impact of high-dose sodium selenite therapy on Bcl-2 expression in adult non-Hodgkin's lymphoma patients: Correlation with response and survival. *Biol. Trace Elem. Res.* **2007**, *120*, 1–10. [CrossRef] [PubMed]
13. Asfour, I.A.; El Shazly, S.; Fayek, M.H.; Hegab, H.M.; Raouf, S.; Moussa, M.A. Effect of high-dose sodium selenite therapy on polymorphonuclear leukocyte apoptosis in non-Hodgkin's lymphoma patients. *Biol. Trace Elem. Res.* **2006**, *110*, 19–32. [CrossRef]
14. Hu, Y.J.; Chen, Y.; Zhang, Y.Q.; Zhou, M.Z.; Song, X.M.; Zhang, B.Z.; Luo, L.; Xu, P.M.; Zhao, Y.N.; Zhao, Y.B.; et al. The protective role of selenium on the toxicity of cisplatin-contained chemotherapy regimen in cancer patients. *Biol. Trace Elem. Res.* **1997**, *56*, 331–341. [CrossRef] [PubMed]
15. Jahangard-Rafsanjani, Z.; Gholami, K.; Hadjibabaie, M.; Shamshiri, A.R.; Alimoghadam, K.; Sarayani, A.; Mojtahedzadeh, M.; Ostadali-Dehaghi, M.; Ghavamzadeh, A. The efficacy of selenium in prevention of oral mucositis in patients undergoing hematopoietic SCT: A randomized clinical trial. *Bone Marrow Transplant.* **2013**, *48*, 832–836. [CrossRef] [PubMed]
16. Muecke, R.; Schomburg, L.; Glatzel, M.; Berndt-Skorka, R.; Baaske, D.; Reichl, B.; Buentzel, J.; Kundt, G.; Prott, F.J.; Devries, A.; et al. Multicenter, phase 3 trial comparing selenium supplementation with observation in gynecologic radiation oncology. *Int. J. Radiat. Oncol. Biol. Phys.* **2010**, *78*, 828–835. [CrossRef] [PubMed]
17. Buntzel, J.; Riesenbeck, D.; Glatzel, M.; Berndt-Skorka, R.; Riedel, T.; Mucke, R.; Kisters, K.; Schonekaes, K.G.; Schafer, U.; Bruns, F.; et al. Limited effects of selenium substitution in the prevention of radiation-associated toxicities. Results of a randomized study in head and neck cancer patients. *Anticancer Res.* **2010**, *30*, 1829–1832. [PubMed]
18. Muecke, R.; Micke, O.; Schomburg, L.; Buentzel, J.; Kisters, K.; Adamietz, I.A. Selenium in Radiation Oncology-15 Years of Experiences in Germany. *Nutrients* **2018**, *10*. [CrossRef] [PubMed]
19. Perry, M.C.; Doll, D.C.; Freter, C.E. *Perry's The Chemotherapy Souce Book*, 5th ed.; Lippincott, Williams & Wilkins: New York, NY, USA, 2012.

20. Mishra, K.; Alsbeih, G. Appraisal of biochemical classes of radioprotectors: Evidence, current status and guidelines for future development. *3Biotech* **2017**, *7*, 292. [CrossRef] [PubMed]

21. Devine, A.; Marignol, L. Potential of Amifostine for Chemoradiotherapy and Radiotherapy-associated Toxicity Reduction in Advanced NSCLC: A Meta-Analysis. *Anticancer Res.* **2016**, *36*, 5–12. [PubMed]

22. Freyer, D.R.; Chen, L.; Krailo, M.D.; Knight, K.; Villaluna, D.; Bliss, B.; Pollock, B.H.; Ramdas, J.; Lange, B.; Van, H.D.; et al. Effects of sodium thiosulfate versus observation on development of cisplatin-induced hearing loss in children with cancer (ACCL0431): A multicentre, randomised, controlled, open-label, phase 3 trial. *Lancet Oncol.* **2017**, *18*, 63–74. [CrossRef]

23. Liu, M.; Hu, C.; Xu, Q.; Chen, L.; Ma, K.; Xu, N.; Zhu, H. Methylseleninic acid activates Keap1/Nrf2 pathway via up-regulating miR-200a in human oesophageal squamous cell carcinoma cells. *Biosci. Rep.* **2015**, *35*. [CrossRef] [PubMed]

24. Lennicke, C.; Rahn, J.; Bukur, J.; Hochgrafe, F.; Wessjohann, L.A.; Lichtenfels, R.; Seliger, B. Modulation of MHC class I surface expression in B16F10 melanoma cells by methylseleninic acid. *Oncoimmunology* **2017**, *6*, e1259049. [CrossRef] [PubMed]

25. Kassam, S.; Goenaga-Infante, H.; Maharaj, L.; Hiley, C.T.; Juliger, S.; Joel, S.P. Methylseleninic acid inhibits HDAC activity in diffuse large B-cell lymphoma cell lines. *Cancer Chemother. Pharmacol.* **2011**, *68*, 815–821. [CrossRef] [PubMed]

26. Ip, C.; Dong, Y.; Ganther, H.E. New concepts in selenium chemoprevention. *Cancer Metastasis Rev.* **2002**, *21*, 281–289. [CrossRef] [PubMed]

27. Ip, C.; Thompson, H.J.; Zhu, Z.; Ganther, H.E. In vitro and in vivo studies of methylseleninic acid: Evidence that a monomethylated selenium metabolite is critical for cancer chemoprevention. *Cancer Res.* **2000**, *60*, 2882–2886. [PubMed]

28. Zakharia, Y.; Bhattacharya, A.; Rustum, Y.M. Selenium targets resistance biomarkers enhancing efficacy while reducing toxicity of anti-cancer drugs: Preclinical and clinical development. *Oncotarget* **2018**, *9*, 10765–10783. [CrossRef] [PubMed]

29. Wu, Y.; Zhang, H.; Dong, Y.; Park, Y.M.; Ip, C. Endoplasmic reticulum stress signal mediators are targets of selenium action. *Cancer Res.* **2005**, *65*, 9073–9079. [CrossRef] [PubMed]

30. Liu, C.; Liu, H.; Li, Y.; Wu, Z.; Zhu, Y.; Wang, T.; Gao, A.C.; Chen, J.; Zhou, Q. Intracellular glutathione content influences the sensitivity of lung cancer cell lines to methylseleninic acid. *Mol. Carcinog.* **2012**, *51*, 303–314. [CrossRef] [PubMed]

31. Shen, H.M.; Ding, W.X.; Ong, C.N. Intracellular glutathione is a cofactor in methylseleninic acid-induced apoptotic cell death of human hepatoma HEPG(2) cells. *Free Radic. Biol. Med.* **2002**, *33*, 552–561. [CrossRef]

32. Caffrey, P.B.; Frenkel, G.D. Selenium compounds prevent the induction of drug resistance by cisplatin in human ovarian tumor xenografts in vivo. *Cancer Chemother. Pharmacol.* **2000**, *46*, 74–78. [CrossRef] [PubMed]

33. Fischer, J.L.; Mihelc, E.M.; Pollok, K.E.; Smith, M.L. Chemotherapeutic selectivity conferred by selenium: A role for p53-dependent DNA repair. *Mol. Cancer Ther.* **2007**, *6*, 355–361. [CrossRef] [PubMed]

34. Seo, Y.R.; Sweeney, C.; Smith, M.L. Selenomethionine induction of DNA repair response in human fibroblasts. *Oncogene* **2002**, *21*, 3663–3669. [CrossRef] [PubMed]

35. Shin, S.H.; Yoon, M.J.; Kim, M.; Kim, J.I.; Lee, S.J.; Lee, Y.S.; Bae, S. Enhanced lung cancer cell killing by the combination of selenium and ionizing radiation. *Oncol. Rep.* **2007**, *17*, 209–216. [CrossRef] [PubMed]

36. Li, Z.; Carrier, L.; Rowan, B.G. Methylseleninic acid synergizes with tamoxifen to induce caspase-mediated apoptosis in breast cancer cells. *Mol. Cancer Ther.* **2008**, *7*, 3056–3063. [CrossRef] [PubMed]

37. Jiang, C.; Wang, Z.; Ganther, H.; Lu, J. Caspases as key executors of methyl selenium-induced apoptosis (anoikis) of DU-145 prostate cancer cells. *Cancer Res.* **2001**, *61*, 3062–3070. [PubMed]

38. Li, S.; Zhou, Y.; Wang, R.; Zhang, H.; Dong, Y.; Ip, C. Selenium sensitizes MCF-7 breast cancer cells to doxorubicin-induced apoptosis through modulation of phospho-Akt and its downstream substrates. *Mol. Cancer Ther.* **2007**, *6*, 1031–1038. [CrossRef] [PubMed]

39. Malhotra, J.D.; Kaufman, R.J. Endoplasmic reticulum stress and oxidative stress: A vicious cycle or a double-edged sword? *Antioxid. Redox Signal.* **2007**, *9*, 2277–2293. [CrossRef] [PubMed]

40. Cai, J.; Jones, D.P. Superoxide in apoptosis. Mitochondrial generation triggered by cytochrome c loss. *J. Biol. Chem.* **1998**, *273*, 11401–11404. [CrossRef] [PubMed]

41. Engel, R.H.; Evens, A.M. Oxidative stress and apoptosis: A new treatment paradigm in cancer. *Front. Biosci.* **2006**, *11*, 300–312. [CrossRef] [PubMed]

42. Voehringer, D.W.; Meyn, R.E. Redox aspects of Bcl-2 function. *Antioxid. Redox Signal.* **2000**, *2*, 537–550. [CrossRef] [PubMed]

43. Biaglow, J.E.; Varnes, M.E.; Epp, E.R.; Clark, E.P.; Tuttle, S.W.; Held, K.D. Role of glutathione and other thiols in cellular response to radiation and drugs. *Drug Metab. Rev.* **1989**, *20*, 1–12. [CrossRef] [PubMed]

44. Bump, E.A.; Brown, J.M. Role of glutathione in the radiation response of mammalian cells in vitro and in vivo. *Pharmacol. Ther.* **1990**, *47*, 117–136. [CrossRef]

45. Coleman, C.N.; Bump, E.A.; Kramer, R.A. Chemical modifiers of cancer treatment. *J. Clin. Oncol.* **1988**, *6*, 709–733. [CrossRef] [PubMed]

46. Mitchell, J.B.; Russo, A. The role of glutathione in radiation and drug induced cytotoxicity. *Br. J. Cancer Suppl.* **1987**, *8*, 96–104. [PubMed]

47. Tew, K.D. Glutathione-associated enzymes in anticancer drug resistance. *Cancer Res.* **1994**, *54*, 4313–4320. [CrossRef] [PubMed]

48. Vadgama, J.V.; Wu, Y.; Shen, D.; Hsia, S.; Block, J. Effect of selenium in combination with Adriamycin or Taxol on several different cancer cells. *Anticancer Res.* **2000**, *20*, 1391–1414. [PubMed]

49. Valdiglesias, V.; Pasaro, E.; Mendez, J.; Laffon, B. In vitro evaluation of selenium genotoxic, cytotoxic, and protective effects: A review. *Arch. Toxicol.* **2010**, *84*, 337–351. [CrossRef] [PubMed]

50. Brodin, O.; Eksborg, S.; Wallenberg, M.; Asker-Hagelberg, C.; Larsen, E.H.; Mohlkert, D.; Lenneby-Helleday, C.; Jacobsson, H.; Linder, S.; Misra, S.; et al. Pharmacokinetics and Toxicity of Sodium Selenite in the Treatment of Patients with Carcinoma in a Phase I Clinical Trial: The SECAR Study. *Nutrients* **2015**, *7*, 4978–4994. [CrossRef] [PubMed]

51. Corcoran, N.M.; Hovens, C.M.; Michael, M.; Rosenthal, M.A.; Costello, A.J. Open-label, phase I dose-escalation study of sodium selenate, a novel activator of PP2A, in patients with castration-resistant prostate cancer. *Br. J. Cancer* **2010**, *103*, 462–468. [CrossRef] [PubMed]

52. Lobb, R.J.; van, A.R.; Wiegmans, A.; Ham, S.; Larsen, J.E.; Moller, A. Exosomes derived from mesenchymal non-small cell lung cancer cells promote chemoresistance. *Int. J. Cancer* **2017**, *141*, 614–620. [CrossRef] [PubMed]

53. Collins, A.R. The comet assay for DNA damage and repair: Principles, applications, and limitations. *Mol. Biotechnol.* **2004**, *26*, 249–261. [CrossRef]

54. Evans, S.O.; Jacobson, G.M.; Goodman, H.J.B.; Bird, S.; Jameson, M.B. Comparative safety and pharmacokinetic evaluation of three oral selenium compounds in cancer patients. *Biol. Trace Elem. Res.* **2018**. [CrossRef] [PubMed]

55. Marshall, J.R.; Burk, R.F.; Payne, O.R.; Hill, K.E.; Perloff, M.; Davis, W.; Pili, R.; George, S.; Bergan, R. Selenomethionine and methyl selenocysteine: Multiple-dose pharmacokinetics in selenium-replete men. *Oncotarget* **2017**, *8*, 26312–26322. [CrossRef] [PubMed]

International Journal of
Molecular Sciences

MDPI

Review

Non-Coding Micro RNAs and Hypoxia-Inducible Factors Are Selenium Targets for Development of a Mechanism-Based Combination Strategy in Clear-Cell Renal Cell Carcinoma—Bench-to-Bedside Therapy

Youcef M. Rustum [1,2,*], Sreenivasulu Chintala [3], Farukh A. Durrani [4] and Arup Bhattacharya [2]

[1]   Department of Internal Medicine, University of Iowa, Iowa City, IA 52242, USA
[2]   Roswell Park Comprehensive Cancer Center, Department of Pharmacology & Therapeutics, Buffalo, NY 14263, USA; Arup.Bhattacharya@Roswellpark.org
[3]   Neurological Surgery, Indiana University, Indianapolis, IN 46202, USA; srchinta@iu.edu
[4]   Roswell Park Comprehensive Cancer Center, Department of Cell Stress Biology, Buffalo, NY 14263, USA; Farukh.Durrani@Roswellpark.org
*   Correspondence: Youcef.Rustum@Roswellpark.org; Tel.: +1-716-208-6667

Received: 24 August 2018; Accepted: 18 October 2018; Published: 29 October 2018

Abstract: Durable response, inherent or acquired resistance, and dose-limiting toxicities continue to represent major barriers in the treatment of patients with advanced clear-cell renal cell carcinoma (ccRCC). The majority of ccRCC tumors are characterized by the loss of Von Hippel–Lindau tumor suppressor gene function, a stable expression of hypoxia-inducible factors 1α and 2α (HIFs), an altered expression of tumor-specific oncogenic microRNAs (miRNAs), a clear cytoplasm with dense lipid content, and overexpression of thymidine phosphorylase. The aim of this manuscript was to confirm that the downregulation of specific drug-resistant biomarkers deregulated in tumor cells by a defined dose and schedule of methylselenocysteine (MSC) or seleno-L-methionine (SLM) sensitizes tumor cells to mechanism-based drug combination. The inhibition of HIFs by selenium was necessary for optimal therapeutic benefit. Durable responses were achieved only when MSC was combined with sunitinib (a vascular endothelial growth factor receptor (VEGFR)-targeted biologic), topotecan (a topoisomerase 1 poison and HIF synthesis inhibitor), and S-1 (a 5-fluorouracil prodrug). The documented synergy was selenium dose- and schedule-dependent and associated with enhanced prolyl hydroxylase-dependent HIF degradation, stabilization of tumor vasculature, downregulation of 28 oncogenic miRNAs, as well as the upregulation of 12 tumor suppressor miRNAs. The preclinical results generated provided the rationale for the development of phase 1/2 clinical trials of SLM in sequential combination with axitinib in ccRCC patients refractory to standard therapies.

Keywords: methylselenocysteine; seleno-L-methionine; clear-cell renal cell carcinoma microRNAs; hypoxia-inducible factor; antitumor activity

1. Introduction

Despite advances in the treatment of patients with advanced clear-cell renal cell carcinoma (ccRCC) with anti-angiogenic agents, checkpoint inhibitors, and mammalian target of rapamycin (mTOR) inhibitors alone and in combination, durable responses are seen in about 30% of treated ccRCC patients [1–13]. A systematic review of the first line for metastatic renal carcinoma reported an average progression-free survival of 8.4 months with a range of 6.5 to 12.3 months, and an average overall survival of 24.4 months with a range of 18.5 to 32.9 months [14]. Based on the clinical data generated in patients with advanced cancer, resistance and the associated dose-limiting toxicities remain major clinical challenges. There is an unmet clinical need to identify a new treatment modality that is patient-centric, selective, and efficacious for metastatic ccRCC patients. Both primary and metastatic ccRCC tumors are uniquely characterized by the expression of altered biomarkers associated with increased angiogenesis, metastasis, and drug resistance, including deletion and/or mutation of the von Hippel–Lindau (VHL) tumor suppressor gene in the majority of ccRCC tumors, resulting in the stable expression of hypoxia-inducible factors 1α and 2α (HIFs), and vascular endothelial growth factor (VEGF) [15–33]. Programmed death 1 (PD-1) is expressed in the membrane and cytoplasm of activated T cells, B cells, and dendritic cells. Programmed death ligand 1 (PD-L1) is expressed in 21–75% of ccRCC tumors, and allows cancer cells to evade immune response [34–47]. Although multiple signaling and epigenetic pathways regulate the expression of PD-L1, interferons γ and α (INF-γ and INF-α) and specific oncogenic micro RNAs (miRNAs) are also known to induce PD-L1 [48–53]. PD-L1 incidence and intensity vary among different tumor types. The analysis of melanoma tumors revealed that 38% of them express both PD-L1 and tumor-infiltrating lymphocyte (TIL), while 41% are negative for both, and 1% are PD-L1-positive, and 20% are TIL-positive [38,54,55]. PD-L1 was expressed in 69 out of 98 (70.9%) ccRCC tumors expressing mutant VHL. In all wild-type VHL tumors, 11.2% express PD-L1 [16]. HIFs and PD-L1 are co-expressed in cancer cells. Under hypoxic conditions, HIFs regulate the expression of PD-L1 by binding to the hypoxia response element in the PD-L1 proximal promoter to activate its transcription [47,48,56]. PD-L1 expression in cancer cells may, therefore, be regulated transcriptionally by HIF and post-transcriptionally by miRNAs. It is likely that the effective downregulation of HIFs would lead to the downregulation of PD-L1, resulting in an increased tumor response to subsequent treatment with anti-PD-1/PD-L1 therapies.

Thymidine phosphorylase (TP), an angiogenic protein and an enzyme required for the activation of several 5-fluorouracil (FU) prodrugs, is overexpressed by approximately 30–40% of cancers [57–63]. TP may function as an independent prognostic factor for increased tumor vascularity, and a target for the activation of 5-FU prodrugs. Utilizing TP to activate 5-FU prodrugs may also reduce its angiogenic activity, and may synergize with VEGF-targeting drugs. The reported overexpression of TP in ccRCC provided the opportunity to evaluate 5-FU prodrugs, such as S-1, in combination with tyrosine kinase inhibitors (TKIs) targeting VEGF/VEGF receptor (VEGFR).

Morphologically, ccRCC tumors are characterized by extensive lipid accumulation. Hypoxia-inducible protein 2 (HIG-2) is highly expressed in tumors expressing HIF1α, but not HIF2α [22,64]. Results generated indicate that HIG-2 is a direct target of HIF1α, but not HIF2α. Carnitine palmitoyltransferase (CPT1A), a fatty-acid transporter in the mitochondria, was recently reported to be a direct target of HIFs [65]. Clear-cell RCC cells transfected with VHL led to the downregulation of CPT1A, resulting in fatty-acid transport into the mitochondria, and forcing the formation of lipid droplets from fatty acids. Recent published reports indicated that ccRCC tumor cells expressing mutant VHL and the stable expression of HIFs participate in lipid deposition. However, HIF2α, but not HIF1α, controls the expression of perilipin 2, resulting in lipid storage [66]. In cells with a co-expression of HIFs, miRNA-155, and miRNA-210, it is possible that HIG-2, CPT1A, and perilipin 2 may also be regulated by miRNAs through HIF-dependent or -independent pathways. Since both HIFs are involved in the regulation of lipid droplets in ccRCC, agents that target HIF2α, but not HIF1α, may express limited antitumor activity. Agents that target both HIFs may have greater therapeutic impact and could avoid the need to regulate or target individual pathways regulated by HIFs.

2. Results

2.1. Hypoxia-Inducible Factors 1α and 2α (HIFs) and VHL Tumor Suppressor Gene

The molecular profiles of ccRCC tumors are summarized in Figure 1 and Table 1. HIFs are transcriptional factors that regulate the expression of over 200 genes involved in angiogenesis, tumor metastasis, and drug resistance. Unlike colorectal and head-and-neck tumors, ccRCC tumors feature a high incidence and intensity of constitutively expressed HIFs, as well as lower levels of VEGF and prolyl hydroxylase 2 (PHD2), with no detectable prolyl hydroxylase 3 (PHD3), as assessed by immunohistochemistry (Table 1).

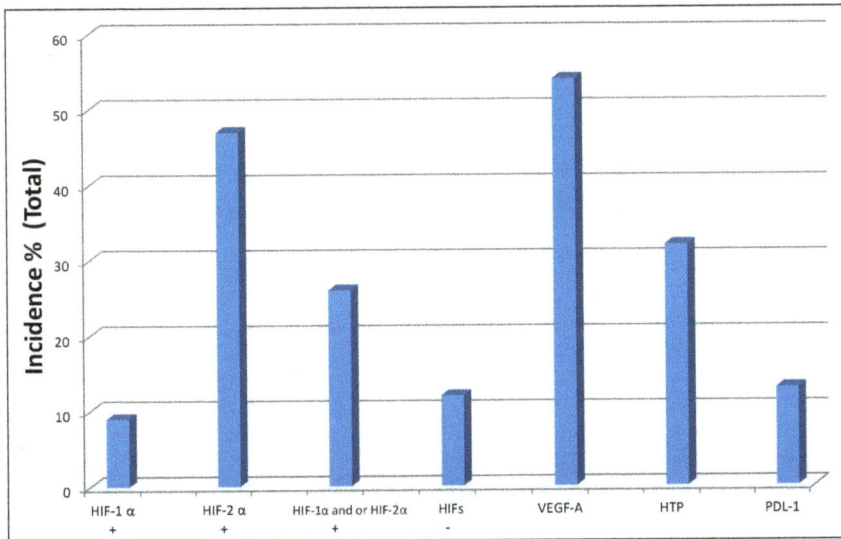

Figure 1. Hypoxia-inducible factors (HIFs, '+' indicates presence and '−' means absence), vascular endothelial growth factor (VEGF), high-intensity thymidine phosphorylase (HTP), and programmed death ligand 1 (PD-L1) biomarker expression in clear-cell renal cell carcinoma (ccRCC) tumors. The data for HIFs and VEGF were generated by our laboratory [20,22], while others are from published reports [13,63].

Our laboratory was the first to report that constitutively expressed HIF1α and HIF2α (Table 1, Figure 2) are selenium targets (adopted from References [20,32]). The data in Figure 2 show that the inhibition of constitutively expressed HIF1α and HIF2α in RC2 and 786.0 Clear-cell RCC cells, and HIF1α in FaDu head and neck [32], A548 lung carcinoma cells, and HT29 colorectal tumor cells is selenium dose-dependent and independent of the disease site/cell type. Unlike other HIF-targeting agents, selenium inhibits HIF expression via PHD-dependent degradation [20,32].

Table 1. Molecular profile of tumor biopsies.

Incidence of HIF-α and PHDs Protein Expression in Primary Human ccRCC, Head & neck (H/N) and Colorectal Cancer (CRC) Tumor Biopsies:					
Tumors	**HIF-1α**	**HIF-2α**	**HIF-1α and/or HIF-2α**	**PHD2**	**PHD3**
ccRCC	45% (40/88)	78% (69/88)	92% (81/88)	35% (31/88)	0% (0/88)
H/N	23% (40/173)	16% (23/146)	38% (46/122)	86% (180/210)	21% (32/153)
CRC	13% (8/62)	15% (10/65)	26% (17/64)	90% (55/61)	50% (31/62)
VEGF(A)					
Tumors	**Incidence of Positions**	**Average Immunoscope**			
ccRCC	54% (48/88)	2.3 (weak)			
H/N	79% (136/173)	4.24 (moderate)			
CRC	97% (60/62)	5.68 (strong)			

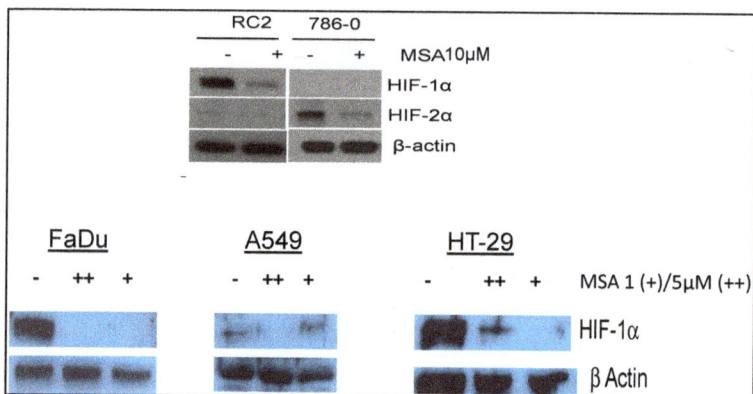

Figure 2. Constitutively expressed HIFs are selenium targets [20,32]. Effects of methylseleninic acid (MSA), the active moiety of methylselenocysteine (MSC) or seleno-L-methionine (SLM), on the expression levels of constitutively expressed HIFs in RC2 and 786.0 renal cell carcinoma, and on HIF1α head and neck in FaDU, lung carcinoma, A549, and colorectal carcinoma cell lines. RC2 and 786.0 cells (adopted from Reference [20]) were exposed to 10 μM MSA for 24 h, while other cells—FaDu (adopted from Reference [32]), A549, and HT29—were exposed to 0.5% O_2 for 24 h and treated with different MSA concentrations. Cells were lysed rapidly on ice and analyzed for HIF expression by Western blot [20,32].

2.2. Tumor Vasculature

To accommodate survival, growth, and metastasis, tumor cells promote the formation and development of new blood vessels [36,39]. Tumor-associated blood vessels within the tumor microenvironment are unstable and leaky, and they could represent a barrier to the delivery of effective therapies to tumor cells [67,68]. Thus, for the development of efficacious therapy, treatment should include drugs targeting biomarkers that induce the normalization of tumor-associated vasculature. Our laboratory was the first to report that the stabilization of tumor vasculature by MSC is dose- and schedule-dependent. We previously reported that the therapeutic dose and schedule of MSC/SLM exert dual effects. Firstly, anti-angiogenic effects were achieved via the inhibition of new vessel formation and a reduction in microvessel density. Secondly, tumor vascular maturation was achieved through an increase in pericyte recruitment. Collectively, these effects were associated with an increase in drug delivery and distribution to tumor cells. As shown in Figure 3, in vivo treatment with therapeutic doses of MSC resulted in a selective increase in vascular maturation index in tumors, but not in normal liver mouse tissues. The data generated demonstrate that tumor cells and their associated vasculature can be successfully and selectively modulated in vivo by a therapeutic, non-toxic dose and schedule of MSC. These results are consistent with the data generated by Jain et al., demonstrating normalization of the tumor microenvironment by Avastin, an anti-angiogenic agent [69].

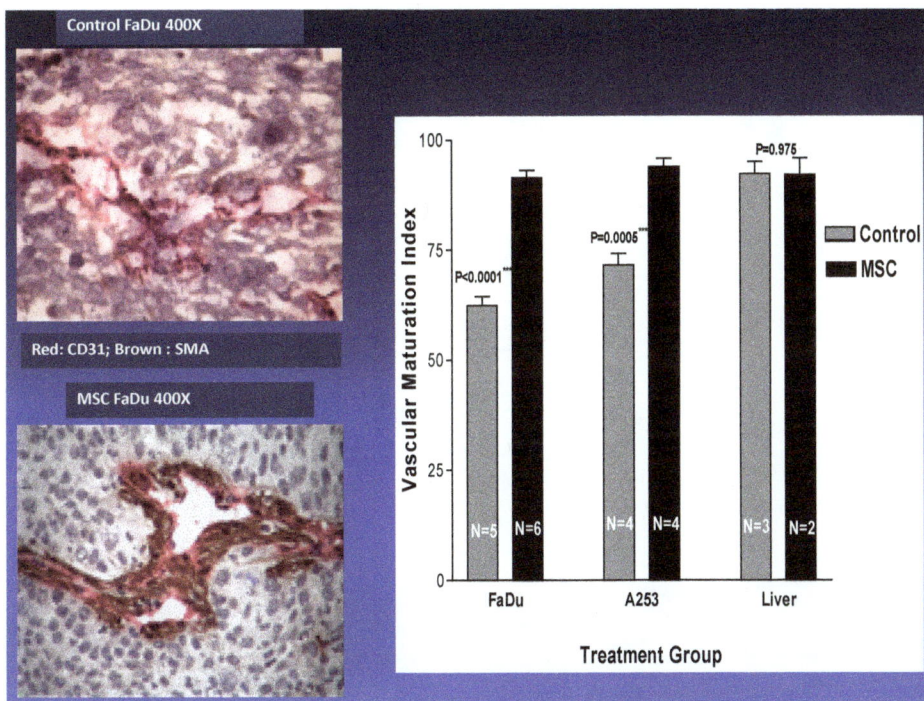

Figure 3. MSC selectively stabilizes tumor vasculature [68,70]. Effects of MSC treatment on the stabilization of tumor vasculature. Xenografts bearing FaDU tumors were treated orally with 10 mg/kg MSC daily for seven days, at which point the vascular maturation in tumor and normal liver tissues was assessed histologically [68,70].

2.3. Oncogenic miRNA-155 and miRNA-210

Non-coding miRNAs are small molecules involved in the post-transcriptional regulation of genes, and are often associated with increased angiogenesis and drug resistance. Micro RNAs function as either tumor suppressors or promoters, and they act by targeting the 3' untranslated region (3'-UTR) of targeted genes [71,72]. Micro RNAs reduce the gene expression of mRNAs by inhibiting translation or via degradation of the transcript. Oncogenic miRNA-155 and miRNA-210 are highly overexpressed in ccRCC tumors expressing HIF1α, HIF2α, VEGF, and PD-L1 [73–81].

To identify a possible link between HIF-α protein expression levels and tumor-associated miRNAs, three primary ccRCC biopsies and two ccRCC cell lines expressing a similar incidence and distribution of HIF-α were analyzed using a microarray for miRNA expression. Microarray analysis using an Exiqon microarray chip of RC2 cells treated with methylselenic acid (MSA), an inhibitor of HIF1α, revealed that 28 miRNAs were downregulated and 12 miRNAs were upregulated (Figure 4A). Although several miRNAs were altered, selected miRNAs which were upregulated and downregulated by MSA treatment are shown in Figure 4B. These results suggest that these miRNAs are likely regulated by HIF1α and can be effectively modulated by therapeutic doses of selenium.

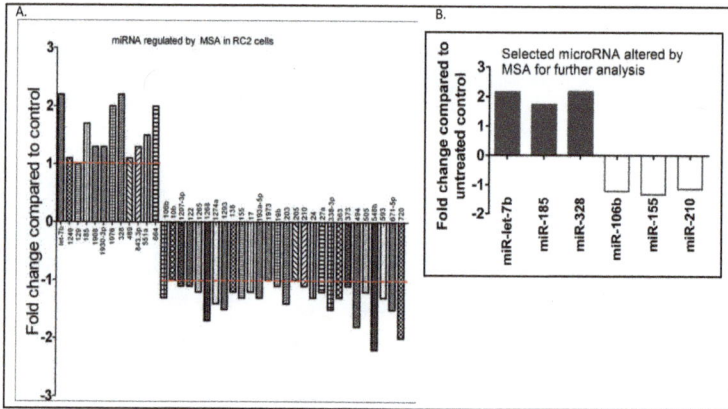

Figure 4. Oncogenic and tumor-suppressor micro RNAs (miRNAs) expressed in ccRCC are selenium targets. Effects of MSA on the expression levels of oncogenic and tumor-suppressor miRNAs altered in RC2 cells expressing HIF1α: (**A**) microarray analysis of miRNAs regulated by the treatment of RC2 cells with 10.0 μM MSA for 24 h, and (**B**) selected miRNAs shortlisted for further analysis.

The data in Figure 5 indicate that the miRNAs that were significantly altered by MSA treatment of RC2 cells expressing HIF1α and of 786.0 cells expressing HIF2α were also altered in primary ccRCC biopsies.

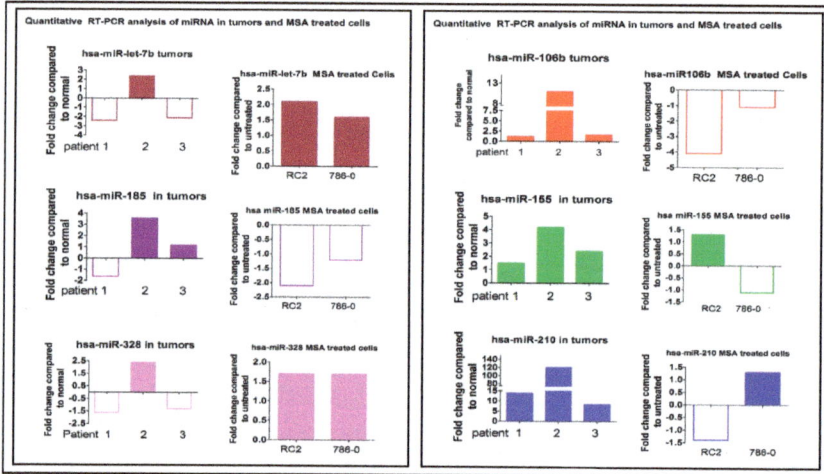

Figure 5. Selected miRNAs expressed in primary ccRCC biopsies are also expressed in ccRCC cell lines and can be modulated by selenium. Modulation of miRNAs expressed in ccRCC patient tumor biopsies, and in RC2 and 786.0 cells lines treated with MSA. Quantitative RT-PCR analysis of selective hypoxia-regulated microRNA in human RCC patient tumors (*n* = 3), and in RC2 and 786.0 cells treated with MSA. MicroRNAs downregulated in human tumors (miR let7b and miR328) (**left panel**) found to be upregulated with MSA treatment in RC2 and 786.0 cells. MicroRNAs which were upregulated (**right panel**: miR106b, miR155, and miR210; **left panel**: miR185) in RCC patients were found to be downregulated with MSA treatment in RC2 and 786.0 cells. Log fold changes are shown compared to matched normal kidney tissues for patients and untreated RC2 and 786.0 cells.

Two miRNAs, Let-7b, and -328, which were upregulated, and miRNA-106b, -155, and -210, which were downregulated by MSA treatment of RC2 and 786.0 cells, were randomly selected to perform qRT-PCR analysis along with four primary ccRCC tumor biopsies and their paired normal kidney cells.

The results presented in Figure 5 confirmed the microarray data that these selected miRNAs which were altered in RC2 and 786.0 cells were similarly altered in the patient biopsies, and their expressions could be modulated in vitro and in vivo by selenium. Collectively, the data generated demonstrate that a defined dose and schedule of selenium can effectively modulate the expression levels of specific oncogenic and tumor-suppressor miRNAs altered in ccRCC tumor cells.

2.4. Selenium: A Selective Modulator of Anticancer Therapies

2.4.1. Nude Mice Bearing HIF1α

The data in Figure 6A demonstrate the antitumor activity of MSC in sequential combination with two representative cytotoxic drugs, irinotecan (an approved drug for the treatment of colorectal cancer) and docetaxel (used in head-and-neck cancers among others), and radiation therapy. Oral daily administration of 10 mg/kg/day MSC for seven days prior to and concurrent with the administration of cytotoxic or radiation therapies beginning on day seven was associated with enhanced therapeutic efficacy.

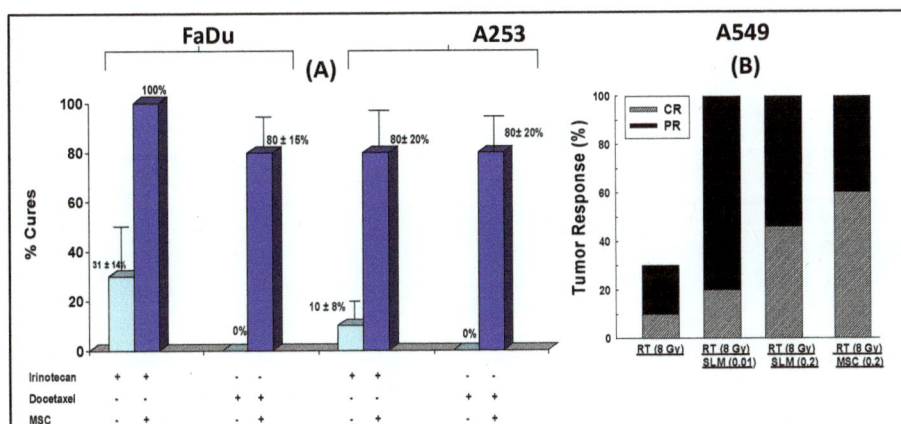

Figure 6. Antitumor activity of MSC in combination with irinotecan and docetaxel in nude mice bearing human head-and-neck cancer cells, FaDU and A253 (**A**), and radiation-treated A549 lung carcinoma (**B**). MSC was administered orally daily for seven days and concurrently with anticancer therapies administered on day seven [82].

The data in Figure 6B demonstrate the antitumor activity of MSC in sequential combination with radiation therapy of mice bearing A549 lung carcinoma tumors expressing HIF. Collectively, MSC was found to significantly enhance the therapeutic efficacy of chemotherapy and radiation in different human cancer xenografts from different disease sites. The results generated suggest that the action of selenium in tumor cells expressing HIFs is a universal phenomenon, irrespective of the cancer type or disease site.

2.4.2. Nude Mice Bearing Tumor Xenografts That Constitutively Expressed HIF2α

Figure 7A,B depict tumor growth inhibition by MSC, SLM, axitinib, sunitinib, and topotecan. The dose and schedule of MSC and SLM that inhibited HIF exhibited limited but similar tumor growth inhibition. Sunitinib exerted greater antitumor activity than Avastin, axitinib, and topotecan [83]. The order of antitumor activity is sunitinib > Avastin ≥ axitinib > topotecan > MSC or SLM. The data in Figure 7C depict the antitumor activity of tyrosine kinase inhibitors (TKIs) that target VEGF/VEGFR, and topotecan alone and in combination with either MSC or SLM. The combination of topotecan and sunitinib in sequential combination with MSC or SLM had the most therapeutic efficacy and achieved long-term and durable responses not observed with these drugs administered individually. The data in Figure 7D indicate that MSC and SLM similarly potentiate the antitumor activity of axitinib, a Food and Drug Administration (FDA)-approved VEGFR-targeting agent for the treatment of relapsed ccRCC patients. The data in Figure 7E confirm that HIFs are a critical therapeutic target of MSC. MSC potentiates the antitumor activity of topotecan, a topoisomerase 1 poison which targets HIF synthesis, as well as that of Avastin, axitinib, and sunitinib, which target VEGF/VEGFR. In comparison, the antitumor activity of irinotecan, a topoisomerase 1 poison with no demonstrable effects on HIF protein expression, was not potentiated by MSC. In this model, S-1 exhibited significant antitumor activity, perhaps due to overexpression of TP. Collectively, the data in Figure 7E indicate that optimal therapeutic benefit was obtained with MSC in sequential combination with topotecan and sunitinib.

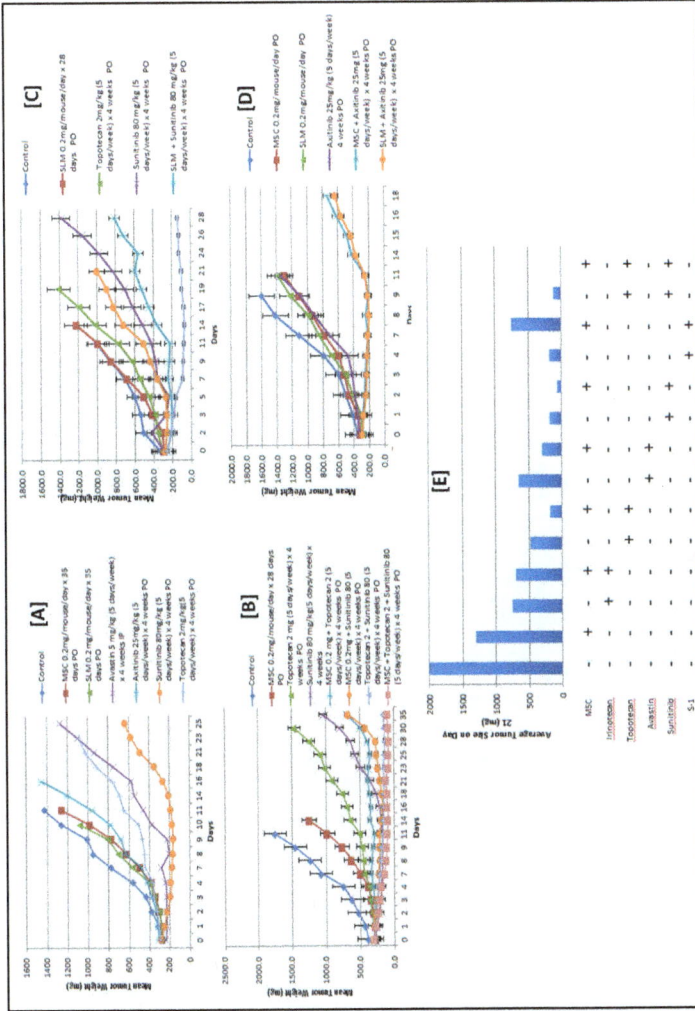

Figure 7. Pretreatment of 786.0 xenografts expressing HIF2α with a nontoxic but molecularly effective dose of MSC sensitizes tumor cells to subsequent treatment with the combination of VEGF/VEGF receptor (VEGFR)-targeted agents and chemotherapy. Assessment of antitumor activity of MSC, SLM, Avastin, axitinib, sunitinib, and topotecan, administered individually to nude mice bearing 786.0 ccRCC tumors (**A**), and in combination with either MSC (**B**) or SLM (**C**), and in combination with axitinib (**D**). (**E**) Summary of the antitumor activity of MSC in sequential combination with anticancer therapies [83].

3. Discussion

Clear-cell RCCs and their associated microenvironment express a unique molecular and morphological profile including a variety of tumor-suppressor and oncogenic miRNAs. However, miRNA-155 and miRNA210 are extensively characterized and overexpressed in multiple tumor types [75–78]. Although VHL may be regulated by multiple biomarkers expressed in tumor cells and their adjacent microenvironment, miRNA-155 and -210 emerged as key modulators of VHL function, and may offer an alternative mechanism for stable expression of HIFs in ccRCC tumors [17,77]. Loss of VHL in ccRCC tumors may mimic the upregulation of HIFs by hypoxia. In recognition of the critical role of VHL in the pathogenesis of ccRCC tumors, efforts are underway to develop anti-VHL chemical agents [84,85]. Similarly, recognizing that HIFs are upregulated by hypoxia-dependent and -independent pathways and that they are critical therapeutic targets, a number of HIF inhibitors are presently under preclinical and clinical development. A recent phase 1 clinical trial of PT2385, a synthetic small-molecule HIF2α antagonist, demonstrated clinical activity in previously treated ccRCC patients [86].

Tumor microarray analysis demonstrated that HIF1α and HIF2α are individually and jointly co-expressed in a majority of primary and metastatic ccRCC biopsies [20]. In addition, it was reported that, although HIF1α and HIF2α are structurally similar, they functionally regulate different target genes in different cell types [25]. Furthermore, under hypoxia, the expression of VEGF is regulated by HIF1α, but not by HIF2α [33]. It is possible that the inhibition of one HIF isoform may induce the activation of the other in support of tumor growth. The data to date suggest that optimal therapeutic benefit may require targeting both HIF1α and HIF2α.

HIFs and PD-L1 are co-expressed in cancer cells. Under hypoxic conditions, HIFs regulate the expression of PD-L1 by binding to the hypoxia response element in the PD-L1 proximal promoter to activate its transcription [42,47]. PD-L1 expression in cancer cells may, therefore, be regulated transcriptionally by HIF and post-transcriptionally by miRNAs. It is likely that effective downregulation of HIFs would lead to the downregulation of PD-L1, resulting in an increased tumor response to subsequent treatment with anti-PD-1/PD-L1 therapies.

Micro RNA-155 and miRNA-210, amongst others, were reported to modulate the tumor microenvironment [74,75], regulate glucose metabolism [87], and target transcription factor E2F2 in ccRCC tumor cells [88]. Neal et al. reported that the VHL/HIF axis regulates the expression of several types of miRNAs in ccRCC tumors, including miRNA155 and miRNA-210 [53]. Increasing evidence suggests that oncogenic miRNA-155 and miRNA-210 are regulators of immune response biomarkers, including forkhead box P3 (FoxP3) regulatory T cell, myeloid-derived suppressor cell (MDSC) T-cells, and immune checkpoint PD-1/PD-L1 [56,80,81,89,90]. Despite the progress made in our understanding of the biology and therapeutic potential of miRNAs, their clinical use as a prognostic and as a predictor of therapeutic outcome is yet to be determined. Efforts to develop miRNA inhibitors fall short of clinical expectations [91–93]. The limited clinical benefits were attributed, in part, to their limited bioavailability, instability, and dose-limiting toxicities, in addition to an inability to demonstrate in vivo modulation of expression of intended targets. Our laboratory was the first to demonstrate that specific types, doses, and a schedule of MSC in ccRCC xenograft models can selectively modulate specific types of miRNAs.

Clear-cell RCC tumors are highly vascular with clear, large cytoplasms expressing perilipin 2, hypoxia-inducible lipid-droplet protein 2, which represses fatty-acid metabolism, and is a target gene of HIF1α [22,64,65]. Molecularly, the majority of ccRCC tumors express high incidence and intensity of HIF1α, HIF2α, and oncogenic miRNA-155 and -210, which target genes involved in ccRCC tumorigenesis, including VEGF and PD-L. The tumor microenvironment associated with ccRCC is leaky and unstable, expressing the common biomarkers that regulate tumor cell growth and metastasis commonly seen in many cancers. Thus, ccRCC tumors provide the opportunity to test the hypothesis and rationale for a mechanism-based treatment combination with selenium that may offer the potential for the development of novel treatment in patients with ccRCC and other cancers with similar expression of Se targets.

Resistance and dose-limiting toxicities continue to represent major clinical challenges for both cytotoxic chemotherapy and biological targeted therapies. In general, in vivo resistance is regulated by multiple molecular and immunological biomarkers expressed in tumor cells and their surrounding microenvironment. These two tumor compartments are functionally interactive. The tumor microenvironment could promote tumor growth while impeding optimal drug delivery and the distribution of effective tumor drug concentrations. Thus, the tumor microenvironment may be considered as the gatekeeper, while tumor cells are the ultimate targets. In order to achieve durable antitumor activity, treatment should include a combination of drugs that enable targeting both the tumor microenvironment and the tumor cells.

In ccRCC, HIFs, miRNA-155, and miRNA-210 are commonly co-expressed and were reported earlier to regulate the expression of gene targets implicated in enhanced angiogenesis, tumor metastasis, and resistance. While considerable efforts are underway to develop miRNA- and HIF-based strategies, in vivo toxicity, tumor instability, and limited drug delivery in effective concentrations continue to plague efforts to have a more clinically effective outcome [93]. In addition, an increased activation of 5-FU prodrugs by TP should result in increased antitumor activity [94–96].

During the last several years, our laboratory determined that SLM, an FDA-approved drug for clinical trials, and MSC (under development) exert several effects that are not shared by other selenium compounds and HIF-targeting compounds that are currently under preclinical and clinical evaluation [20,23,70,83–90]. We were the first to demonstrate [97,98], in several tumor xenograft models, that (1) therapeutic and nontoxic doses and a schedule of organic selenium compounds, SLM and MSC, potently enhance constitutively expressed HIF1α and HIF2α degradation; (2) SLM and MSC downregulate VEGF, which is regulated by HIF1α, but not by HIF2α; (3) SLM and MSC stabilize tumor vasculature resulting in the selective enhancement of drug delivery to tumor cells, consistent with results reported by Jain [69]; (4) SLM and MSC modulate the expression of a number of tumor-suppressor and oncogenic miRNAs altered in ccRCC tumors; (5) SLM and MSC offer selective protection against toxicity induced by toxic and often lethal doses of cytotoxic drugs in preclinical models [83]; and (6) treatment with MSC and SLM was associated with significant enhancement of the efficacy and selectivity of anticancer therapies in head-and-neck, colorectal, and renal cancer xenografts. The antitumor activity of VEFG/VEGFR-targeted therapies alone and in combination with topotecan and S-1 can be further enhanced by MSC in mice bearing VHL-deficient 786.0 ccRCC tumors expressing HIF2α, VEGF, miRNA-155, and miRNA-210. Taken together, non-toxic doses of selenium may offer the potential for the development of novel therapeutic modality. Chart 1 is an outline of the approach used in the translational development of selenium in combination with anticancer drugs in preclinical models to phase 1 and 2 clinical trials. The data generated in several xenograft models provided the rationale for the development of a phase 1 clinical trial in ccRCC patients. The aim was to confirm that the SLM dose used to yield blood selenium concentrations similar to those determined therapeutically, synergistic with anticancer drugs in the preclinical model, could be achieved clinically without toxicity. The optimal SLM dose defined in the phase 1 trial [99] was used to design a phase 2 trial of SLM in sequential combination with axitinib, aimed at assessing the efficacy and modulation of relevant molecular correlates.

Xenograft Models	Phase 1 Clinical	Phase 2 Clinical
-Dose optimizations	-Dose escalation	-Efficacy
-Antitumor activity	-Blood/Plasma Levels	-Toxicity
-Toxicity	-Safety	-Blood/plasma levels
-Mechanisms	-Efficacy?	-Correlative mechanisms

Chart 1. "Bench-to-bedside" therapeutic development of SLM from preclinical models to phase 1 and 2 clinical trials in patients with advanced clear-cell renal cell carcinoma (ccRCC).

Based on the preclinical results generated, a mechanism-based combination therapy is proposed, as outlined in Chart 2. In order to achieve optimal therapeutic benefit with the proposed mechanism-based drug combination, the dose, schedule, and sequence of MSC and SLM are critical parameters. Pretreatment with selenium prior to and concurrent with the administration of anticancer therapy is necessary for the optimal modulation of relevant selenium biomarkers in tumor cells and for the optimal stabilization of tumor vasculature. To maintain the optimal and sustained inhibition of HIFs and associated gene targets, it is recommended that topotecan be administered in combination with MSC or SLM. Since therapeutic doses and the schedule of selenium partially downregulate the expression levels of VEGF in tumor cells expressing HIF1α but not HIF2α [20,23], we propose, therefore, adding TKI inhibitors to the combination regimen in order for maximum downregulation of VEGF/VEGFR. This proposed mechanism-based combination was evaluated in 786.0 xenografts and was determined to be highly selective and therapeutically effective. The dose and schedule of the SLM/MSC used were selected based on their molecularly effective dose instead of the maximum tolerated dose. Furthermore, since the expression level of PD-L1 is regulated by HIFs and miRNAs, it is reasonable to expect that SLM/MSC will also modulate the therapeutic efficacy of checkpoint inhibitors. Proof of principle in ccRCC could provide the basis for the verification of this mechanism-based treatment combination in other tumors expressing these molecular targets similarly affected by SLM/MSC.

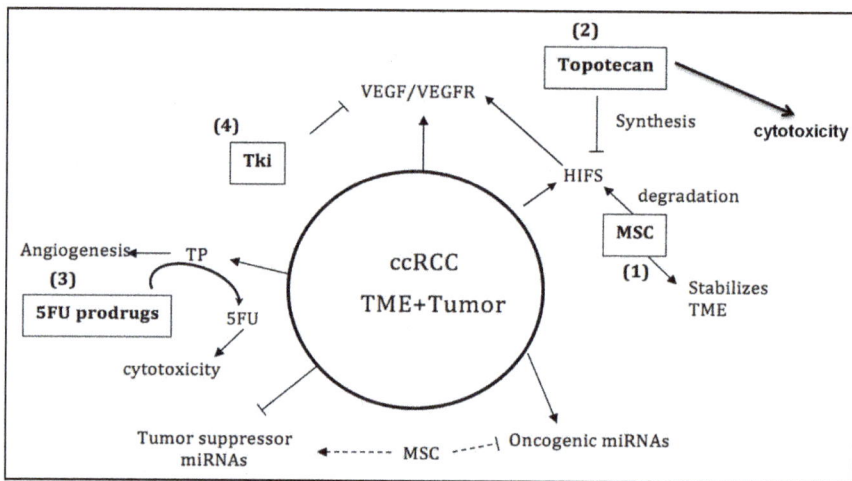

Chart 2. Schematic representation of targetable markers expressed in ccRCC. Methylselenocysteine (MSC) targets hypoxia-inducible factors (HIFs) and micro RNAs (miRNAs). Topotecan targets HIF synthesis, while tyrosine kinase inhibitors (TKIs) target vascular endothelial growth factor (VEGF)/VEGF receptor (VEGFR) and 5-fluorouracil (5-FU) prodrugs are the substrate for activation by thymidine phosphorylase.

4. Conclusions and Future Perspectives

The aim of this paper was to determine that the levels of specific biomarkers altered in the majority of ccRCC tumors, such as HIFs, oncogenic miRNA-155 and miRNA-210, and VEGF, can be selectively downregulated by therapeutic nontoxic doses and a schedule of MSC and SLM. In addition, the aim was also to confirm that downregulation of these biomarkers would translate into therapeutic synergy with anticancer therapies. The results in several xenograft models and with multiple cytotoxic and biologic agents demonstrated that the dose- and time-dependent downregulation of constitutively expressed HIFs, miRNA-155 and -210, and VEGF-A by selenium was associated with enhanced therapeutic efficacy and selectivity of anticancer therapies. Preclinical data generated provided the rationale for the development of a phase 1 clinical trial in ccRCC patients treated with escalating doses of SLM in sequential combination with a fixed dose of axitinib [99,100]. Unlike the 200 µg/day SLM dose used in prevention clinical trials, the SLM doses used in combination therapy were 10 mg/kg in nude mice, and 8000 µg/day in ccRCC patients, which was the dose recommended for the ongoing phase 2 clinical trial for efficacy assessment and for the monitoring of the effects of SLM on relevant biomarkers. The plasma selenium concentrations achieved clinically with the recommended SLM dose were comparable with those achieved with SLM doses determined therapeutically synergistic with anticancer drugs in preclinical models. The mechanism-based drug combination proposed in Chart 2 warrants expanded preclinical investigation and clinical verification. Proof of concept that enhanced therapeutic efficacy and selectivity of axitinib in refractory ccRCC patients are SLM dose- and schedule-dependent will be highly innovative and significant. Furthermore, the ability of selenium to downregulate specific biomarkers associated with drug resistance may provide the opportunity for the clinical development of SLM in sequential combination with other clinically available targeted therapies.

5. Material and Methods

5.1. Cell Culture and Drug Treatments

Clear-cell RCC cell lines 786.0 and RC2 were cultured in Rosewell Memorial Park Institute (RMPI-1640) medium with 10% fetal bovine serum (FBS) and 1% penicillin/streptomycin (PenStrep, Sigma-Aldrich, St. Louis, MO, USA) at 37 °C in an incubator with 5% CO_2. Cells were routinely tested for mycoplasma contamination. Cells were seeded in T75 and/or T150 flasks, and were allowed to grow overnight. Cells were treated with MSA for 24 to 48 h, and were processed to isolate total RNA. Untreated control cells were maintained without treatment.

5.2. Animals

Female athymic nude mice (Envigo, nu/nu, 20–25 g body weight), 8–12 weeks of age, were used for the tumor xenograft experiment as previously described [97]. All studies were carried out as approved by the Institutional Roswell Park Comprehensive Cancer Center Animal Care and Use Committee (207M, 2009).

Tumor Xenografts

Clear-cell RCC 786.0 cells were cultured in RMPI-1640 and transplanted into nude mice to establish xenografts. Tumors were harvested, and ~50 mg of non-necrotic tumor tissue was transplanted into nude mice and randomized to groups of 5–10 mice each. Treatment with drugs alone or in combination was started when tumors reached ~200 mg, and the tumor volume and response were measured as described previously [97]. Drug toxicity was evaluated by measuring the weight loss of the mice biweekly.

5.3. Drugs

MSC and SLM (Sigma-Aldrich, St. Louis, MO, USA) were given at 0.2 mg/kg for 35 days starting seven days prior to the start of drug treatment. Axitinib (AdooQ Bioscience, Irvine, CA, USA), sunitinib (LC laboratories, Woburn, MA, USA), and topotecan (Selleckchem, Houston, TX, USA) were administered orally at 25 mg/kg, 80 mg/kg, or 2 mg/kg five days per week for four weeks, either as a single drug or in combination. Avastin (Genentech, South San Francisco, CA, USA), was given at 5 mg/kg via intraperitoneal injection for five days/week for four weeks either by itself or in combination with selenium.

5.4. Total RNA Isolation from ccRCC Cells Treated with and without MSA

Cells were treated with MSA for 24–48 h and processed for isolation of total RNA using Trizol reagent as per the instructions of the manufacturer (Invitrogen, Liverpool, NY, USA). RNA quantity and quality was measured using Nanodrop (Thermo-Fisher Scientific, Liverpool, NY, USA), and then used for microRNA microarray analysis and quantitative PCR analysis of microRNA.

5.5. Total RNA from ccRCC Patient Tumors and Their Matched Normal Tissues

Total RNA of de-identified ccRCC patient tumors and their matched normal kidneys were obtained from the RPCI Pathology core facility. RNA samples were isolated using Trizol reagent (Thermo-Fisher Scientific, Liverpool, NY, USA) from the non-necrotic tissues selected by the pathologist, and purity was determined before use for detecting microRNA expression by qRT-PCR.

5.6. Reverse Transcription (RT) and miRNA qPCR

Complementary DNA (cDNA) was prepared using the following quantities of each reagent and RNA: 4 μL (20 ng) of RNA, 9 μL of H_2O, 1 μL of Spike-In, 4 μL of reverse transcription (RT) buffer, and 2 μL of enzyme in a total volume of 20 μL. Immediately after the RT reaction was finished, a 1:80 dilution was made on the cDNA, and ROX was added. The reaction mix for qRT-PCR was prepared using 400 μL of SYBR® Green Master Mix (Thermo-Fisher Scientific, Liverpool, NY, USA) and 320 μL of cDNA (from the above diluted RT reaction). Then, 9 μL of this mix was added to a 384-well plate pre-loaded with specific miR primers in triplicate using an electronic multichannel pipette. Plates were sealed with optical tape and shaken on a plate shaker for 30 s, before being centrifuged for one minute and run on the ABI7900 qPCR machine (Applied Biosystem, Foster City, CA, USA). Quantitative PCR machine cycling conditions and parameters were set exactly the same for every plate.

Normalization of Exiqon miRNA Panels (http://www.exiqon.com/mirna-pcr-panels) Excerpt from Exiqon Manual: Inter-Plate Calibrator (IPC). Since each assay was present only once on each plate, replicates were performed using separate plates. This raises the issue of run-to-run differences. To allow for simple inter-plate calibration, we designed a calibration assay with an accompanying template (annotated as UniSp3 or IPC in the plate layout files). Three wells were assigned for inter-plate calibration to provide triplicate values with the possibility for outlier removal. In each of these wells, both the primers and the DNA template were present, giving high reproducibility. The inter-plate calibrator requires only the addition of the SYBR® Green master mix in order to give a signal and can, therefore, be used for quality control of each plate run.

GenEx Software (ver 6.1, Thermo-Fisher Scientific, Liverpool, NY, USA: http://www.exiqon.com/qpcr-software.

Plates were imported into the GenEx software (ver 6.1, Thermo-Fisher Scientific, Liverpool, NY, USA) and the IPCs (in triplicate on each plate) were used to normalize the plates helping to eliminate run-to-run variation when comparing multiple plates. All Ct values above 38 were set to 38 as the maximum value (this is arbitrary and may even be left blank to denoted non-amplification). All miRNAs were listed in an excel file regardless of whether or not they were expressed in the samples, with normalized Ct values for each sample. Data were represented as individual triplicate runs and as averages of triplicates (with outliers excluded). Expressions of miRNA were normalized to untreated controls, and fold changes with the selenium treatment were determined. In ccRCC patient tumors, microRNA expression was normalized to normal tissue and fold changes were determined.

Author Contributions: All authors were involved in conceptualization, data collection, writing, editing, and reviewing the manuscript.

Funding: This work was supported by the National cancer institute grant P30CA016056 involving the use of Roswell Park Cancer Institute's Pathology Network, Genomic, and Clinical Data Network Shared Resources.

Acknowledgments: The authors would like to recognize the important and valuable contributions provided by Ms. Tara Rustum in editing the figures, tables, and the text.

Conflicts of Interest: The authors declare no conflicts of interest.

Abbreviations

ccRCC	Clear-cell renal cell carcinoma
HIF	Hypoxia-inducible factor
HIG-2	Hypoxia-inducible protein 2
IFN	Interferon
MSC	Se-methylselenocysteine
PD-1	Programmed death 1 receptor
PD-L1	Program death ligand 1
SLM	Seleno-L-methionine
TP	Thymidine phosphorylase
TNF	Tumor necrosis factor
TKI	Tyrosine kinase inhibitor
TIL	Tumor-infiltrating lymphocyte
VEGF	Vascular endothelial growth factor
VHL	von Hippel–Lindau

References

1. Albiges, L.; Choueiri, T.K. Renal-cell carcinoma in 2016: Advances in treatment—Jostling for pole position. *Nat. Rev. Clin. Oncol.* **2017**, *14*, 82–84. [CrossRef] [PubMed]
2. Armstrong, A.J.; Halabi, S.; Eisen, T.; Broderick, S.; Stadler, W.M.; Jones, R.J.; Garcia, J.A.; Vaishampayan, U.N.; Picus, J.; Hawkins, R.E.; et al. Everolimus versus sunitinib for patients with metastatic non-clear cell renal cell carcinoma (ASPEN): A multicentre, open-label, randomised phase 2 trial. *Lancet Oncol.* **2016**, *17*, 378–388. [CrossRef]
3. Escudier, B.; Powles, T.; Motzer, R.J.; Olencki, T.; Aren Frontera, O.; Oudard, S.; Rolland, F.; Tomczak, P.; Castellano, D.; Appleman, L.J.; et al. Cabozantinib, a New Standard of Care for Patients With Advanced Renal Cell Carcinoma and Bone Metastases? Subgroup Analysis of the METEOR Trial. *J. Clin. Oncol.* **2018**, *36*, 765–772. [CrossRef] [PubMed]
4. Fernandez-Pello, S.; Hofmann, F.; Tahbaz, R.; Marconi, L.; Lam, T.B.; Albiges, L.; Bensalah, K.; Canfield, S.E.; Dabestani, S.; Giles, R.H.; et al. A Systematic Review and Meta-analysis Comparing the Effectiveness and Adverse Effects of Different Systemic Treatments for Non-clear Cell Renal Cell Carcinoma. *Eur. Urol.* **2017**, *71*, 426–436. [CrossRef] [PubMed]
5. Greef, B.; Eisen, T. Medical treatment of renal cancer: New horizons. *Br. J. Cancer* **2016**, *115*, 505–516. [CrossRef] [PubMed]
6. Hsieh, J.J.; Chen, D.; Wang, P.I.; Marker, M.; Redzematovic, A.; Chen, Y.B.; Selcuklu, S.D.; Weinhold, N.; Bouvier, N.; Huberman, K.H.; et al. Genomic Biomarkers of a Randomized Trial Comparing First-line Everolimus and Sunitinib in Patients with Metastatic Renal Cell Carcinoma. *Eur. Urol.* **2017**, *71*, 405–414. [CrossRef] [PubMed]
7. Larkin, J.; Hodi, F.S.; Wolchok, J.D. Combined Nivolumab and Ipilimumab or Monotherapy in Untreated Melanoma. *N. Engl. J. Med.* **2015**, *373*, 1270–1271. [CrossRef] [PubMed]
8. Motzer, R.J.; Escudier, B.; McDermott, D.F.; George, S.; Hammers, H.J.; Srinivas, S.; Tykodi, S.S.; Sosman, J.A.; Procopio, G.; Plimack, E.R.; et al. Nivolumab versus Everolimus in Advanced Renal-Cell Carcinoma. *N. Engl. J. Med.* **2015**, *373*, 1803–1813. [CrossRef] [PubMed]
9. Powles, T.; Albiges, L.; Staehler, M.; Bensalah, K.; Dabestani, S.; Giles, R.H.; Hofmann, F.; Hora, M.; Kuczyk, M.A.; Lam, T.B.; et al. Updated European Association of Urology Guidelines Recommendations for the Treatment of First-line Metastatic Clear Cell Renal Cancer. *Eur. Urol.* **2017**. [CrossRef] [PubMed]
10. Robert, C.; Long, G.V.; Brady, B.; Dutriaux, C.; Maio, M.; Mortier, L.; Hassel, J.C.; Rutkowski, P.; McNeil, C.; Kalinka-Warzocha, E.; et al. Nivolumab in previously untreated melanoma without BRAF mutation. *N. Engl. J. Med.* **2015**, *372*, 320–330. [CrossRef] [PubMed]

11. Tannir, N.M.; Jonasch, E.; Albiges, L.; Altinmakas, E.; Ng, C.S.; Matin, S.F.; Wang, X.; Qiao, W.; Dubauskas Lim, Z.; Tamboli, P.; et al. Everolimus Versus Sunitinib Prospective Evaluation in Metastatic Non-Clear Cell Renal Cell Carcinoma (ESPN): A Randomized Multicenter Phase 2 Trial. *Eur. Urol.* **2016**, *69*, 866–874. [CrossRef] [PubMed]

12. Xu, K.Y.; Wu, S. Update on the treatment of metastatic clear cell and non-clear cell renal cell carcinoma. *Biomark Res.* **2015**, *3*, 5. [CrossRef] [PubMed]

13. Choueiri, T.K.; Figueroa, D.J.; Fay, A.P.; Signoretti, S.; Liu, Y.; Gagnon, R.; Deen, K.; Carpenter, C.; Benson, P.; Ho, T.H.; et al. Correlation of PD-L1 tumor expression and treatment outcomes in patients with renal cell carcinoma receiving sunitinib or pazopanib: Results from COMPARZ, a randomized controlled trial. *Clin. Cancer Res.* **2015**, *21*, 1071–1077. [CrossRef] [PubMed]

14. Wallis, C.J.D.; Klaassen, Z.; Bhindi, B.; Ye, X.Y.; Chandrasekar, T.; Farrell, A.M.; Goldberg, H.; Boorjian, S.A.; Leibovich, B.; Kulkarni, G.S.; et al. First-line Systemic Therapy for Metastatic Renal Cell Carcinoma: A Systematic Review and Network Meta-analysis. *Eur. Urol.* **2018**, *74*, 309–321. [CrossRef] [PubMed]

15. Kammerer-Jacquet, S.F.; Brunot, A.; Pladys, A.; Bouzille, G.; Dagher, J.; Medane, S.; Peyronnet, B.; Mathieu, R.; Verhoest, G.; Bensalah, K.; et al. Synchronous Metastatic Clear-Cell Renal Cell Carcinoma: A Distinct Morphologic, Immunohistochemical, and Molecular Phenotype. *Clin. Genitourin Cancer* **2017**, *15*, e1–e7. [CrossRef] [PubMed]

16. Kammerer-Jacquet, S.F.; Crouzet, L.; Brunot, A.; Dagher, J.; Pladys, A.; Edeline, J.; Laguerre, B.; Peyronnet, B.; Mathieu, R.; Verhoest, G.; et al. Independent association of PD-L1 expression with noninactivated VHL clear cell renal cell carcinoma-A finding with therapeutic potential. *Int. J. Cancer* **2017**, *140*, 142–148. [CrossRef] [PubMed]

17. Kong, W.; He, L.; Richards, E.J.; Challa, S.; Xu, C.X.; Permuth-Wey, J.; Lancaster, J.M.; Coppola, D.; Sellers, T.A.; Djeu, J.Y.; et al. Upregulation of miRNA-155 promotes tumour angiogenesis by targeting VHL and is associated with poor prognosis and triple-negative breast cancer. *Oncogene* **2014**, *33*, 679–689. [CrossRef] [PubMed]

18. Nickerson, M.L.; Jaeger, E.; Shi, Y.; Durocher, J.A.; Mahurkar, S.; Zaridze, D.; Matveev, V.; Janout, V.; Kollarova, H.; Bencko, V.; et al. Improved identification of von Hippel-Lindau gene alterations in clear cell renal tumors. *Clin. Cancer Res.* **2008**, *14*, 4726–4734. [CrossRef] [PubMed]

19. Melendez-Rodriguez, F.; Roche, O.; Sanchez-Prieto, R.; Aragones, J. Hypoxia-Inducible Factor 2-Dependent Pathways Driving Von Hippel-Lindau-Deficient Renal Cancer. *Front. Oncol.* **2018**, *8*, 214. [CrossRef] [PubMed]

20. Chintala, S.; Najrana, T.; Toth, K.; Cao, S.; Durrani, F.A.; Pili, R.; Rustum, Y.M. Prolyl hydroxylase 2 dependent and Von-Hippel-Lindau independent degradation of Hypoxia-inducible factor 1 and 2α by selenium in clear cell renal cell carcinoma leads to tumor growth inhibition. *BMC Cancer* **2012**, *12*, 293. [CrossRef] [PubMed]

21. Schodel, J.; Grampp, S.; Maher, E.R.; Moch, H.; Ratcliffe, P.J.; Russo, P.; Mole, D.R. Hypoxia, Hypoxia-inducible Transcription Factors, and Renal Cancer. *Eur. Urol.* **2016**, *69*, 646–657. [CrossRef] [PubMed]

22. Toth, K.; Chintala, S.; Rustum, Y.M. Constitutive expression of HIF-alpha plays a major role in generation of clear-cell phenotype in human primary and metastatic renal carcinoma. *Appl. Immunohistochem. Mol. Morphol.* **2014**, *22*, 642–647. [CrossRef] [PubMed]

23. Yin, M.B.; Li, Z.R.; Toth, K.; Cao, S.; Durrani, F.A.; Hapke, G.; Bhattacharya, A.; Azrak, R.G.; Frank, C.; Rustum, Y.M. Potentiation of irinotecan sensitivity by Se-methylselenocysteine in an in vivo tumor model is associated with downregulation of cyclooxygenase-2, inducible nitric oxide synthase, and hypoxia-inducible factor 1-alpha expression, resulting in reduced angiogenesis. *Oncogene* **2006**, *25*, 2509–2519. [PubMed]

24. Kondo, K.; Klco, J.; Nakamura, E.; Lechpammer, M.; Kaelin, W.G., Jr. Inhibition of HIF is necessary for tumor suppression by the von Hippel-Lindau protein. *Cancer Cell* **2002**, *1*, 237–246. [CrossRef]

25. Raval, R.R.; Lau, K.W.; Tran, M.G.; Sowter, H.M.; Mandriota, S.J.; Li, J.L.; Pugh, C.W.; Maxwell, P.H.; Harris, A.L.; Ratcliffe, P.J. Contrasting properties of hypoxia-inducible factor 1 (HIF-1) and HIF-2 in von Hippel-Lindau-associated renal cell carcinoma. *Mol. Cell. Boil.* **2005**, *25*, 5675–5686. [CrossRef] [PubMed]

26. del Peso, L.; Castellanos, M.C.; Temes, E.; Martin-Puig, S.; Cuevas, Y.; Olmos, G.; Landazuri, M.O. The von Hippel Lindau/hypoxia-inducible factor (HIF) pathway regulates the transcription of the HIF-proline hydroxylase genes in response to low oxygen. *J. Boil. Chem.* **2003**, *278*, 48690–48695. [CrossRef] [PubMed]

27. Gossage, L.; Eisen, T.; Maher, E.R. VHL, the story of a tumour suppressor gene. *Nat. Rev. Cancer* **2015**, *15*, 55–64. [CrossRef] [PubMed]

28. Kaelin, W.G., Jr. Molecular basis of the VHL hereditary cancer syndrome. *Nat. Rev. Cancer* **2002**, *2*, 673–682. [CrossRef] [PubMed]

29. Linehan, W.M.; Lerman, M.I.; Zbar, B. Identification of the von Hippel-Lindau (VHL) gene. Its role in renal cancer. *JAMA* **1995**, *273*, 564–570. [CrossRef] [PubMed]

30. Yao, X.; Tan, J.; Lim, K.J.; Koh, J.; Ooi, W.F.; Li, Z.; Huang, D.; Xing, M.; Chan, Y.S.; Qu, J.Z.; et al. VHL Deficiency Drives Enhancer Activation of Oncogenes in Clear Cell Renal Cell Carcinoma. *Cancer Discov.* **2017**, *7*, 1284–1305. [CrossRef] [PubMed]

31. Razafinjatovo, C.; Bihr, S.; Mischo, A.; Vogl, U.; Schmidinger, M.; Moch, H.; Schraml, P. Characterization of VHL missense mutations in sporadic clear cell renal cell carcinoma: Hotspots, affected binding domains, functional impact on pVHL and therapeutic relevance. *BMC Cancer* **2016**, *16*, 638. [CrossRef] [PubMed]

32. Chintala, S.; Toth, K.; Cao, S.; Durrani, F.A.; Vaughan, M.M.; Jensen, R.L.; Rustum, Y.M. Se-methylselenocysteine sensitizes hypoxic tumor cells to irinotecan by targeting hypoxia-inducible factor 1 alpha. *Cancer Chemother. Pharmacol.* **2010**, *66*, 899–911. [CrossRef] [PubMed]

33. Hu, C.J.; Wang, L.Y.; Chodosh, L.A.; Keith, B.; Simon, M.C. Differential roles of hypoxia-inducible factor 1 alpha (HIF-1alpha) and HIF-2alpha in hypoxic gene regulation. *Mol. Cell. Boil.* **2003**, *23*, 9361–9374. [CrossRef]

34. Alsaab, H.O.; Sau, S.; Alzhrani, R.; Tatiparti, K.; Bhise, K.; Kashaw, S.K.; Iyer, A.K. PD-1 and PD-L1 Checkpoint Signaling Inhibition for Cancer Immunotherapy: Mechanism, Combinations, and Clinical Outcome. *Front. Pharmacol.* **2017**, *8*, 561. [CrossRef] [PubMed]

35. Senthebane, D.A.; Rowe, A.; Thomford, N.E.; Shipanga, H.; Munro, D.; Mazeedi, M.; Almazyadi, H.A.M.; Kallmeyer, K.; Dandara, C.; Pepper, M.S.; et al. The Role of Tumor Microenvironment in Chemoresistance: To Survive, Keep Your Enemies Closer. *Int. J. Mol. Sci.* **2017**, *18*, 1586. [CrossRef] [PubMed]

36. Forster, J.C.; Harriss-Phillips, W.M.; Douglass, M.J.; Bezak, E. A review of the development of tumor vasculature and its effects on the tumor microenvironment. *Hypoxia* **2017**, *5*, 21–32. [CrossRef] [PubMed]

37. Noman, M.Z.; Chouaib, S. Targeting hypoxia at the forefront of anticancer immune responses. *Oncoimmunology* **2014**, *3*, e954463. [CrossRef] [PubMed]

38. Baine, M.K.; Turcu, G.; Zito, C.R.; Adeniran, A.J.; Camp, R.L.; Chen, L.; Kluger, H.M.; Jilaveanu, L.B. Characterization of tumor infiltrating lymphocytes in paired primary and metastatic renal cell carcinoma specimens. *Oncotarget* **2015**, *6*, 24990–25002. [CrossRef] [PubMed]

39. Schaaf, M.B.; Garg, A.D.; Agostinis, P. Defining the role of the tumor vasculature in antitumor immunity and immunotherapy. *Cell Death Dis.* **2018**, *9*, 115. [CrossRef] [PubMed]

40. Abbas, M.; Steffens, S.; Bellut, M.; Eggers, H.; Grosshennig, A.; Becker, J.U.; Wegener, G.; Schrader, A.J.; Grunwald, V.; Ivanyi, P. Intratumoral expression of programmed death ligand 1 (PD-L1) in patients with clear cell renal cell carcinoma (ccRCC). *Med. Oncol.* **2016**, *33*, 80. [CrossRef] [PubMed]

41. Callea, M.; Albiges, L.; Gupta, M.; Cheng, S.C.; Genega, E.M.; Fay, A.P.; Song, J.; Carvo, I.; Bhatt, R.S.; Atkins, M.B.; et al. Differential Expression of PD-L1 between Primary and Metastatic Sites in Clear-Cell Renal Cell Carcinoma. *Cancer Immunol. Res.* **2015**, *3*, 1158–1164. [CrossRef] [PubMed]

42. Chen, J.; Jiang, C.C.; Jin, L.; Zhang, X.D. Regulation of PD-L1: A novel role of pro-survival signalling in cancer. *Ann. Oncol.* **2016**, *27*, 409–416. [CrossRef] [PubMed]

43. Choueiri, T.K.; Fay, A.P.; Gray, K.P.; Callea, M.; Ho, T.H.; Albiges, L.; Bellmunt, J.; Song, J.; Carvo, I.; Lampron, M.; et al. PD-L1 expression in nonclear-cell renal cell carcinoma. *Ann. Oncol.* **2014**, *25*, 2178–2184. [CrossRef] [PubMed]

44. Jilaveanu, L.B.; Shuch, B.; Zito, C.R.; Parisi, F.; Barr, M.; Kluger, Y.; Chen, L.; Kluger, H.M. PD-L1 Expression in Clear Cell Renal Cell Carcinoma: An Analysis of Nephrectomy and Sites of Metastases. *J. Cancer* **2014**, *5*, 166–172. [CrossRef] [PubMed]

45. Joseph, R.W.; Millis, S.Z.; Carballido, E.M.; Bryant, D.; Gatalica, Z.; Reddy, S.; Bryce, A.H.; Vogelzang, N.J.; Stanton, M.L.; Castle, E.P.; et al. PD-1 and PD-L1 Expression in Renal Cell Carcinoma with Sarcomatoid Differentiation. *Cancer Immunol. Res.* **2015**, *3*, 1303–1307. [CrossRef] [PubMed]

46. Leite, K.R.; Reis, S.T.; Junior, J.P.; Zerati, M.; Gomes Dde, O.; Camara-Lopes, L.H.; Srougi, M. PD-L1 expression in renal cell carcinoma clear cell type is related to unfavorable prognosis. *Diagn. Pathol.* **2015**, *10*, 189. [CrossRef] [PubMed]

47. Messai, Y.; Gad, S.; Noman, M.Z.; Le Teuff, G.; Couve, S.; Janji, B.; Kammerer, S.F.; Rioux-Leclerc, N.; Hasmim, M.; Ferlicot, S.; et al. Renal Cell Carcinoma Programmed Death-ligand 1, a New Direct Target of Hypoxia-inducible Factor-2 Alpha, is Regulated by von Hippel-Lindau Gene Mutation Status. *Eur. Urol.* **2016**, *70*, 623–632. [CrossRef] [PubMed]

48. Ruf, M.; Moch, H.; Schraml, P. PD-L1 expression is regulated by hypoxia inducible factor in clear cell renal cell carcinoma. *Int. J. Cancer* **2016**, *139*, 396–403. [CrossRef] [PubMed]

49. Sun, C.; Mezzadra, R.; Schumacher, T.N. Regulation and Function of the PD-L1 Checkpoint. *Immunity* **2018**, *48*, 434–452. [CrossRef] [PubMed]

50. Lastwika, K.J.; Wilson, W., 3rd; Li, Q.K.; Norris, J.; Xu, H.; Ghazarian, S.R.; Kitagawa, H.; Kawabata, S.; Taube, J.M.; Yao, S.; et al. Control of PD-L1 Expression by Oncogenic Activation of the AKT-mTOR Pathway in Non-Small Cell Lung Cancer. *Cancer Res.* **2016**, *76*, 227–238. [CrossRef] [PubMed]

51. Mimura, K.; Teh, J.L.; Okayama, H.; Shiraishi, K.; Kua, L.F.; Koh, V.; Smoot, D.T.; Ashktorab, H.; Oike, T.; Suzuki, Y.; et al. PD-L1 expression is mainly regulated by interferon gamma associated withJAK-STAT pathway in gastric cance. *Cancer Sci.* **2018**, *109*, 43–53. [CrossRef] [PubMed]

52. Wang, X.; Yang, L.; Huang, F.; Zhang, Q.; Liu, S.; Ma, L.; You, Z. Inflammatory cytokines IL-17 and TNF-alpha up-regulate PD-L1 expression in human prostate and colon cancer cells. *Immunol. Lett.* **2017**, *184*, 7–14. [CrossRef] [PubMed]

53. Neal, C.S.; Michael, M.Z.; Rawlings, L.H.; Van der Hoek, M.B.; Gleadle, J.M. The VHL-dependent regulation of microRNAs in renal cancer. *BMC Med.* **2010**, *8*, 64. [CrossRef] [PubMed]

54. Teng, M.W.; Ngiow, S.F.; Ribas, A.; Smyth, M.J. Classifying Cancers Based on T-cell Infiltration and PD-L1. *Cancer Res.* **2015**, *75*, 2139–2145. [CrossRef] [PubMed]

55. Kaunitz, G.J.; Cottrell, T.R.; Lilo, M.; Muthappan, V.; Esandrio, J.; Berry, S.; Xu, H.; Ogurtsova, A.; Anders, R.A.; Fischer, A.H.; et al. Melanoma subtypes demonstrate distinct PD-L1 expression profiles. *Lab. Investig.* **2017**, *97*, 1063–1071. [CrossRef] [PubMed]

56. Noman, M.Z.; Janji, B.; Hu, S.; Wu, J.C.; Martelli, F.; Bronte, V.; Chouaib, S. Tumor-Promoting Effects of Myeloid-Derived Suppressor Cells Are Potentiated by Hypoxia-Induced Expression of miR-210. *Cancer Res.* **2015**, *75*, 3771–3787. [CrossRef] [PubMed]

57. Miscoria, M.; Di Loreto, C.; Puglisi, F.; Murray, P.G.; Deroma, L.; Atmadini, M. Thymidine phosphorylase expression in metastatic kidney cancer as a potential predictor of outcome in patients treated with sunitinib. *J. Clin. Oncol.* **2012**. [CrossRef]

58. Eda, H.; Fujimoto, K.; Watanabe, S.; Ura, M.; Hino, A.; Tanaka, Y.; Wada, K.; Ishitsuka, H. Cytokines induce thymidine phosphorylase expression in tumor cells and make them more susceptible to 5′-deoxy-5-fluorouridine. *Cancer Chemother. Pharmacol.* **1993**, *32*, 333–338. [CrossRef] [PubMed]

59. Huang, X.; Wang, L.; Chen, Y.; Zheng, X.; Wang, X. Poor Prognosis Associated with High Levels of Thymidine Phosphorylase and Thrombocytosis in Patients with Renal Cell Carcinoma. *Urol. Int.* **2017**, *98*, 162–168. [CrossRef] [PubMed]

60. Lin, S.; Lai, H.; Qin, Y.; Chen, J.; Lin, Y. Thymidine phosphorylase and hypoxia-inducible factor 1-alpha expression in clinical stage II/III rectal cancer: Association with response to neoadjuvant chemoradiation therapy and prognosis. *Int. J. Clin. Exp. Pathol.* **2015**, *8*, 10680–10688. [PubMed]

61. Atrih, A.; Mudaliar, M.A.V.; Zakikhani, P.; Lamont, D.J.; Huang, J.T.-J.; Bray, S.E.; Barton, G.; Fleming, S.; Nabi, G. Quantitative proteomics in resected renal cancer tissue for biomarker discovery and profiling. *Br. J. Cancer* **2014**, *110*, 1622–1633. [CrossRef] [PubMed]

62. Padrik, P.; Saar, H. Thymidine phosphorylase as a prognostic factor in renal cell carcinoma. *Int. Urol. Nephrol.* **2010**, *42*, 295–298. [CrossRef] [PubMed]

63. Takayama, T.; Mugiya, S.; Sugiyama, T.; Aoki, T.; Furuse, H.; Liu, H.; Hirano, Y.; Kai, F.; Ushiyama, T.; Ozono, S. High levels of thymidine phosphorylase as an independent prognostic factor in renal cell carcinoma. *Jpn. J. Clin. Oncol.* **2006**, *36*, 564–569. [CrossRef] [PubMed]

64. Gimm, T.; Wiese, M.; Teschemacher, B.; Deggerich, A.; Schodel, J.; Knaup, K.X.; Hackenbeck, T.; Hellerbrand, C.; Amann, K.; Wiesener, M.S.; et al. Hypoxia-inducible protein 2 is a novel lipid droplet protein and a specific target gene of hypoxia-inducible factor-1. *FASEB J.* **2010**, *24*, 4443–4458. [CrossRef] [PubMed]

65. Du, W.; Zhang, L.; Brett-Morris, A.; Aguila, B.; Kerner, J.; Hoppel, C.L.; Puchowicz, M.; Serra, D.; Herrero, L.; Rini, B.I.; et al. HIF drives lipid deposition and cancer in ccRCC via repression of fatty acid metabolism. *Nat. Commun.* **2017**, *8*, 1769. [CrossRef] [PubMed]

66. Cao, Q.; Ruan, H.; Wang, K.; Song, Z.; Bao, L.; Xu, T.; Xiao, H.; Wang, C.; Cheng, G.; Tong, J.; et al. Overexpression of PLIN2 is a prognostic marker and attenuate tumor progression in clear cell renal cell carcinoma. *Int J. Oncol.* **2018**, *53*, 137–147. [CrossRef] [PubMed]

67. Azrak, R.G.; Cao, S.; Durrani, F.A.; Toth, K.; Bhattacharya, A.; Rustum, Y.M. Augmented therapeutic efficacy of irinotecan is associated with enhanced drug accumulation. *Cancer Lett.* **2011**, *311*, 219–229. [CrossRef] [PubMed]

68. Bhattacharya, A.; Seshadri, M.; Oven, S.D.; Toth, K.; Vaughan, M.M.; Rustum, Y.M. Tumor vascular maturation and improved drug delivery induced by methylselenocysteine leads to therapeutic synergy with anticancer drugs. *Clin. Cancer Res.* **2008**, *14*, 3926–3932. [CrossRef] [PubMed]

69. Jain, R.K. Normalizing tumor microenvironment to treat cancer: Bench to bedside to biomarkers. *J. Clin. Oncol.* **2013**, *31*, 2205–2218. [CrossRef] [PubMed]

70. Rustum, Y.M.; Toth, K.; Seshadri, M.; Sen, A.; Durrani, F.A.; Stott, E.; Morrison, C.D.; Cao, S.; Bhattacharya, A. Architectural heterogeneity in tumors caused by differentiation alters intratumoral drug distribution and affects therapeutic synergy of antiangiogenic organoselenium compound. *J. Oncol.* **2010**, *2010*, 396286. [CrossRef] [PubMed]

71. Lytle, J.R.; Yario, T.A.; Steitz, J.A. Target mRNAs are repressed as efficiently by microRNA-binding sites in the 5′ UTR as in the 3′ UTR. *Proc. Natl. Acad. Sci. USA* **2007**, *104*, 9667–9672. [CrossRef] [PubMed]

72. Li, Z.; Rana, T.M. Therapeutic targeting of microRNAs: Current status and future challenges. *Nat. Rev. Drug Discov.* **2014**, *13*, 622–638. [CrossRef] [PubMed]

73. Gulyaeva, L.F.; Kushlinskiy, N.E. Regulatory mechanisms of microRNA expression. *J. Transl. Med.* **2016**, *14*, 143. [CrossRef] [PubMed]

74. Kuninty, P.R.; Schnittert, J.; Storm, G.; Prakash, J. MicroRNA Targeting to Modulate Tumor Microenvironment. *Front. Oncol.* **2016**, *6*, 3. [CrossRef] [PubMed]

75. Li, M.; Wang, Y.; Song, Y.; Bu, R.; Yin, B.; Fei, X.; Guo, Q.; Wu, B. MicroRNAs in renal cell carcinoma: A systematic review of clinical implications (Review). *Oncol. Rep.* **2015**, *33*, 1571–1578. [CrossRef] [PubMed]

76. McCormick, R.I.; Blick, C.; Ragoussis, J.; Schoedel, J.; Mole, D.R.; Young, A.C.; Selby, P.J.; Banks, R.E.; Harris, A.L. miR-210 is a target of hypoxia-inducible factors 1 and 2 in renal cancer, regulates ISCU and correlates with good prognosis. *Br. J. Cancer* **2013**, *108*, 1133–1142. [CrossRef] [PubMed]

77. Schanza, L.M.; Seles, M.; Stotz, M.; Fosselteder, J.; Hutterer, G.C.; Pichler, M.; Stiegelbauer, V. MicroRNAs Associated with Von Hippel-Lindau Pathway in Renal Cell Carcinoma: A Comprehensive Review. *Int. J. Mol. Sci.* **2017**, *18*, 2495. [CrossRef] [PubMed]

78. Shen, G.; Li, X.; Jia, Y.F.; Piazza, G.A.; Xi, Y. Hypoxia-regulated microRNAs in human cancer. *Acta Pharmacol. Sin.* **2013**, *34*, 336–341. [CrossRef] [PubMed]

79. Tang, K.; Xu, H. Prognostic value of meta-signature miRNAs in renal cell carcinoma: An integrated miRNA expression profiling analysis. *Sci. Rep.* **2015**, *5*, 10272. [CrossRef] [PubMed]

80. Wang, Q.; Lin, W.; Tang, X.; Li, S.; Guo, L.; Lin, Y.; Kwok, H.F. The Roles of microRNAs in Regulating the Expression of PD-1/PD-L1 Immune Checkpoint. *Int. J. Mol. Sci.* **2017**, *18*, 2540. [CrossRef] [PubMed]

81. Yee, D.; Shah, K.M.; Coles, M.C.; Sharp, T.V.; Lagos, D. MicroRNA-155 induction via TNF-alpha and IFN-gamma suppresses expression of programmed death ligand-1 (PD-L1) in human primary cells. *J. Boil. Chem.* **2017**, *292*, 20683–20693. [CrossRef] [PubMed]

82. Durrani, F.; Cao, S.; Park, Y.-M.; Thiompson, G.; Martin, J.; Yang, G.; Kuettel, M.; Rustum, Y.M. Synergistic effect of selenium compounds with radiation therapy in human A549 lung xenografts. *Cancer Res.* **2007**, *67* (Suppl. 9), 750.

83. Durrani, F.A.; Chintala, S.; Toth, K.; Cao, S.; Rustum, Y.M. Mechanism-based drug combination targeting HIF-2α and VEGF in renal cancer xenografts. *Trends Cell Mol. Boil.* **2015**, *10*, 12.

84. Frost, J.; Galdeano, C.; Soares, P.; Gadd, M.S.; Grzes, K.M.; Ellis, L.; Epemolu, O.; Shimamura, S.; Bantscheff, M.; Grandi, P.; et al. Potent and selective chemical probe of hypoxic signalling downstream of HIF-alpha hydroxylation via VHL inhibition. *Nat. Commun.* **2016**, *7*, 13312. [CrossRef] [PubMed]

85. Soares, P.; Gadd, M.S.; Frost, J.; Galdeano, C.; Ellis, L.; Epemolu, O.; Rocha, S.; Read, K.D.; Ciulli, A. Group-Based Optimization of Potent and Cell-Active Inhibitors of the von Hippel-Lindau (VHL) E3 Ubiquitin Ligase: Structure-Activity Relationships Leading to the Chemical Probe (2*S*,4*R*)-1-((*S*)-2-(1-Cyanocyclopropanecarboxamido)-3,3-dimethylbutanoyl)-4-hydroxy-*N*-(4-(4-methylthiazol-5-yl)benzyl)pyrrolidine-2-carboxamide (VH298). *J. Med. Chem.* **2018**, *61*, 599–618. [PubMed]

86. Courtney, K.D.; Infante, J.R.; Lam, E.T.; Figlin, R.A.; Rini, B.I.; Brugarolas, J.; Zojwalla, N.J.; Lowe, A.M.; Wang, K.; Wallace, E.M.; et al. Phase I Dose-Escalation Trial of PT2385, a First-in-Class Hypoxia-Inducible Factor-2alpha Antagonist in Patients with Previously Treated Advanced Clear Cell Renal Cell Carcinoma. *J. Clin. Oncol.* **2018**, *36*, 867–874. [CrossRef] [PubMed]

87. Kim, S.; Lee, E.; Jung, J.; Lee, J.W.; Kim, H.J.; Kim, J.; Yoo, H.J.; Lee, H.J.; Chae, S.Y.; Jeon, S.M.; et al. microRNA-155 positively regulates glucose metabolism via PIK3R1-FOXO3a-cMYC axis in breast cancer. *Oncogene* **2018**, *37*, 2982–2991. [CrossRef] [PubMed]

88. Gao, Y.; Ma, X.; Yao, Y.; Li, H.; Fan, Y.; Zhang, Y.; Zhao, C.; Wang, L.; Ma, M.; Lei, Z.; et al. miR-155 regulates the proliferation and invasion of clear cell renal cell carcinoma cells by targeting E2F2. *Oncotarget* **2016**, *7*, 20324–20337. [CrossRef] [PubMed]

89. Yao, R.; Ma, Y.L.; Liang, W.; Li, H.H.; Ma, Z.J.; Yu, X.; Liao, Y.H. MicroRNA-155 modulates Treg and Th17 cells differentiation and Th17 cell function by targeting SOCS1. *PLoS ONE* **2012**, *7*, e46082. [CrossRef] [PubMed]

90. Ivan, M.; Harris, A.L.; Martelli, F.; Kulshreshtha, R. Hypoxia response and microRNAs: No longer two separate worlds. *J. Cell. Mol. Med.* **2008**, *12*, 1426–1431. [CrossRef] [PubMed]

91. Nguyen, D.D.; Chang, S. Development of Novel Therapeutic Agents by Inhibition of Oncogenic MicroRNAs. *Int. J. Mol. Sci.* **2017**, *19*, 65. [CrossRef] [PubMed]

92. Schmidt, M.F. Drug target miRNAs: Chances and challenges. *Trends Biotechnol.* **2014**, *32*, 578–585. [CrossRef] [PubMed]

93. Wigerup, C.; Pahlman, S.; Bexell, D. Therapeutic targeting of hypoxia and hypoxia-inducible factors in cancer. *Pharmacol. Ther.* **2016**, *164*, 152–169. [CrossRef] [PubMed]

94. Jedeszko, C.; Paez-Ribes, M.; Di Desidero, T.; Man, S.; Lee, C.R.; Xu, P.; Bjarnason, G.A.; Bocci, G.; Kerbel, R.S. Postsurgical adjuvant or metastatic renal cell carcinoma therapy models reveal potent antitumor activity of metronomic oral topotecan with pazopanib. *Sci. Transl. Med.* **2015**, *7*, 282ra250. [CrossRef] [PubMed]

95. Oevermann, K.; Buer, J.; Hoffmann, R.; Franzke, A.; Schrader, A.; Patzelt, T.; Kirchner, H.; Atzpodien, J. Capecitabine in the treatment of metastatic renal cell carcinoma. *Br. J. Cancer* **2000**, *83*, 583–587. [CrossRef] [PubMed]

96. Tannir, N.M.; Thall, P.F.; Ng, C.S.; Wang, X.; Wooten, L.; Siefker-Radtke, A.; Mathew, P.; Pagliaro, L.; Wood, C.; Jonasch, E. A phase II trial of gemcitabine plus capecitabine for metastatic renal cell cancer previously treated with immunotherapy and targeted agents. *J. Urol.* **2008**, *180*, 867–872; discussion 872. [CrossRef] [PubMed]

97. Cao, S.; Durrani, F.A.; Rustum, Y.M. Selective modulation of the therapeutic efficacy of anticancer drugs by selenium containing compounds against human tumor xenografts. *Clin. Cancer Res.* **2004**, *10*, 2561–2569. [CrossRef] [PubMed]

98. Cao, S.; Durrani, F.A.; Toth, K.; Rustum, Y.M. Se-methylselenocysteine offers selective protection against toxicity and potentiates the antitumour activity of anticancer drugs in preclinical animal models. *Br. J. Cancer* **2014**, *110*, 1733–1743. [CrossRef] [PubMed]

99. Zakharia, Y.; Garje, R.; Brown, J.A.; Nepple, K.G.; Dahmoush, L.; Gibson-Corley, K. Phase1 clinical trial of high doses of Seleno-L-methionine (SLM), in sequential combination with axitinib in previously treated and relapsed clear cell renal cell carcinoma (ccRCC) patients. *J. Clin. Oncol.* **2018**, *36* (Suppl. 6), 630. [CrossRef]

100. Zakharia, Y.; Bhattacharya, A.; Rustum, Y.M. Selenium targets resistance biomarkers enhancing efficacy while reducing toxicity of anti-cancer drugs: Preclinical and clinical development. *Oncotarget* **2018**, *9*, 10765–10783. [CrossRef] [PubMed]

International Journal of
Molecular Sciences

MDPI

Review

Selenium-Binding Protein 1 in Human Health and Disease

Mostafa Elhodaky and Alan M. Diamond *

Department of Pathology, University of Illinois at Chicago, Chicago, IL 60612, USA; melhod2@uic.edu
* Correspondence: adiamond@uic.edu; Tel.: +1-312-413-8747

Received: 20 August 2018; Accepted: 31 October 2018; Published: 2 November 2018

Abstract: Selenium-binding protein 1 (SBP1) is a highly conserved protein that covalently binds selenium. SBP1 may play important roles in several fundamental physiological functions, including protein degradation, intra-Golgi transport, cell differentiation, cellular motility, redox modulation, and the metabolism of sulfur-containing molecules. SBP1 expression is often reduced in many cancer types compared to the corresponding normal tissues and low levels of SBP1 are frequently associated with poor clinical outcome. In this review, the transcriptional regulation of *SBP1*, the different physiological roles reported for SBP1, as well as the implications of SBP1 function in cancer and other diseases are presented.

Keywords: selenium-binding protein 1; SBP1; SELENBP1; hSP56; cancer; disease

1. Introduction

Selenium (Se) is a non-metallic, essential trace element for many organisms, including humans. Se has long been recognized for its potential in cancer prevention as evidenced by multiple animal, and human epidemiological studies that have reported an inverse association between Se status and cancer risk [1–8]. Many mechanisms have been suggested for the chemopreventive effect of Se [9–11], including DNA hypomethylation [12], blocked cell cycle progression, enhanced cell death, decreased cell proliferation, increased glutathione peroxidase or thioredoxin reductases activities [13], modulated ER stress response [14], and enhanced DNA repair [15]. Furthermore, Se has been found to play a key role in mammalian development [16] and immune function [17,18]. Low levels of Se may be a contributing factor to several pathologies, including male infertility [19], heart disease [20], inflammation [21,22], and neuromuscular disorders [23]. It is generally recognized that important cellular and organismal functions of Se are likely mediated by the action of selenoproteins constituting the mammalian selenoproteome [24]. While the functions of many selenoproteins are still unknown, they likely have a significant role in human health and disease. Human selenoproteins are generally classified into three categories [11,25]. The first category includes proteins in which Se is cotranslationally incorporated into the elongating peptide as the amino acid selenocysteine in response to an in-frame UGA codon in the corresponding messenger RNA [26]. The human selenoproteome contains 25 genes [25]. The second category consists of proteins in which Se is incorrectly substituted for sulfur in sulfur-containing amino acids due to the similarity in structure between these two elements. The third category is composed of selenium-binding proteins which bind Se by an unknown mechanism. This review will be primarily focused on one member of the latter category, selenium-binding protein 1 (SBP1, SELENBP1, hSP56).

2. SBP1 Discovery

SBP1 was first discovered in mouse liver in 1989 by Bansal et al. using ^{75}Se labelling. Normal 6-week old female BALB/c mice were given a single intraperitoneal injection of ^{75}Se in the form of

Na$_2$SeO$_3$. After 40 h, animals were euthanized and livers were harvested for preparation of liver cytosols which were then used for a combination of gel filtration, ion-exchange chromatography and SDS-PAGE techniques. This led to identification of four selenium-binding proteins of apparent molecular weights of 12, 14, 24, and 56 kDa [27]. The 56-kDa protein was designated as SBP1, whereas the 24-kDa protein was identified as glutathione peroxidase 1 (GPX1), an enzyme that detoxifies hydroperoxides using reducing equivalents from glutathione [28]. The full-length human *SBP1* cDNA clone was first described by Chang et al. in 1997 and determined to be 1668 base pair (bp) long with an open reading frame encoding 472 amino acids [29]. *SBP1* is abundantly expressed in various human tissues, including liver, lung, prostate, colon, and pancreas, while moderate levels were detected in spleen, heart, and ovary. In contrast, its expression was barely detectable in thymus, testis, and peripheral blood leukocytes [30]. SBP1 is a highly-conserved protein. Flemetakis et al. reported that the predicted amino acid sequence of SBP1 is conserved in both plants and animals, ranging from 77 to 88% in plants, while the identity between the plants and mammalian proteins ranged from 57 to 60% [31]. By comparison, this degree of homology is higher than other conserved proteins, such as HSP60, γ-tubulin, apoptotic cell death 1 protein, and eIF4E whose identities of the plant and human proteins are 44, 49, 48, and 52%, respectively [31]. The homology between the mammalian *SBP1* of mice and humans is 86% [31], indicating that the potential fundamental cellular and molecular functions for SBP1 are also conserved across different species. SBP1 is very similar to another selenium-associated protein, selenium liver binding protein (AP-56, SBP2), whose sequence differs by only 14 residues from SBP1 and is encoded by a distinct gene [32]. AP-56 is implicated in the detoxification of acetaminophen in the liver [32]. Although these genes are regulated differently, their similarity may indicate a role for SBP1 in detoxification.

3. The Role of Se in SBP1

The form of Se in SBP1 is currently unknown. Se is stably associated with SBP1, probably through a selenosulfide bond (perselenide), as indicated by the binding of Se to SBP1 being reversed by the addition of a reducing agent during SDS-PAGE [33]. Based on structural and functional studies, it was suggested that one cysteine in SBP1 was the likely binding site for the Se molecule, the cysteine at position 57 [34]. Converting cysteine 57 in SBP1 to a glycine and ectopically expressing that protein in human HCT116 cells that do not express detectable SBP1 levels indicated that the loss of the cysteine reduced the half-life of the protein, induced mitochondrial damage, and attenuated the degree of phosphorylation of signaling proteins such as p53 and GSK3β compared to the native protein expressed at similar levels [35].

The Se in SBP1 may facilitate its interaction with other proteins. SBP1 physically interacts with von Hippel–Lindau protein–interacting deubiquitinating enzyme 1 (VDU1), which plays a role in proteasomal protein degradation [33,36]. This indicates that SBP1, via its interaction with VDU1, may have a role in ubiquitination/deubiquitination-mediated protein degradation and detoxification pathways. When the Se moiety was dissociated from SBP1 by the addition of ß-mercaptoethanol, the interaction with VDU1 was completely blocked, indicating that Se may be essential for the interaction of these two proteins [33]. While the Se moiety is likely required for its interaction with VDU1, the inclusion of Se in SBP1 does not appear to be essential for functioning as methanethiol oxidase (MTO), a recently-discovered novel human SBP1 enzyme activity that metabolizes sulfur-containing molecules [37].

As a non-selenocysteine containing protein, SBP1 is not considered as a part of the "selenium hierarchy" that describes the relative response of selenoproteins to the availability of Se [38]. Initial studies feeding rats varying amounts of Se led to the conclusion that SBP1 levels were not likely dependent upon dietary Se supplementation [39]. However, there may be indirect regulation of SBP1 by Se due to its interaction with GPX1, a member of the selenocysteine-containing selenoproteins. GPX1 is a highly conserved and ubiquitously expressed enzyme that detoxifies hydrogen and lipid peroxides and is implicated in several diseases by human genetics [40]. There is a reciprocal regulatory relationship between SBP1 and GPX1. Ectopically expressing SBP1 in HCT116 human colon cancer cells that do not express endogenous SBP1 resulted in the inhibition of GPX1 enzyme activity without affecting protein levels [28], indicating a likely physical interaction. Consistent with this possibility was data indicating that knocking down *SBP1* in human liver cells resulted in a 4–5 fold increase in GPX activity, also without altering protein levels [41].

Expressing GPX1 in MCF7 human breast cancer cells that do not exhibit detectable GPX1 levels resulted in a decline in both SBP1 mRNA and protein levels [28]. The reciprocal relationship between SBP1 and GPX1 has also been established in mouse colon and duodenum epithelial cells [28], as well as human prostate and liver tissues [41,42]. This raises the possibility that SBP1 can be indirectly downregulated by Se because GPX1 is high on the Se hierarchy, being among the selenoproteins most responsive to Se availability. Support for the indirect regulation of SBP1 by GPX1 comes from experiments showing that increasing Se in the culture media of MCF7 cells caused a dramatic reduction in SBP1 levels only when GPX1 was present and GPX1 levels were increased by the Se supplementation [28].

4. SBP1 Levels Are Reduced in Cancer and Low Levels Are Predictive of Clinical Outcome

One of the striking observations about SBP1 is the diversity of the types of cancers in which SBP1 was found to be reduced compared to normal or benign tissues (reviewed in [43]), including cancers of the thyroid [44], lung [45], stomach [46,47], liver [41], kidney [48], ovary [49–51], breast [52], prostate [53,54], colon [55,56], head and neck [57], and malignant melanoma [58]. In addition to being lower in cancers, the degree of reduction of SBP1 in resected tissues is often predictive of how long a patient will be cancer free and survive their disease [43]. Reduced SBP1 levels have been correlated with poor survival in several types of carcinomas, including colorectal [55,59], gastric [47], nasopharyngeal [57], pulmonary [45], renal [48], and prostate [53] cancers. Recently, a search for genetic variations in selenoprotein genes revealed that a polymorphism in the gene for SBP1, along with variations in the genes of selenocysteine encoding genes, were associated with prostate cancer aggressiveness at diagnosis [60]. The exception to this pattern is ovarian cancer where higher levels of SBP1 were associated with poor survival [50].

In addition to its levels, the distribution of SBP1 between cellular compartments may be relevant to cancer etiology. The associations between prostatic SBP1 levels, tumor grade, and disease recurrence following prostatectomy were investigated using a tissue microarray containing tissue from more than 200 prostate cancer patients who experienced biochemical (PSA) recurrence after prostatectomy and matched control patients whose cancer did not recur [53]. Reduced SBP1 levels were associated with a higher likelihood of prostate cancer recurrence, as has been seen in other cancer types. The subcellular localization of SBP1 was both nuclear and cytoplasmic, with nuclear staining being sporadic (Figure 1). However, a lower nuclear-to-cytoplasmic distribution of SBP1 was associated with a higher tumor grade (Gleason score) [53]. These results indicate that sequestration of SBP1 in a particular cellular compartment may restrict access to relevant substrates or the protein has different functions at these locations.

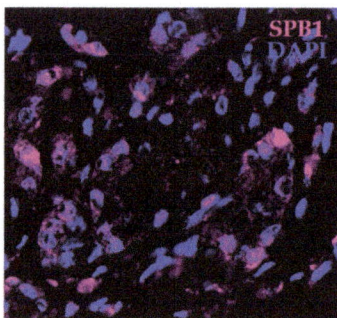

Figure 1. Localization of SBP1 in prostate cancer. Human prostate cancer tissue showing cells that express SBP1 (magenta) mostly in the cytoplasm and several cells that express SBP1 in the nucleus. Nuclei are highlighted with DAPI (blue).

4.1. Is SBP1 a Tumor Suppressor?

The frequent loss of SBP1 in cancer and the association of reduced SBP1 levels with greater mortality could imply that SBP1 is a tumor suppressor. Alternatively, its loss or downregulation may be a consequence of cancer development and progression, and the reduced levels represent a mere "bystander effect". Data supporting the direct role of SBP1 in cancer comes from studies where its levels are altered in cells and consequences relevant to transformation and tumorigenesis are revealed. Over-expressing SBP1 in colon, gastric, and prostate cancer cells have generally yielded results consistent with a tumor suppressor function, including reduced growth in semi-solid media and decreased tumorigenicity in xenograft studies using immune-deficient mice [46,53,54,61,62]. When over-expressed in lung cancer cells, SBP1 reduced proliferation and induced greater apoptosis compared to control cells only when the cells were challenged with H_2O_2 [41]. Some of the phenotypic consequences of over-expressing SBP1 may be due to the downstream activation of the p53 tumor suppressor protein. Over-expression of SBP1 in HCT116 human colon cancer cells resulted in the increased phosphorylation of p53 [53]. In addition to the phosphorylation of p53, SBP1 over-expression in the same cells resulted in the differential expression of 132 proteins, many are associated with energy metabolism and MAPK, Wnt, NF-κB, and Notch signaling [61]. This same study reported that the expression of SBP1 resulted in the reduction of TWIST1, a critical regulator of the epithelial-mesenchymal transition and metastasis.

Consistent with over-expression data, either knocking down *SBP1* or inactivating the gene using CRISPR/Cas9 editing in mouse lung cancer cells and injecting these cells into syngeneic hosts increased the size of tumors obtained compared to controls, although the number of tumors was not increased [63]. Knockout mice that are null for SBP1 exhibit very limited pathology and do not develop tumors [64]. However, examining the ovaries from these animals by gene expression microarrays indicated the increased expression of several genes associated with ovarian carcinogenesis, including *Notch1* and *Fas1* [64]. Less clear is why tumor suppressor genes such as *Apc*, *RB1*, and *Trp53* were also enhanced in the ovaries from these mice. Collectively, studies altering the levels of SBP1 provide substantial evidence that SBP1 serves as a tumor suppressor and its loss or downregulation during cancer development contributes to disease development or progression.

4.2. Is SBP1 Downregulation an Early or Late Event in the Process of Tumorigenesis?

Given the data presented above indicating the frequent downregulation of SBP1 in cancers and its association with poor outcomes, it raises the issue of whether SBP1 loss occurs early in cancer development or late in the process, contributing to cancer progression. This issue was investigated by Zhang et al. who examined SBP1 levels in tissues classified as gastric cancer, precursor lesions, and matched controls of corresponding non-neoplastic epithelial tissues [65]. SBP1 was reduced in most of the gastric cancer tissues compared to its abundant expression in matched non-neoplastic controls and precursor lesions, including tissues obtained from gastric ulcers, gastric polyps, as well as tissues presenting with chronic atrophic gastritis, intestinal metaplasia and dysplasia [65]. SBP1 expression was similar in tissues with different levels of intestinal metaplasia or dysplasia indicating that the reduction of SBP1 levels may be a late event associated with gastric carcinoma progression from normal gastric epithelium or premalignant lesions [65]. These results are consistent with those of Kim et al. who observed much lower levels of SBP1 in colorectal carcinomas compared to matched controls of normal tissues and colon adenomas, supporting the notion that SBP1 loss is a late event during tumorigenesis [59]. In addition, changes in the levels of proteins that occur during the progression of human squamous lung cancer were investigated using isobaric tags for relative and absolute quantitation labeling combined with 2D LC-MS/MS [66]. SBP1 levels were determined by western blotting and immunohistochemistry and shown to be progressively lost during bronchial epithelial cancer progression [66].

In contrast to the data obtained examining gastric, bronchial, and colonic tissues, results have also been reported indicating that the reduction in SBP1 expression may be an early event in the evolution of some tumor types, including ovarian cancer [50] and uterine leiomyoma [67]. Huang et al. identified SBP1 to be the most significantly reduced protein in ovarian cancer cell lines, including DOV13, OVCA429, OVCA882, TOV112D, and SKOV3 using a membrane proteome profiling analysis [50]. However, relatively low levels of SBP1 were also observed in the immortalized human ovarian cell line, HOSE2089, indicating that the reduction of SBP1 may have occurred during the immortalization process [50]. SBP1 expression was also approximately 4-fold lower in leiomyoma samples compared to normal myometrium, as determined by western blotting and immunohistochemistry [67]. However, SBP1 levels were similar in tissues obtained from patients with proliferative secretory and atrophic endometrium in either leiomyoma or normal myometrium. These contrasting results may indicate distinct roles of SBP1 in the development of cancers of different origins.

5. Physiological Roles for SBP1

The impact of SBP1 on normal biological processes and pathologies other than cancer may be due to roles in the modulation of cellular redox homeostasis. The SBP1 amino acid sequence contains two bis (cysteinyl) sequence motifs, Cys-X-X-Cys, at Cys5-X-X-Cys8 and Cys80-X-X-Cys83 shown to be a characteristic feature among several proteins which are involved in modulating the cellular redox state in vivo [68]. In addition, SBP1 may also modulate the redox state of the extracellular environment. Experimental data in support of this comes from a study where the knockdown of *SBP1* in MCF-7 breast cancer and HC116 colon cancer cells by siRNA resulted in increased levels of H_2O_2 and superoxide ion, leading to enhanced apoptosis when cells were exposed to selenite [69]. The authors attributed this effect to the significant increase in extracellular glutathione in the culture media. Changes in either the intracellular or extracellular environment can potentially impact a broad range of biological processes responsive to reactive oxygen in signaling pathways and contribute to the pathology associated with SBP1 dysregulation.

SBP1 has also been implicated in the late stages of intra-Golgi transport. Using an in vitro intra-Golgi cell-free transport assay, both endogenous and recombinant SBP1 (rSBP1) exhibited transport activity in the cell-free assay and the addition of antibodies directed against SBP1 abolished this activity [70]. This data indicates that SBP1 may be regulating vesicular intra-Golgi transport, particularly at the docking or fusions stages [70]. The reported roles for SBP1 are summarized in Figure 2.

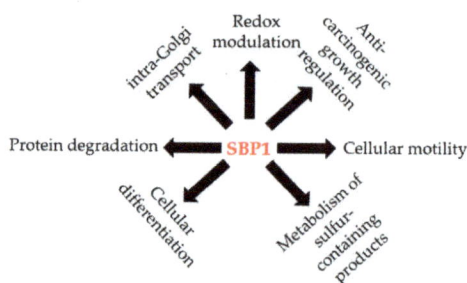

Figure 2. The potential roles of SBP1 in human health and disease. Illustration of the different potential functions reported for SBP1 in the published literature.

Tissue-Specific Roles for SBP1

Several studies have indicated a potential role for SBP1 in neurobiology. SBP1 has been localized at the tips of rapidly extending protrusions in T98G glioblastoma multiforme cells in vitro [71]. Cell protrusive motility, which is tightly associated with actin filament polymerization, is an essential function for multiple cellular processes, including cell proliferation and migration. Monomeric G-actin, but not filamentous F-actin, was shown to be recruited to the SBP1-positive tip, indicating that the recruitment of SBP1 and G-actin at the cell margin precedes actin polymerization [71]. In addition, SBP1 recruitment to the cell margin was observed to precede that of G-actin. The extension of the protrusion will stop when G-actin polymerizes to F-actin at the protruding edges, hence, SBP1 and G-actin disappear from these margins. SBP1 also localized to the growing tips of neurites in SH-SY5Y neuroblastoma cells in vitro [71], possibly indicating a role for SBP1 in neuronal cell outgrowth.

Changes in the levels of SBP1 in neuronal tissues may implicate the protein in several neuropathologies. *SBP1* mRNA has reported to be elevated in the frontal cortex of patients with schizophrenia, indicating a potential specialized role in the pathophysiology of schizophrenia and the central nervous system [72–74]. Genetic data has also implicated SBP1 in the risk of schizophrenia as two single nucleotide polymorphisms in the *SBP1* gene (rs2800953 and rs10788804) have been identified as susceptibility loci for schizophrenia in a family-wide association study [75]. This, and a report of plasma SBP1 protein levels being decreased in patients with recent-onset schizophrenia [76], collectively indicate a potential specialized role for SBP1 in the pathophysiology of this disease. Whether these data indicate a role for SBP1 in the proper functioning of the central nervous system or the potential neuroprotective effect of Se against oxidative and excitatory brain damage [77] remains to be determined.

SBP1 may also be involved with the pathogenesis of glaucoma. Elevated levels of SBP1 have been associated with elevated ocular pressure [78]. It was also identified as a differentially expressed gene in datasets comparing transcripts in glaucoma to normal control tissues, which has been verified in a rat model of acute elevated intraocular pressure [79]. SBP1 was also identified as a novel target antigen in patients with Behçet's disease (BD) with uveitis, where an autoimmune response to retinal antigens is considered to be involved in the pathogenesis of the uveitis in those patients [80,81]. What if any role SBP1 plays in these diseases has not yet been investigated.

6. The Transcriptional Regulation of SBP1

A greater understanding of the biological roles of SBP1 could be gained by examining how its expression is regulated. A subtractive hybridization approach was used to identify transcripts that were more abundant in the relatively fast growing PC-3 human prostate cancer cells compared to slow growing LNCaP cells [30]. The low levels of *SBP1* mRNA in LNCaP cells was shown to be due to the down regulation of *SBP1* transcription as treatment with of androgen-sensitive LNCaP cells with dihydrotestosterone (DHT, active form of androgen) reduced the levels of *SBP1* mRNA in a reversible, concentration-dependent manner [30]. A more complicated picture was revealed by the analysis of the effect of androgen on normal ovarian epithelial cells obtained from the scraping of the ovary surface of patients with benign disease, an immortalized cell line, and ovarian cancer cell lines [50]. Treatment of the primary and immortalized cells with DHT reduced the levels of *SBP1* mRNA while SBP1 levels were increased in four tumor-derived cell lines by DHT treatment. The mechanism accounting for the differential response of these cell lines to DHT has not been resolved.

In addition to androgens, *SBP1* expression is also downregulated by estrogen treatment (17-β estradiol) in estrogen receptor (ER)-positive breast cancer cells, but not in ER-negative cells [52]. The suppression of SBP1 expression by transforming growth factor beta (TGF-β) was also observed using a rhesus monkey renal allograft model to identify molecules involved in the pathogenesis of chronic allograft nephropathy (CAN) [82]. SBP1 was absent or markedly reduced in vascular smooth muscle cells in monkey kidney allografts with CAN. Testing growth factors previously associated with graft rejection, including IFNγ, TNFα, and PDGF, only TGF-β blocked the expression of SBP1 in the normal human vascular smooth muscle cell line, CRL-1999 [82]. It is unlikely that the effects of androgens or estrogens on *SBP1* transcription is a direct consequence of the binding of the corresponding receptor to the *SBP1* promoter as there does not appear to be a consensus binding sequence for the receptor/transcription factor.

The mouse *Sbp1* gene has been identified as a direct target gene of the hypoxia-inducible factor-1α (HIF-1α) transcription factor in primary keratinocyte cell cultures [83]. Scortegagna et al. examined HIF-1α gain of function during multistage murine skin chemical carcinogenesis in K14-HIF-1α^{Pro402A564G} transgenic mice. They concluded that HIF-1α was functioning as a tumor suppressor, most likely by upregulating target genes, including *Sbp1*. Four hypoxia response elements were located within 1400 bp of the transcription start site of the human promoter region of *SBP1*, although the demonstration that these were bona fide response elements was not provided [83]. HIF-1α is a central mediator of the cellular response to environmental stresses, such as hypoxia [84]. It is overexpressed in many types of human cancer [85–87] and its overexpression is associated with treatment failure and increased mortality in some cancers including cancers of the cervix [88,89], breast [90,91], ovary [92], uterus [93], stomach [94], and brain [95]. It is also associated with decreased mortality in other cancers, including those of the head and neck [96] and non-small-cell lung cancer [97]. The consequences of the changes of HIF-1α levels are cancer-type specific and the accompanying molecular alterations, such as SBP1 reduction/loss, can affect the balance between pro- and anti-apoptotic factors. A study by Huang et al. demonstrated that the decreased expression of SBP1 could lead to a higher GPX1 activity and reduced HIF-1α expression in hepatocellular carcinoma, indicating that SBP1 might exert its tumor suppressive function as a regulator of the tumor redox microenvironment [41].

In addition to the putative HIF-1α response elements in the *SBP1* promoter, two potential antioxidant response elements (ARE) with strong homology to the consensus ARE recognition motif are present in promoter region of *SBP1* [28], although the functionality of these sequences as AREs has yet to be established. The presence of functional AREs in the promoter region of *SBP1* may account for the repression of transcription observed when the anti-oxidant selenoprotein GPX1 is ectopically expressed in colon carcinoma cell lines [28], as well as the reciprocal relationship observed in cells and tissues [43]. The factors potentially regulating *SBP1* and some downstream targets are summarized in Figure 3.

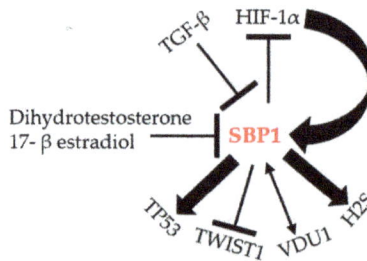

Figure 3. The molecular regulation of SBP1. Illustration of the different proteins that can potentially interact, be regulated by, or regulate SBP1 according to the literature. TGF-β: transforming growth factor beta; HIF-1α: hypoxia-inducible factor-1α; H2S: hydrogen sulfide; VDU1: von Hippel–Lindau protein–interacting deubiquitinating enzyme 1; TWIST1: Twist Family BHLH Transcription Factor 1; TP53: tumor protein p53.

In some cases, epigenetic silencing by promoter methylation may be a mechanism by which the expression of *SBP1* is reduced in human colon cancers. Comparing DNA obtained from colon cancer samples to DNA obtained from matched normal tissue indicated significantly more methylation in the promoter region of samples from the cancers [56]. Hypermethylation of the *SBP1* promoter region was demonstrated in the human colon cell lines SW480, Caco-2, HT-29, and HCT1161 in which the extent of promoter methylation was associated with the degree of SBP1 protein levels. Moreover, treatment of these cells with 5-aza-deoxycytidine, a demethylation agent, decreased promoter methylation and resulted in increased promoter activity and protein levels [56]. In contrast, treatment of three different human colon cancer cell lines, LOVO, SNU-C4, and A549, with 5-aza-deoxycytidine did not result in increased SBP1 expression, nor was there any evidence of genetic loss at the *SBP1* locus [59]. There was also a lack of evidence for either hypermethylation or genetic deletion accounting for the low levels of *SBP1* observed in lung cancers [98]. While there is consistent loss of SBP1 in many cancer types, there may be a multitude of ways in which tumor cells can achieve the reduction in SBP1 expression.

7. SBP1 Is a Methanethiol Oxidase

The enzymatic function of SBP1 was recently revealed by investigators examining the genetic determinants of extraoral halitosis, bad breath [37]. The authors analyzed breaths and body fluids of five affected individuals with extraoral halitosis from three unrelated families using NMR spectroscopy and gas chromatography with a sulfur-specific detector. All patients exhibited elevated levels of methanethiol (MT), dimethylsulfide (DMS), dimethylsufoxide, and dimethylsulfone in breaths and body fluids [37]. The authors postulated that the accumulation of these compounds was due to a defect in a protein which oxidizes MT, leading to its accumulation in affected individuals. Methanethiol oxidases (MTOs) have not previously been reported in humans, but *SBP1* was identified as a candidate gene for extraoral halitosis by searching for human sequences that were similar to the gene encoding an MT-metabolizing protein previously recognized in methylotropic bacteria, the *mtoX* gene. This effort revealed a 26% identity and a 54% sequence similarity between the two genes [37]. Subsequent sequencing of *SBP1* in patients DNAs revealed four different biallelic mutations in the five patients (1039G>T, 481+1G>A, 673G>T, and 985C>T) that were predicted to be pathogenic. Fibroblasts from these patients had significantly reduced SBP1 protein levels and undetectable MTO enzymatic activity, compared to the control cells [37].

MTO converts MT to H_2O_2, formaldehyde, and hydrogen sulfide (H_2S), the latter is a gaseous signaling molecule with distinct functions at different cellular concentrations [99,100]. At low concentrations, H_2S stimulates mitochondrial electron transport in mammalian cells, increasing oxygen consumption [101]. At high concentrations, H_2S is toxic through the inhibition of mitochondrial respiratory-chain complex IV, and consequently reduces oxygen consumption [101]. H_2S has been proposed as a therapy for multiple disorders by suppressing inflammation, affecting apoptotic pathways, increasing anti-oxidant defenses, and vasodilatation [99,102,103]. It is quite conceivable that many of the consequences of SBP1 expression can be due to the effects on H_2S levels as well as the other products of the MTO-mediated reaction, on a broad spectrum of physiological endpoints.

8. Conclusions

Among the selenium-associated proteins, SBP1 is relatively less studied, but is a highly-conserved protein that may be critical for a variety of physiological functions, including cell differentiation, protein degradation, intra-Golgi vesicular transport, cell motility, and redox modulation. The variety of processes where SBP1 has been implicated to have a role is suggestive that there may be cell type-specific functions that are yet to be identified. The only enzymatic function of SBP1 identified to date is MTO activity and it is possible that different levels of both its substrates and products provide differential signals, resulting in distinct intracellular and extracellular environments for the SBP1-expressing cells. With a better understanding of the function of SBP1 in these tissues, its role in diseases such as cancer may be resolved and SBP1 may become a novel therapeutic target for interventions to control it levels and/or activity.

Author Contributions: Both authors contributed to the writing of this manuscript.

Funding: This work was supported by grants from the National Institutes of Health (grant nos. R21CA182103, R01CA193497) to AMD.

Conflicts of Interest: The authors declare no conflict of interest.

References

1. Clark, L.C.; Combs, G.F., Jr.; Turnbull, B.W.; Slate, E.H.; Chalker, D.K.; Chow, J.; Davis, L.S.; Glover, R.A.; Graham, G.F.; Gross, E.G.; et al. Effects of selenium supplementation for cancer prevention in patients with carcinoma of the skin. A randomized controlled trial. Nutritional prevention of cancer study group. *JAMA* **1996**, *276*, 1957–1963. [CrossRef] [PubMed]

2. Clark, L.C.; Dalkin, B.; Krongrad, A.; Combs, G.F., Jr.; Turnbull, B.W.; Slate, E.H.; Witherington, R.; Herlong, J.H.; Janosko, E.; Carpenter, D.; et al. Decreased incidence of prostate cancer with selenium supplementation: Results of a double-blind cancer prevention trial. *Br. J. Urol.* **1998**, *81*, 730–734. [CrossRef] [PubMed]

3. Schrauzer, G.N.; White, D.A.; Schneider, C.J. Cancer mortality correlation studies. Iii. Statistical association with dietary selenium intakes. *Bioinorg. Chem.* **1977**, *7*, 23–34. [CrossRef]

4. Shamberger, R.J.; Frost, D.V. Possible protective effect of selenium against human cancer. *Can. Med. Assoc. J.* **1969**, *100*, 682. [PubMed]

5. Brinkman, M.; Reulen, R.C.; Kellen, E.; Buntinx, F.; Zeegers, M.P. Are men with low selenium levels at increased risk of prostate cancer? *Eur. J. Cancer* **2006**, *42*, 2463–2471. [CrossRef] [PubMed]

6. Fortmann, S.P.; Burda, B.U.; Senger, C.A.; Lin, J.S.; Whitlock, E.P. Vitamin and mineral supplements in the primary prevention of cardiovascular disease and cancer: An updated systematic evidence review for the u.S. Preventive services task force. *Ann. Intern. Med.* **2013**, *159*, 824–834. [CrossRef] [PubMed]

7. Steinbrenner, H.; Speckmann, B.; Sies, H. Toward understanding success and failures in the use of selenium for cancer prevention. *Antioxid. Redox Signal.* **2013**, *19*, 181–191. [CrossRef] [PubMed]

8. Vinceti, M.; Crespi, C.M.; Malagoli, C.; Del Giovane, C.; Krogh, V. Friend or foe? The current epidemiologic evidence on selenium and human cancer risk. *J. Environ. Sci. Health Part C Environ. Carcinog. Ecotoxicol. Rev.* **2013**, *31*, 305–341. [CrossRef] [PubMed]

9. Ip, C. Lessons from basic research in selenium and cancer prevention. *J. Nutr.* **1998**, *128*, 1845–1854. [CrossRef] [PubMed]

10. Klein, E.A. Selenium: Epidemiology and basic science. *J. Urol.* **2004**, *171*, S50–S53. [CrossRef] [PubMed]

11. Behne, D.; Kyriakopoulos, A. Mammalian selenium-containing proteins. *Annu. Rev. Nutr.* **2001**, *21*, 453–473. [CrossRef] [PubMed]

12. Davis, C.D.; Uthus, E.O. Dietary folate and selenium affect dimethylhydrazine-induced aberrant crypt formation, global DNA methylation and one-carbon metabolism in rats. *J. Nutr.* **2003**, *133*, 2907–2914. [CrossRef] [PubMed]

13. Ip, C.; Dong, Y.; Ganther, H. New concepts in selenium chemoprevention. *Cancer Metastasis Rev.* **2002**, *21*, 281–289. [CrossRef] [PubMed]

14. Wu, Y.; Zhang, H.; Dong, Y.; Park, Y.M.; Ip, C. Endoplasmic reticulum stress signal mediators are targets of selenium action. *Cancer Res.* **2005**, *65*, 9073–9079. [CrossRef] [PubMed]

15. Bera, S.; De Rosa, V.; Rachidi, W.; Diamond, A.M. Does a role for selenium in DNA damage repair explain apparent controversies in its use in chemoprevention? *Mutagenesis* **2013**, *28*, 127–134. [CrossRef] [PubMed]

16. Kohrle, J. The deiodinase family: Selenoenzymes regulating thyroid hormone availability and action. *Cell. Mol. Life Sci.* **2000**, *57*, 1853–1863. [CrossRef] [PubMed]

17. Rayman, M.P. Selenium and human health. *Lancet* **2012**, *379*, 1256–1268. [CrossRef]

18. Avery, J.C.; Hoffmann, P.R. Selenium, selenoproteins, and immunity. *Nutrients* **2018**, *10*, 1203. [CrossRef] [PubMed]

19. Turanov, A.A.M.M.; Gladyshev, V.N. Selenium and male reproduction. In *Selenium: Its Molecular Biology and Role in Human Health*, 3rd ed.; Hatfield, D.L.B.M., Gladyshev, V.N., Eds.; Springer: New York, NY, USA, 2012; pp. 409–417.

20. Handy, D.E.L.J. Selenoproteins in cardiovascular redox pathology. In *Selenium: Its Molecular Biology and Role in Human Health*, 3rd ed.; Hatfield, D.L.B.M., Gladyshev, V.N., Eds.; Springer: New York, NY, USA, 2012; pp. 249–259.

21. Kaushal, N.G.U.; Nelson, S.M.; Narayan, V.; Prabhu, K.S. Selenium and inflamation. In *Selenium: Its Molecular Biology and Role in Human Health*, 3rd ed.; Hatfield, D.L.B.M., Gladyshev, V.N., Eds.; Springer: New York, NY, USA, 2012; pp. 443–456.

22. Kudva, A.K.; Shay, A.E.; Prabhu, K.S. Selenium and inflammatory bowel disease. *Am. J. Physiol. Gastrointest. Liver Physiol.* **2015**, *309*, G71–G77. [CrossRef] [PubMed]

23. Moghadaszadeh, B.; Petit, N.; Jaillard, C.; Brockington, M.; Quijano Roy, S.; Merlini, L.; Romero, N.; Estournet, B.; Desguerre, I.; Chaigne, D.; et al. Mutations in sepn1 cause congenital muscular dystrophy with spinal rigidity and restrictive respiratory syndrome. *Nat. Genet.* **2001**, *29*, 17–18. [CrossRef] [PubMed]

24. Labunskyy, V.M.; Hatfield, D.L.; Gladyshev, V.N. Selenoproteins: Molecular pathways and physiological roles. *Physiol. Rev.* **2014**, *94*, 739–777. [CrossRef] [PubMed]

25. Kryukov, G.V.; Castellano, S.; Novoselov, S.V.; Lobanov, A.V.; Zehtab, O.; Guigo, R.; Gladyshev, V.N. Characterization of mammalian selenoproteomes. *Science* **2003**, *300*, 1439–1443. [CrossRef] [PubMed]

26. Hatfield, D.L.; Gladyshev, V.N. How selenium has altered our understanding of the genetic code. *Mol. Cell. Biol.* **2002**, *22*, 3565–3576. [CrossRef] [PubMed]

27. Bansal, M.P.; Oborn, C.J.; Danielson, K.G.; Medina, D. Evidence for two selenium-binding proteins distinct from glutathione peroxidase in mouse liver. *Carcinogenesis* **1989**, *10*, 541–546. [CrossRef] [PubMed]

28. Fang, W.; Goldberg, M.L.; Pohl, N.M.; Bi, X.; Tong, C.; Xiong, B.; Koh, T.J.; Diamond, A.M.; Yang, W. Functional and physical interaction between the selenium-binding protein 1 (sbp1) and the glutathione peroxidase 1 selenoprotein. *Carcinogenesis* **2010**, *31*, 1360–1366. [CrossRef] [PubMed]

29. Chang, P.W.; Tsui, S.K.; Liew, C.; Lee, C.C.; Waye, M.M.; Fung, K.P. Isolation, characterization, and chromosomal mapping of a novel cdna clone encoding human selenium binding protein. *J. Cell. Biochem.* **1997**, *64*, 217–224. [CrossRef]

30. Yang, M.; Sytkowski, A.J. Differential expression and androgen regulation of the human selenium-binding protein gene hsp56 in prostate cancer cells. *Cancer Res.* **1998**, *58*, 3150–3153. [PubMed]

31. Flemetakis, E.; Agalou, A.; Kavroulakis, N.; Dimou, M.; Martsikovskaya, A.; Slater, A.; Spaink, H.P.; Roussis, A.; Katinakis, P. Lotus japonicus gene ljsbp is highly conserved among plants and animals and encodes a homologue to the mammalian selenium-binding proteins. *Mol. Plant Microbe Interact.* **2002**, *15*, 313–322. [CrossRef] [PubMed]

32. Lanfear, J.; Fleming, J.; Walker, M.; Harrison, P. Different patterns of regulation of the genes encoding the closely related 56 kda selenium- and acetaminophen-binding proteins in normal tissues and during carcinogenesis. *Carcinogenesis* **1993**, *14*, 335–340. [CrossRef] [PubMed]

33. Jeong, J.Y.; Wang, Y.; Sytkowski, A.J. Human selenium binding protein-1 (hsp56) interacts with vdu1 in a selenium-dependent manner. *Biochem. Biophys. Res. Commun.* **2009**, *379*, 583–588. [CrossRef] [PubMed]

34. Raucci, R.; Colonna, G.; Guerriero, E.; Capone, F.; Accardo, M.; Castello, G.; Costantini, S. Structural and functional studies of the human selenium binding protein-1 and its involvement in hepatocellular carcinoma. *Biochim. Biophys. Acta* **2011**, *1814*, 513–522. [CrossRef] [PubMed]

35. Ying, Q.; Ansong, E.; Diamond, A.M.; Yang, W. A critical role for cysteine 57 in the biological functions of selenium binding protein-1. *Int. J. Mol. Sci.* **2015**, *16*, 27599–27608. [CrossRef] [PubMed]

36. Li, Z.; Na, X.; Wang, D.; Schoen, S.R.; Messing, E.M.; Wu, G. Ubiquitination of a novel deubiquitinating enzyme requires direct binding to von hippel-lindau tumor suppressor protein. *J. Biol. Chem.* **2002**, *277*, 4656–4662. [CrossRef] [PubMed]

37. Pol, A.; Renkema, G.H.; Tangerman, A.; Winkel, E.G.; Engelke, U.F.; de Brouwer, A.P.M.; Lloyd, K.C.; Araiza, R.S.; van den Heuvel, L.; Omran, H.; et al. Mutations in SELENBP1, encoding a novel human methanethiol oxidase, cause extraoral halitosis. *Nat. Genet.* **2018**, *50*, 120–129. [CrossRef] [PubMed]

38. Driscoll, D.M.; Copeland, P.R. Mechanism and regulation of selenoprotein synthesis. *Annu. Rev. Nutr.* **2003**, *23*, 17–40. [CrossRef] [PubMed]

39. Bansal, M.P.; Ip, C.; Medina, D. Levels and 75se-labeling of specific proteins as a consequence of dietary selenium concentration in mice and rats. *Proc. Soc. Exp. Biol. Med.* **1991**, *196*, 147–154. [CrossRef] [PubMed]

40. Lubos, E.; Loscalzo, J.; Handy, D.E. Glutathione peroxidase-1 in health and disease: From molecular mechanisms to therapeutic opportunities. *Antioxid. Redox Signal.* **2011**, *15*, 1957–1997. [CrossRef] [PubMed]

41. Huang, C.; Ding, G.; Gu, C.; Zhou, J.; Kuang, M.; Ji, Y.; He, Y.; Kondo, T.; Fan, J. Decreased selenium-binding protein 1 enhances glutathione peroxidase 1 activity and downregulates hif-1alpha to promote hepatocellular carcinoma invasiveness. *Clin. Cancer Res.* **2012**, *18*, 3042–3053. [CrossRef] [PubMed]

42. Jerome-Morais, A.; Wright, M.E.; Liu, R.; Yang, W.; Jackson, M.I.; Combs, G.F., Jr.; Diamond, A.M. Inverse association between glutathione peroxidase activity and both selenium-binding protein 1 levels and gleason score in human prostate tissue. *Prostate* **2012**, *72*, 1006–1012. [CrossRef] [PubMed]

43. Ansong, E.; Yang, W.; Diamond, A.M. Molecular cross-talk between members of distinct families of selenium containing proteins. *Mol. Nutr. Food Res.* **2014**, *58*, 117–123. [CrossRef] [PubMed]

44. Brown, L.M.; Helmke, S.M.; Hunsucker, S.W.; Netea-Maier, R.T.; Chiang, S.A.; Heinz, D.E.; Shroyer, K.R.; Duncan, M.W.; Haugen, B.R. Quantitative and qualitative differences in protein expression between papillary thyroid carcinoma and normal thyroid tissue. *Mol. Carcinog.* **2006**, *45*, 613–626. [CrossRef] [PubMed]

45. Tan, X.; Liao, L.; Wan, Y.P.; Li, M.X.; Chen, S.H.; Mo, W.J.; Zhao, Q.L.; Huang, L.F.; Zeng, G.Q. Downregulation of selenium-binding protein 1 is associated with poor prognosis in lung squamous cell carcinoma. *World J. Surg. Oncol.* **2016**, *14*, 70. [CrossRef] [PubMed]

46. Zhang, C.; Xu, W.; Pan, W.; Wang, N.; Li, G.; Fan, X.; Xu, X.; Shen, S.; Das, U.N. Selenium-binding protein 1 may decrease gastric cellular proliferation and migration. *Int. J. Oncol.* **2013**, *42*, 1620–1629. [CrossRef] [PubMed]

47. Xia, Y.J.; Ma, Y.Y.; He, X.J.; Wang, H.J.; Ye, Z.Y.; Tao, H.Q. Suppression of selenium-binding protein 1 in gastric cancer is associated with poor survival. *Hum. Pathol.* **2011**, *42*, 1620–1628. [CrossRef] [PubMed]

48. Ha, Y.S.; Lee, G.T.; Kim, Y.H.; Kwon, S.Y.; Choi, S.H.; Kim, T.H.; Kwon, T.G.; Yun, S.J.; Kim, I.Y.; Kim, W.J. Decreased selenium-binding protein 1 mrna expression is associated with poor prognosis in renal cell carcinoma. *World J. Surg. Oncol.* **2014**, *12*, 288. [CrossRef] [PubMed]

49. Zhang, C.; Wang, Y.E.; Zhang, P.; Liu, F.; Sung, C.J.; Steinhoff, M.M.; Quddus, M.R.; Lawrence, W.D. Progressive loss of selenium-binding protein 1 expression correlates with increasing epithelial proliferation and papillary complexity in ovarian serous borderline tumor and low-grade serous carcinoma. *Hum. Pathol.* **2010**, *41*, 255–261. [CrossRef] [PubMed]

50. Huang, K.C.; Park, D.C.; Ng, S.K.; Lee, J.Y.; Ni, X.; Ng, W.C.; Bandera, C.A.; Welch, W.R.; Berkowitz, R.S.; Mok, S.C.; et al. Selenium binding protein 1 in ovarian cancer. *Int. J. Cancer* **2006**, *118*, 2433–2440. [CrossRef] [PubMed]

51. Stammer, K.; Edassery, S.L.; Barua, A.; Bitterman, P.; Bahr, J.M.; Hales, D.B.; Luborsky, J.L. Selenium-binding protein 1 expression in ovaries and ovarian tumors in the laying hen, a spontaneous model of human ovarian cancer. *Gynecol. Oncol.* **2008**, *109*, 115–121. [CrossRef] [PubMed]

52. Zhang, S.; Li, F.; Younes, M.; Liu, H.; Chen, C.; Yao, Q. Reduced selenium-binding protein 1 in breast cancer correlates with poor survival and resistance to the anti-proliferative effects of selenium. *PLoS ONE* **2013**, *8*, e63702. [CrossRef] [PubMed]

53. Ansong, E.; Ying, Q.; Ekoue, D.N.; Deaton, R.; Hall, A.R.; Kajdacsy-Balla, A.; Yang, W.; Gann, P.H.; Diamond, A.M. Evidence that selenium binding protein 1 is a tumor suppressor in prostate cancer. *PLoS ONE* **2015**, *10*, e0127295. [CrossRef] [PubMed]

54. Jeong, J.Y.; Zhou, J.R.; Gao, C.; Feldman, L.; Sytkowski, A.J. Human selenium binding protein-1 (hsp56) is a negative regulator of hif-1alpha and suppresses the malignant characteristics of prostate cancer cells. *BMB Rep.* **2014**, *47*, 411–416. [CrossRef] [PubMed]

55. Wang, N.; Chen, Y.; Yang, X.; Jiang, Y. Selenium-binding protein 1 is associated with the degree of colorectal cancer differentiation and is regulated by histone modification. *Oncol. Rep.* **2014**, *31*, 2506–2514. [CrossRef] [PubMed]

56. Pohl, N.M.; Tong, C.; Fang, W.; Bi, X.; Li, T.; Yang, W. Transcriptional regulation and biological functions of selenium-binding protein 1 in colorectal cancer in vitro and in nude mouse xenografts. *PLoS ONE* **2009**, *4*, e7774. [CrossRef] [PubMed]

57. Chen, F.; Chen, C.; Qu, Y.; Xiang, H.; Ai, Q.; Yang, F.; Tan, X.; Zhou, Y.; Jiang, G.; Zhang, Z. Selenium-binding protein 1 in head and neck cancer is low-expression and associates with the prognosis of nasopharyngeal carcinoma. *Medicine* **2016**, *95*, e4592. [CrossRef] [PubMed]

58. Schott, M.; de Jel, M.M.; Engelmann, J.C.; Renner, P.; Geissler, E.K.; Bosserhoff, A.K.; Kuphal, S. Selenium-binding protein 1 is down-regulated in malignant melanoma. *Oncotarget* **2018**, *9*, 10445–10456. [CrossRef] [PubMed]

59. Kim, H.; Kang, H.J.; You, K.T.; Kim, S.H.; Lee, K.Y.; Kim, T.I.; Kim, C.; Song, S.Y.; Kim, H.J.; Lee, C.; et al. Suppression of human selenium-binding protein 1 is a late event in colorectal carcinogenesis and is associated with poor survival. *Proteomics* **2006**, *6*, 3466–3476. [CrossRef] [PubMed]

60. Xie, W.; Yang, M.; Chan, J.; Sun, T.; Mucci, L.A.; Penney, K.L.; Lee, G.S.; Kantoff, P.W. Association of genetic variations of selenoprotein genes, plasma selenium levels, and prostate cancer aggressiveness at diagnosis. *Prostate* **2016**, *76*, 691–699. [CrossRef] [PubMed]

61. Ying, Q.; Ansong, E.; Diamond, A.M.; Lu, Z.; Yang, W.; Bie, X. Quantitative proteomic analysis reveals that anti-cancer effects of selenium-binding protein 1 in vivo are associated with metabolic pathways. *PLoS ONE* **2015**, *10*, e0126285. [CrossRef] [PubMed]

62. Gao, P.T.; Ding, G.Y.; Yang, X.; Dong, R.Z.; Hu, B.; Zhu, X.D.; Cai, J.B.; Ji, Y.; Shi, G.M.; Shen, Y.H.; et al. Invasive potential of hepatocellular carcinoma is enhanced by loss of selenium-binding protein 1 and subsequent upregulation of cxcr4. *Am. J. Cancer Res.* **2018**, *8*, 1040–1049. [PubMed]

63. Caswell, D.R.; Chuang, C.H.; Ma, R.K.; Winters, I.P.; Snyder, E.L.; Winslow, M.M. Tumor suppressor activity of SELENBP1, a direct nkx2-1 target, in lung adenocarcinoma. *Mol. Cancer Res.* **2018**. [CrossRef] [PubMed]

64. Tsujimoto, S.; Ishida, T.; Takeda, T.; Ishii, Y.; Onomura, Y.; Tsukimori, K.; Takechi, S.; Yamaguchi, T.; Uchi, H.; Suzuki, S.O.; et al. Selenium-binding protein 1: Its physiological function, dependence on aryl hydrocarbon receptors, and role in wasting syndrome by 2,3,7,8-tetrachlorodibenzo-p-dioxin. *Biochim. Biophys. Acta* **2013**, *1830*, 3616–3624. [CrossRef] [PubMed]

65. Zhang, J.; Zhan, N.; Dong, W.G. Altered expression of selenium-binding protein 1 in gastric carcinoma and precursor lesions. *Med. Oncol.* **2011**, *28*, 951–957. [CrossRef] [PubMed]

66. Zeng, G.Q.; Yi, H.; Zhang, P.F.; Li, X.H.; Hu, R.; Li, M.Y.; Li, C.; Qu, J.Q.; Deng, X.; Xiao, Z.Q. The function and significance of SELENBP1 downregulation in human bronchial epithelial carcinogenic process. *PLoS ONE* **2013**, *8*, e71865. [CrossRef] [PubMed]

67. Zhang, P.; Zhang, C.; Wang, X.; Liu, F.; Sung, C.J.; Quddus, M.R.; Lawrence, W.D. The expression of selenium-binding protein 1 is decreased in uterine leiomyoma. *Diagn. Pathol.* **2010**, *5*, 80. [CrossRef] [PubMed]

68. Hiroshi, I.; Masayuki, I.; Shosuke, K. Early decreases in pulmonary, hepatic and renal glutathione levels in response to cadmium instillation into rat trachea. *J. Appl. Toxicol.* **1991**, *11*, 211–217. [CrossRef]

69. Wang, Y.; Fang, W.; Huang, Y.; Hu, F.; Ying, Q.; Yang, W.; Xiong, B. Reduction of selenium-binding protein 1 sensitizes cancer cells to selenite via elevating extracellular glutathione: A novel mechanism of cancer-specific cytotoxicity of selenite. *Free Radic. Biol. Med.* **2015**, *79*, 186–196. [CrossRef] [PubMed]

70. Porat, A.; Sagiv, Y.; Elazar, Z. A 56-kda selenium-binding protein participates in intra-golgi protein transport. *J. Biol. Chem.* **2000**, *275*, 14457–14465. [CrossRef] [PubMed]

71. Miyaguchi, K. Localization of selenium-binding protein at the tips of rapidly extending protrusions. *Histochem. Cell Biol.* **2004**, *121*, 371–376. [CrossRef] [PubMed]

72. Glatt, S.J.; Everall, I.P.; Kremen, W.S.; Corbeil, J.; Sasik, R.; Khanlou, N.; Han, M.; Liew, C.C.; Tsuang, M.T. Comparative gene expression analysis of blood and brain provides concurrent validation of SELENBP1 up-regulation in schizophrenia. *Proc. Natl. Acad. Sci. USA* **2005**, *102*, 15533–15538. [CrossRef] [PubMed]

73. Kanazawa, T.; Chana, G.; Glatt, S.J.; Mizuno, H.; Masliah, E.; Yoneda, H.; Tsuang, M.T.; Everall, I.P. The utility of SELENBP1 gene expression as a biomarker for major psychotic disorders: Replication in schizophrenia and extension to bipolar disorder with psychosis. *Am. J. Med. Genet. Part B Neuropsychiatr. Genet.* **2008**, *147B*, 686–689. [CrossRef] [PubMed]

74. Udawela, M.; Money, T.T.; Neo, J.; Seo, M.S.; Scarr, E.; Dean, B.; Everall, I.P. SELENBP1 expression in the prefrontal cortex of subjects with schizophrenia. *Transl. Psychiatry* **2015**, *5*, e615. [CrossRef] [PubMed]

75. Kanazawa, T.; Glatt, S.J.; Faraone, S.V.; Hwu, H.G.; Yoneda, H.; Tsuang, M.T. Family-based association study of SELENBP1 in schizophrenia. *Schizophr. Res.* **2009**, *113*, 268–272. [CrossRef] [PubMed]

76. Chau, E.J.; Mostaid, M.S.; Cropley, V.; McGorry, P.; Pantelis, C.; Bousman, C.A.; Everall, I.P. Downregulation of plasma SELENBP1 protein in patients with recent-onset schizophrenia. *Prog. Neuro-Psychopharmacol. Biol. Psychiatry* **2018**, *85*, 1–6. [CrossRef] [PubMed]

77. Brauer, A.U.; Savaskan, N.E. Molecular actions of selenium in the brain: Neuroprotective mechanisms of an essential trace element. *Rev. Neurosci.* **2004**, *15*, 19–32. [CrossRef] [PubMed]

78. Xue, W.; Du, P.; Lin, S.; Dudley, V.J.; Hernandez, M.R.; Sarthy, V.P. Gene expression changes in retinal muller (glial) cells exposed to elevated pressure. *Curr. Eye Res.* **2011**, *36*, 754–767. [CrossRef] [PubMed]

79. Wang, J.; Qu, D.; An, J.; Yuan, G.; Liu, Y. Integrated microarray analysis provided novel insights to the pathogenesis of glaucoma. *Mol. Med. Rep.* **2017**, *16*, 8735–8746. [CrossRef] [PubMed]

80. Okunuki, Y.; Usui, Y.; Takeuchi, M.; Kezuka, T.; Hattori, T.; Masuko, K.; Nakamura, H.; Yudoh, K.; Usui, M.; Nishioka, K.; et al. Proteomic surveillance of autoimmunity in behcet's disease with uveitis: Selenium binding protein is a novel autoantigen in behcet's disease. *Exp. Eye Res.* **2007**, *84*, 823–831. [CrossRef] [PubMed]

81. Krause, I.; Weinberger, A. Behcet's disease. *Curr. Opin. Rheumatol.* **2008**, *20*, 82–87. [CrossRef] [PubMed]

82. Torrealba, J.R.; Colburn, M.; Golner, S.; Chang, Z.; Scheunemann, T.; Fechner, J.H.; Roenneburg, D.; Hu, H.; Alam, T.; Kim, H.T.; et al. Selenium-binding protein-1 in smooth muscle cells is downregulated in a rhesus monkey model of chronic allograft nephropathy. *Am. J. Transplant.* **2005**, *5*, 58–67. [CrossRef] [PubMed]

83. Scortegagna, M.; Martin, R.J.; Kladney, R.D.; Neumann, R.G.; Arbeit, J.M. Hypoxia-inducible factor-1alpha suppresses squamous carcinogenic progression and epithelial-mesenchymal transition. *Cancer Res.* **2009**, *69*, 2638–2646. [CrossRef] [PubMed]

84. Semenza, G.L. Targeting hif-1 for cancer therapy. *Nat. Rev. Cancer* **2003**, *3*, 721–732. [CrossRef] [PubMed]

85. Zhong, H.; De Marzo, A.M.; Laughner, E.; Lim, M.; Hilton, D.A.; Zagzag, D.; Buechler, P.; Isaacs, W.B.; Semenza, G.L.; Simons, J.W. Overexpression of hypoxia-inducible factor 1alpha in common human cancers and their metastases. *Cancer Res.* **1999**, *59*, 5830–5835. [PubMed]

86. Feldser, D.; Agani, F.; Iyer, N.V.; Pak, B.; Ferreira, G.; Semenza, G.L. Reciprocal positive regulation of hypoxia-inducible factor 1alpha and insulin-like growth factor 2. *Cancer Res.* **1999**, *59*, 3915–3918. [PubMed]

87. Talks, K.L.; Turley, H.; Gatter, K.C.; Maxwell, P.H.; Pugh, C.W.; Ratcliffe, P.J.; Harris, A.L. The expression and distribution of the hypoxia-inducible factors hif-1alpha and hif-2alpha in normal human tissues, cancers, and tumor-associated macrophages. *Am. J. Pathol.* **2000**, *157*, 411–421. [CrossRef]

88. Birner, P.; Schindl, M.; Obermair, A.; Plank, C.; Breitenecker, G.; Oberhuber, G. Overexpression of hypoxia-inducible factor 1alpha is a marker for an unfavorable prognosis in early-stage invasive cervical cancer. *Cancer Res.* **2000**, *60*, 4693–4696. [PubMed]

89. Burri, P.; Djonov, V.; Aebersold, D.M.; Lindel, K.; Studer, U.; Altermatt, H.J.; Mazzucchelli, L.; Greiner, R.H.; Gruber, G. Significant correlation of hypoxia-inducible factor-1alpha with treatment outcome in cervical cancer treated with radical radiotherapy. *Int. J. Radiat. Oncol. Biol. Phys.* **2003**, *56*, 494–501. [CrossRef]

90. Bos, R.; van der Groep, P.; Greijer, A.E.; Shvarts, A.; Meijer, S.; Pinedo, H.M.; Semenza, G.L.; van Diest, P.J.; van der Wall, E. Levels of hypoxia-inducible factor-1alpha independently predict prognosis in patients with lymph node negative breast carcinoma. *Cancer* **2003**, *97*, 1573–1581. [CrossRef] [PubMed]

91. Schindl, M.; Schoppmann, S.F.; Samonigg, H.; Hausmaninger, H.; Kwasny, W.; Gnant, M.; Jakesz, R.; Kubista, E.; Birner, P.; Oberhuber, G.; et al. Overexpression of hypoxia-inducible factor 1alpha is associated with an unfavorable prognosis in lymph node-positive breast cancer. *Clin. Cancer Res.* **2002**, *8*, 1831–1837. [PubMed]

92. Birner, P.; Schindl, M.; Obermair, A.; Breitenecker, G.; Oberhuber, G. Expression of hypoxia-inducible factor 1alpha in epithelial ovarian tumors: Its impact on prognosis and on response to chemotherapy. *Clin. Cancer Res.* **2001**, *7*, 1661–1668. [PubMed]

93. Sivridis, E.; Giatromanolaki, A.; Gatter, K.C.; Harris, A.L.; Koukourakis, M.I.; Tumor and Angiogenesis Research Group. Association of hypoxia-inducible factors 1alpha and 2alpha with activated angiogenic pathways and prognosis in patients with endometrial carcinoma. *Cancer* **2002**, *95*, 1055–1063. [CrossRef] [PubMed]

94. Takahashi, R.; Tanaka, S.; Hiyama, T.; Ito, M.; Kitadai, Y.; Sumii, M.; Haruma, K.; Chayama, K. Hypoxia-inducible factor-1alpha expression and angiogenesis in gastrointestinal stromal tumor of the stomach. *Oncol. Rep.* **2003**, *10*, 797–802. [PubMed]

95. Birner, P.; Gatterbauer, B.; Oberhuber, G.; Schindl, M.; Rossler, K.; Prodinger, A.; Budka, H.; Hainfellner, J.A. Expression of hypoxia-inducible factor-1 alpha in oligodendrogliomas: Its impact on prognosis and on neoangiogenesis. *Cancer* **2001**, *92*, 165–171. [CrossRef]

96. Beasley, N.J.; Leek, R.; Alam, M.; Turley, H.; Cox, G.J.; Gatter, K.; Millard, P.; Fuggle, S.; Harris, A.L. Hypoxia-inducible factors hif-1alpha and hif-2alpha in head and neck cancer: Relationship to tumor biology and treatment outcome in surgically resected patients. *Cancer Res.* **2002**, *62*, 2493–2497. [PubMed]

97. Volm, M.; Koomagi, R. Hypoxia-inducible factor (hif-1) and its relationship to apoptosis and proliferation in lung cancer. *Anticancer Res.* **2000**, *20*, 1527–1533. [PubMed]

98. Chen, G.; Wang, H.; Miller, C.T.; Thomas, D.G.; Gharib, T.G.; Misek, D.E.; Giordano, T.J.; Orringer, M.B.; Hanash, S.M.; Beer, D.G. Reduced selenium-binding protein 1 expression is associated with poor outcome in lung adenocarcinomas. *J. Pathol.* **2004**, *202*, 321–329. [CrossRef] [PubMed]

99. Barr, L.A.; Calvert, J.W. Discoveries of hydrogen sulfide as a novel cardiovascular therapeutic. *Circ. J.* **2014**, *78*, 2111–2118. [CrossRef] [PubMed]

100. Kohl, J.B.; Mellis, A.T.; Schwarz, G. Homeostatic impact of sulfite and hydrogen sulfide on cysteine catabolism. *Br. J. Pharmacol.* **2018**. [CrossRef] [PubMed]

101. Szabo, C.; Ransy, C.; Modis, K.; Andriamihaja, M.; Murghes, B.; Coletta, C.; Olah, G.; Yanagi, K.; Bouillaud, F. Regulation of mitochondrial bioenergetic function by hydrogen sulfide. Part i. Biochemical and physiological mechanisms. *Br. J. Pharmacol.* **2014**, *171*, 2099–2122. [CrossRef] [PubMed]

102. Bos, E.M.; van Goor, H.; Joles, J.A.; Whiteman, M.; Leuvenink, H.G. Hydrogen sulfide: Physiological properties and therapeutic potential in ischaemia. *Br. J. Pharmacol.* **2015**, *172*, 1479–1493. [CrossRef] [PubMed]

103. Wallace, J.L.; Wang, R. Hydrogen sulfide-based therapeutics: Exploiting a unique but ubiquitous gasotransmitter. *Nat. Rev. Drug Discov.* **2015**, *14*, 329–345. [CrossRef] [PubMed]

International Journal of
Molecular Sciences

MDPI

Review

Current Landscape and the Potential Role of Hypoxia-Inducible Factors and Selenium in Clear Cell Renal Cell Carcinoma Treatment

Rohan Garje [1,2], Josiah J. An [1], Kevin Sanchez [3], Austin Greco [3], Jeffrey Stolwijk [4], Eric Devor [5], Youcef Rustum [1,6] and Yousef Zakharia [1,2,*]

1   Department of Internal Medicine, Division of Hematology and Oncology, University of Iowa, Iowa City, IA 52242, USA; rohan-garje@uiowa.edu; josiah-an@uiowa.edu (R.G.); josiah-an@uiowa.edu (J.J.A.); Youcef.Rustum@RoswellPark.org (Y.R.)
2   Holden Comprehensive Cancer Center, University of Iowa, Iowa City, IA 52242, USA
3   Department of Internal Medicine, University of Iowa, Iowa City, IA 52242, USA; kevin-sanchez@uiowa.edu (K.S.); austin-greco@uiowa.edu (A.G.)
4   Interdisciplinary Graduate Program in Human Toxicology, Department of Occupational and Environmental Health, College of Public Health, University of Iowa, Iowa City, IA 52242, USA; jeffrey-stolwijk@uiowa.edu
5   Department of Obstetrics and Gynecology, University of Iowa, Iowa City, IA 52242, USA; eric-devor@uiowa.edu
6   Roswell Park Cancer Institute, Buffalo, NY 14203, USA
*   Correspondence: yousef-zakharia@uiowa.edu

Received: 18 September 2018; Accepted: 28 November 2018; Published: 1 December 2018

Abstract: In the last two decades, the discovery of various pathways involved in renal cell carcinoma (RCC) has led to the development of biologically-driven targeted therapies. Hypoxia-inducible factors (HIFs), angiogenic growth factors, von Hippel–Lindau (*VHL*) gene mutations, and oncogenic microRNAs (miRNAs) play essential roles in the pathogenesis and drug resistance of clear cell renal cell carcinoma. These insights have led to the development of vascular endothelial growth factor (VEGF) inhibitors, Mechanistic target of rapamycin (mTOR) inhibitors, and immunotherapeutic agents, which have significantly improved the outcomes of patients with advanced RCC. HIF inhibitors will be a valuable asset in the growing therapeutic armamentarium of RCC. Various histone deacetylase (HDAC) inhibitors, selenium, and agents like PT2385 and PT2977 are being explored in various clinical trials as potential HIF inhibitors, to ameliorate the outcomes of RCC patients. In this article, we will review the current treatment options and highlight the potential role of selenium in the modulation of drug resistance biomarkers expressed in clear cell RCC (ccRCC) tumors.

Keywords: clear cell renal cell carcinoma; hypoxia-inducible factors (HIFs); selenium; PD-L1; miRNA; VEGF; mTOR inhibitors

1. Introduction

Clear cell renal cell carcinoma (ccRCC) is the most common malignancy in the kidney. Over 65,000 new kidney cancer cases and 14,000 deaths were estimated in the United States in 2018 [1]. Renal cell carcinoma (RCC) is the most lethal genitourinary cancer, given that its disease course is largely asymptomatic and incidentally found in more than half of new cases [2,3]. Established modifiable risk factors for RCC include obesity, smoking, and hypertension [4]. Other studies link alcohol use, type 2 diabetes, and occupational or environmental exposures to increased risk of RCC [5].

RCC is categorized into three major histological subtypes: ccRCC, comprising 70% of cases; papillary and chromophobe RCC, which together comprise 25% of cases; and tumors of the medullary and collecting systems, which comprise 5% of cases [6,7]. These subtypes arise from distinct genetics and therefore are treated differently [8].

Localized RCC is often managed surgically with a partial or radical nephrectomy, with tumor ablation or active surveillance for small tumors. Systemic therapy is primarily reserved for metastatic RCC. Current evidence for adjuvant systemic therapy after complete resection of the tumor has shown no survival benefit [9]. For stage IV disease, cytoreductive nephrectomy in addition to systemic therapy has not shown improvement in overall survival [9,10]. In the last two decades, there has been significant improvement in our knowledge of renal cell carcinogenesis that has, in turn, led to the development of biologically-driven targeted therapies.

2. Role of Hypoxia-Inducible Factors in Renal Cell Carcinogenesis

Adaptation to a hypoxic environment is a key attribute of cancer cells. This is mediated via transcription factors called hypoxia-inducible factors (HIFs). These factors are heterodimers with an α-subunit (HIF1α, HIF2α, or HIF3α) and a β-subunit (HIF1β) [11]. Previously, HIF1α was considered to be a predominant oncogenic driver, but recent evidence shows HIF2α as a key player in renal cancer progression [12]. Along with hypoxia, the von Hippel–Lindau (*VHL*) gene and other oncogenic signaling pathways (e.g., PI3K, RAS) are known to regulate HIF activation. Once activated, HIF transcription factors translocate to the nucleus and bind to the hypoxia response elements, which leads to transcription of several target genes involved in angiogenesis (vascular endothelial growth factor (VEGF)), oxygen transport and metabolism (erythropoietin), glycolysis (LDH), glucose transport (GLUT1), cell proliferation, and migration, which eventually leads to carcinogenesis (Figure 1) [13,14]. VEGF plays a vital role in tumor angiogenesis, and is a key target of anti-cancer therapeutic agents. Regulation of the HIF pathway is vital for cellular homeostasis.

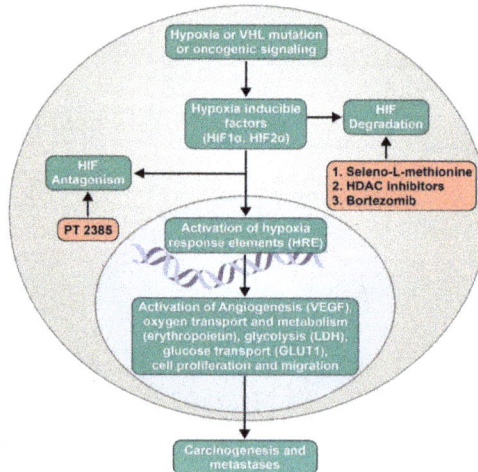

Figure 1. Inhibitors of the hypoxia-inducible factor (HIF) pathway currently being evaluated in clinical trials. VHL: von Hippel–Lindau; HIF: hypoxia-inducible factors; HDAC: histone deacetylase; VEGF: vascular endothelial growth factor; LDH: lactate dehydrogenase; GLUT1: glucose transporter 1.

3. Regulation of HIF Pathway by VHL Gene

Von Hippel–Lindau (*VHL*) is a tumor suppressor gene located on the short arm of chromosome 3 that is commonly mutated in both hereditary and sporadic renal cell carcinoma. The *VHL* gene encodes two isoforms of VHL proteins (pVHL) that play a crucial role in cellular oxygen sensing and regulation of HIFs. In normoxic conditions, the pVHL form a ubiquitin ligase complex and bind to the hydroxylated HIF1α and HIF2α, which subsequently undergoes ubiquitination and proteasomal degradation. However, in cellular hypoxic conditions, the pVHL ubiquitin ligase complex cannot bind to HIFα and lead to its degradation, as they are not hydroxylated (an oxygen-dependent process). Hence, this leads to HIFα accumulation and formation of heterodimer complex with HIF1β and subsequent induction of several target genes in the nucleus [15,16]. In ccRCC, a wide range of intragenic mutations, deletions (complete or partial) and splicing defects have been identified that derange normal function of the *VHL* gene, which creates a situation similar to cellular hypoxia and the accumulation of HIFs [17].

In addition to the *VHL* gene, multitudinous genetic and enzymatic derangements have been identified that predispose a region to various histologies of renal cell carcinoma. These include folliculin (*FLCN*; chromophobe RCC/oncocytoma in Birt–Hogg–Dubé syndrome), papillary type 1 RCC (*MET*), fumarate hydratase (FH; papillary type 2 RCC in hereditary leiomyomatosis and renal cell cancer syndrome), SDHB/SDHD/SDHC/SDHA (succinate dehydrogenase subunit-related RCC), chromosome 3 translocations-associated clear cell RCC, papillary RCC (PTEN), and BAP1 (clear cell RCC) [18].

This interdependency on biological pathways by cancer cells has laid the foundation for the development of several targeted therapeutic agents for the treatment of advanced renal cell carcinoma.

4. Angiogenesis (Vascular Endothelial Growth Factor Pathway) Inhibitors

Current first-line therapy for stage IV, unresectable, or relapsed disease of clear cell histology includes the oral VEGF tyrosine kinase inhibitors (TKIs) sunitinib and pazopanib [9]. Additionally, for intermediate to poor risk groups, based on the international metastatic renal cell carcinoma database consortium (IMDC) criteria [19], either the combination of ipilimumab and nivolumab or cabozantinib are options.

Sunitinib is a multi-kinase inhibitor targeting several tyrosine kinase receptors, including platelet-derived growth factor receptors (PDGFR-α and -β), VEGF receptors (VEGFR-1, -2, and -3), stem cell factor receptor (c-KIT), FMS-like tyrosine kinase (FLT-3), colony-stimulating factor (CSF-1R), and neurotropic factor receptor (RET) [9]. In the landmark phase III, multicenter clinical trial by Motzer et al., sunitinib was compared with interferon-α in patients with previously untreated metastatic renal-cell carcinoma [20]. Progression-free survival (PFS) in the sunitinib arm was 11 months, and in the interferon-α arm the PFS was 5 months. The overall survival (OS) with sunitinib was 26 months.

Pazopanib is another oral angiogenesis inhibitor targeting VEGFR-1, -2, and -3, PDGFR-α and -β, and c-KIT. In a phase III, open-label study of pazopanib in patients with no prior treatment or one prior cytokine-based treatment, PFS was prolonged significantly with pazopanib versus a placebo. For the treatment naïve group, PFS was 11.1 months, compared to 2.8 months for pazopanib and placebo, respectively [21]. In a phase III non-inferiority trial, pazopanib was compared to sunitinib in patients with advanced renal cell carcinoma. The study was positive for non-inferiority, with a progression-free survival of 8.4 and 9.5 months for pazopanib and sunitinib, respectively [22]. In addition, the median OS with pazopanib was 28.3 and 29.1 months for sunitinib. In subgroup analysis for patients with favorable-risk disease, the median OS for pazopanib and sunitinib was found to be 52.5 and 43.6 months, respectively [23]. Both of these medications had similar rates of adverse events that led to dose reduction, and had no differences in grades 3/4 adverse events. Symptoms associated with discomfort, such as fatigue, hand–foot syndrome, and mouth sores occurred more frequently with sunitinib, while pazopanib was associated with elevations in liver-function tests, weight loss, and changes in hair color. The study also showed lower monthly use of medical resources with pazopanib than with sunitinib [22].

Cabozantinib is a small molecule inhibitor of tyrosine kinases, which include VEGF receptors, MET, and AXL [9]. Cabozantinib was compared to sunitinib in a phase II study of intermediate to poor IMDC risk, treatment naïve patients with metastatic RCC [24]. In this study, PFS was 8.6 months versus 5.3 months for cabozantinib and sunitinib, respectively, and the median OS was found to be 34.5 months

Int. J. Mol. Sci. **2018**, *19*, 3834

and 26.6 months, respectively. Based on these results, cabozantinib has been approved by the United States Food and Drug Administration (FDA) as a first-line agent. Cabozantinib has also been studied in a phase III trial (METEOR trial) of patients with disease progression after previous TKI therapy [25]. The study compared second-line therapy with cabozantinib versus everolimus. The results showed a median PFS of 7.4 months compared to 3.8 months for cabozantinib and everolimus, respectively. Thus, in addition to first-line therapy, cabozantinib is a viable option as a second-line therapy for patients with disease progression after other TKI therapy.

Axitinib is a selective, second-generation tyrosine kinase inhibitor targeting VEGFR-1, -2, and -3 [9]. The phase III AXIS trial compared axitinib and sorafenib as second-line therapy, following other systemic therapy. PFS was 6.7 for axitinib versus 4.7 months for sorafenib. PFS was favored in both subgroups of patients treated with axitinib whose prior systemic therapy was sunitinib or cytokine therapy. Median OS was 20.1 months with axitinib, as compared to 19.2 months with sorafenib, although this was not statistically significant [26].

Bevacizumab along with interferon (IFN) α-2b also has a category 1 level of evidence for first-line therapy. It is a recombinant humanized monoclonal antibody that binds to circulating VEGF-A. A double-blind phase III trial (AVOREN) compared bevacizumab plus IFN-α-2b versus placebo plus IFN-α-2b [27]. With the addition of bevacizumab, PFS was significantly increased (10.4 versus 5.4 months), with a tumor response rate of 30.6% in the bevacizumab group compared to 12.4% in the placebo group. This was achieved without significant increase in adverse events. OS was improved in the bevacizumab group versus the placebo group (23.3 versus 21.3 months); however, this was not statistically significant.

5. Mechanistic Target of Rapamycin Inhibitors

Mechanistic target of rapamycin (mTOR) proteins are known to regulate cellular metabolism, growth, apoptosis, and angiogenesis through protein expression. Cellular growth factors stimulate the PI3K/Akt/mTOR pathway and eventually lead to HIF accumulation [28]. These discoveries led to the evaluation of temsirolimus and everolimus, which are both mTOR inhibitors in the management of renal cell carcinoma. Temsirolimus was compared to interferon-α in previously untreated patients with poor risk prognostic risk factors per, the MSKCC prognostic model [29]. The group receiving temsirolimus alone demonstrated significant improvement in median OS compared to IFN-α alone (10.9 versus 7.3 months, respectively). Similarly, PFS was shown to have improved from 3.1 months with IFN-α to 5.5 months with temsirolimus. Based on these study results, temsirolimus was the first FDA-approved mTOR inhibitor for patients with advanced renal cell carcinoma [30]. Currently, temsirolimus is the only mTOR inhibitor that is FDA-approved as a monotherapy.

Everolimus in combination with lenvatinib, a TKI, is utilized in patients who progress after prior therapy. In a phase II clinical trial, lenvatinib plus everolimus was compared to single-agent everolimus in previously-treated metastatic RCC (mRCC) patients [31]. The combination therapy showed increased median OS of 25.5 months, compared to 15.4 months for the monotherapy. Similarly, median PFS improved to 14.6 months in the combination group, compared to 5.5 months for the everolimus-only group [31].

6. Immunotherapy

Until late 2005, medical treatment options for RCC involved cytokine-based immunotherapy with the use of high-dose interleukin-2 (IL-2) and IFN-α. Though high-dose IL-2 is associated with significant toxicity, long-term durable response rates were seen in a small fraction of patients. High-dose IL-2 therapy is utilized in highly selected patients with excellent performance status and normal organ function [32]. IFN-α as a monotherapy has fallen out of favor, as a phase III multinational trial between sunitinib and IFN-α demonstrated a strong trend toward a median overall survival advantage of sunitinib over IFN-α [33].

Checkmate-214, a randomized phase III clinical trial, evaluated the combination of two immune checkpoint inhibitors, nivolumab and ipilimumab, in comparison to sunitinib in treatment-naïve patients with metastatic ccRCC. In patients with IMDC intermediate and poor risk, PFS was found to be 11.6 and 8.4 months for the nivolumab/ipilimumab combination and sunitinib, respectively. However, discontinuation due to adverse events was 24% in the combination group, as compared to 12% in the sunitinib group. The median OS with sunitinib was 26 months, whereas the median OS was not reached with the combination therapy [34]. Nivolumab was also shown to be effective as second-line therapy. In a phase III trial that studied patients previously treated with at least one line of therapy excluding mTOR inhibitors, nivolumab demonstrated an increase in OS of 5.4 months in comparison to everolimus monotherapy (25 versus 19.6 months, respectively). Median PFS, however, was not statistically significant, with 4.6 months for nivolumab and 4.4 months for everolimus [35].

7. Strategies to Inhibit the Hypoxia-Inducible Factor Pathway: A Plausible Therapeutic Avenue

VEGF inhibitors target one of the myriad oncogenic pathways that are activated by HIF. Hence, the cancer eventually develops resistance and progresses, despite an initial good response to various oral TKIs. Inhibiting the HIF pathway and subsequently its translational activity is an attractive treatment modality, as it blocks the activation of all downstream genes. The mTOR inhibitors inhibit HIF activation, but the responses are limited, as noted above. Recently, further strategies have been explored targeting the HIF pathway in combination with VEGF inhibitors, with variable success.

HIF regulation, either by blocking its production, antagonizing it effects, or by enhancing its degradation, has provided multiple opportunities to expand the current therapeutic armamentarium of renal cell carcinoma. In a small study of mRCC, HIF expression was predictive of increased response to sunitinib treatment [36]. In this study, 26 of 49 patients had high HIF1α and HIF2α expression on the tumor cells (based on immunoblot analysis). These patients had a higher rate of complete or partial response when compared to patients with low or absent HIF1α/HIF2α expression.

8. Role of Selenium in Cancer Therapeutics and the Hypoxia-Inducible Factor Pathway

Selenium is an essential micronutrient; in the human body, it is involved with the regulation of cell metabolism, DNA, and RNA, as well as protein synthesis, and is at the active site of several enzymes of the antioxidant network [37]. Inorganic forms of selenium, such as selenide and selenite, are converted by plants into organic forms, such as selenomethionine (SLM) and Se-methylselenocysteine (MSC), which are retained in the human body [37]. Epidemiologic studies have suggested that dietary selenium intake is a protective factor for some forms of cancers, such as colorectal, prostate, lung, and bladder cancer [38,39]. However, additional studies in healthy men did not show benefit of selenium in the prevention of prostate cancer [40].

SLM and MSC are forms of selenium that are currently being investigated as possible anti-tumor agents. In their natural form, these agents have a relatively low toxicity profile. They are converted via β-lyase into the active form methylselenol (MSA). HIF1α appears to be a target of selenium. In pre-clinical studies of head and neck squamous cell carcinoma cells that express HIF1α, it was found that in the setting of hypoxia, where HIF1α expression is increased, the cytotoxicity of SN38, the active metabolite of irinotecan, was enhanced with the addition of MSA [41]. This is possibly due to the inhibition of HIF1α by MSA, and demonstrates the potential for reversal of chemoresistance by MSA. Moreover, selenium has been found to target β-catenin, and increases drug cytotoxicity through the reduction of β-catenin's drug-resistant effects [42]. Selenium compounds may also improve efficaciousness of other anti-tumor agents through a reduction in treatment-induced toxicities, allowing for higher tolerated doses. In one study of A253 and HT 29 xenografts, coadministration of MSC with irinotecan at two to three times the maximum tolerated dose of irinotecan led to a response without intolerable toxicity [43].

Selenium can also affect the tumor microenvironment (TME), and may be able to stabilize the TME to improve drug delivery. MSC has an anti-vascular effect, and can increase the antitumor effect of irinotecan through the inhibition of HIF1α, which leads to decreased microvessel density, lowered tumor interstitial pressure, and increased pericyte coverage of blood vessels [41].

Selenium may also be able to act through its effects in the expression of miRNAs. Non-coding miRNAs are small molecules involved in post-transcriptional regulation of genes, which are often associated with increased angiogenesis and drug resistance.

9. Base-Line Transcription and Translation Biomarkers Expressed in Clear Cell Renal Cell Carcinoma Cell Lines with Disentail Expression of HIF1α and HIF2α

The expression levels of oncogenic, tumor-suppressor miRNAs, as well as hypoxia-inducible factors 1α and 2α, and program death ligand1 (PD-L1), are altered in many advanced cancers and implicated in multi-drug resistance, angiogenesis, and tumor growth and metastasis. Specifically, the oncogenic miRNA-155 and miRNA-210 are highly expressed in ccRCC tumors with differential expression of HIFs [44,45].

We demonstrated that these miRNAs and HIF proteins are targets of therapeutic doses and a schedule of selenomethionine and methylselenocysteine (selenium). It was not fully understood, however, whether selenium exerts its effects at the transcription or at the translation levels. Studies were carried out in ccRCC cell lines expressing differential levels of HIFs. We have determined that the base-line expression of three genes and the two oncogenic miRNAs in four cell lines: 786-O, RC2, RCC4, and RCC4-VHL. The three genes are *HIF-1α*, *HIF-2α*, and *PD-L1*. The two miRNAs are hsa-miR-155 and hsa-miR-210, Figure 2A. Results shown in Figure 2B indicate that there is robust mRNA transcription of each locus in all four of the cell lines. Note that lower normalized transcription levels (ΔCt) indicate higher expression levels. In general, the 786-O cell line expressing *HIF2α* displays the highest transcription. However, translation of these messages into a protein, as shown in Figure 2C, reveals a very different pattern. For example, in the 786-O cell line, both *HIF-1α* and *HIF-2α* are equally expressed at the RNA level, but reveal very different protein levels. Similarly, the RC2 cell line has consistently lower levels of RNA transcripts than 786-O, but higher levels of protein for *HIF-2α* and *PD-L1*. These inconsistencies suggest that there may be post-transcriptional targeting of these mRNAs by the miRNAs or another, as yet unknown post-transcriptional process at work.

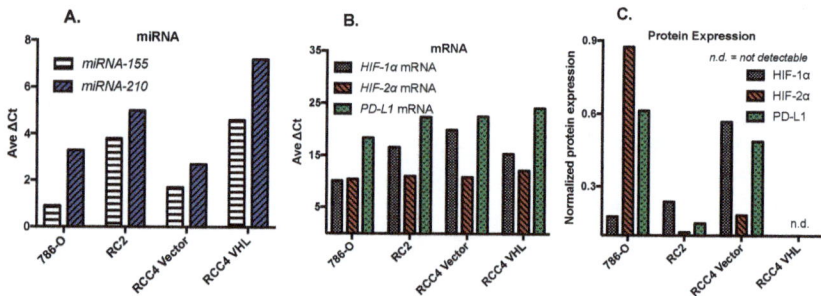

Figure 2. Base-line levels of constitutively expressed miRNAs (**A**), mRNAs (**B**), and HIFs and PD-L1 proteins (**C**) in normoxic clear cell renal cell carcinoma (ccRCC) cell lines. These cell lines show minimal differences in mRNA levels of *PD-L1*, *HIF-1α*, and *HIF-2α* (**B**), but express differential levels of the PD-L1, HIF-1α, and HIF-2α proteins (**C**). Ave ΔCt: average cycle threshold change compared to the reference sequence.

In normal cells, selenium has been shown to have a selective protective effect against chemotherapy-induced DNA damage via p53 mediated DNA repair. However, it did not confer a similar benefit to cancer cells [46].

SLM has been FDA-approved for clinical trials. A phase Ib dose-escalation trial in patients with metastatic ccRCC after failure of prior treatment is ongoing (NCT02535533). Preliminary results of nine evaluable patients demonstrated two patients achieving complete response, and three patients achieving partial response. No dose-limiting toxicities have been noted. The most common side effects included anorexia, fatigue, cough, diarrhea, and proteinuria. There were no grade 4 toxicities or deaths associated with this combination therapy. The phase II part of the clinical trial with SLM and axitinib 5 mg twice daily is planned [47]. The multiple avenues in which selenium interacts with other chemotherapies, the tumor microenvironment, and its interaction with miRNA and transcription factors make it a very favorable target for further research.

10. Studies on Hypoxia-Inducible Factor Inhibitors in Advanced Renal Cell Carcinoma

PT2385 is an HIF-2α antagonist that was evaluated in a phase I, standard "3 + 3" dose escalation study of heavily pretreated metastatic ccRCC [48]. In this study, 51 patients were treated with oral PT2385 twice a day, and the recommended phase-II dose (RP2D) was 800 mg BID. One patient had complete response (2%), six had partial responses (12%), and the rest had either stable disease or progression. No dose-limiting toxicities were noted. The most common treatment-related side effects included anemia, peripheral edema, fatigue, and nausea. Considering the promising response signals of a single agent in a heavily pretreated patient population, further studies are ongoing, with a combination of PT2385 with nivolumab and cabozantinib, respectively (NCT02293980).

In a multi-institutional, phase I/II clinical trial, vorinostat (histone deacetylase inhibitor) was evaluated in combination with bevacizumab in ccRCC patients [49]. HDAC inhibitors modulate the HIF pathway by affecting Hsp90 acetylation and HIF-α nuclear translocation [50,51]. In this study, 33 evaluable patients were treated with vorinostat 200 mg orally, twice daily for two weeks, in combination with bevacizumab 15 mg/kg administered intravenously every three weeks. There were no dose-limiting toxicities. Two patients had grade 4 thrombocytopenia. The most common adverse events included fatigue, nausea, pain, anorexia, diarrhea, and elevated creatinine. About 10 patients discontinued therapy due to toxicities, but there were no treatment-related deaths. One patient achieved complete response, and five patients had partial responses. Currently, a phase I/Ib study of pembrolizumab with vorinostat is in progress for patients with advanced renal or urothelial cell carcinoma (NCT02619253).

The safety and efficacy of another HDAC inhibitor, abexinostat, as an epigenetic downregulator of HIF-1α and VEGF expression was evaluated in combination with pazopanib by Aggarwal and colleagues, in a study of advanced solid tumor malignancies [52]. The RCC cohort included 22 patients. The dosing schedule of abexinostat was modified, due to five dose-limiting toxicities that included grade 3 thrombocytopenia ($n = 2$), grade 3 fatigue ($n = 2$), and grade 3 AST/ALT elevation ($n = 1$). There were no treatment-related deaths. The objective response rate in the RCC cohort was 27%, including patients who were previously refractory to pazopanib.

Bortezomib is a proteasome inhibitor currently approved for the treatment of multiple myeloma and mantle cell lymphoma. It is a reversible inhibitor of the chymotrypsin-like activity of the 26S proteasome in mammalian cells. By inhibiting proteasomes, it causes protein buildup and then leads to cell cytotoxicity. In preclinical models, Shin and colleagues have shown its role in *HIF-1α* repression by inhibiting the recruitment of the p300 coactivator [53]. In a phase II clinical trial of treatment-naïve, metastatic ccRCC, 17 patients were treated with sorafenib 200 mg orally twice daily, in combination with bortezomib 1 mg/m² intravenously administered on days 1, 4, 8, and 11, and then every 21 days [54]. The combination was safe, but the study was negative, as it did not meet the pre-specified endpoint of median progression-free survival of 70 weeks. Further studies are not planned with this combination.

The clinical efficacy of bortezomib in combination with bevacizumab was evaluated in 91 patients with treatment-refractory advanced cancers [55]. In the RCC cohort, 5 of 20 patients had a partial response or stable disease. No treatment-related deaths were noted. Common toxicities included thrombocytopenia, fatigue, nausea/vomiting, diarrhea, neuropathy, anemia, neutropenia, and hypertension. Table 1 summarizes the concluded clinical trials.

Table 1. Summary of reported clinical trials exploring HIF inhibitors in metastatic renal cell carcinoma.

Investigational Agent(s)	Phase	N	Trial Design	Dose-Limiting Toxicities (DLTs)	ORR	PFS, OS
HIF Antagonist						
PT2385	Phase 1	51	PT2385 administered twice daily orally from 100 to 1800 mg, followed by RP2D expansion phase	No DLTs reported	CR: 2% PR: 12% SD: 52% PD: 34%	PFS, OS: N/A
HIF Degradation						
Seleno-L-methionine (SLM) + axitinib	Phase 1b	9	SLM administered at 2500, 3000, or 4000 µg twice daily orally for 14 days, followed by once daily in combination with axitinib	No DLTs reported	CR: 22% PR: 33% SD: 11% PD: 33%	PFS, OS: N/A
HIF Degradation via Proteasomes						
Vorinostat + bevacizumab	Phase 1/2	36	Vorinostat administered at 200 mg twice daily orally for 14 days, in combination with bevacizumab at 15 mg/kg intravenously every 3 weeks	No DLTs reported in phase 1; 2 patients with grade 4 thrombocytopenia and grade 3 thromboembolic events	CR: 2.7% PR: 13.8%	mPFS: 5.7 months mOS: 12.9 months
Abexinostat + pazopanib	Phase 1	RCC cohort: 22 Total: 51	Pazopanib administered once daily on days 1 to 28, and abexinostat orally twice daily on days 1 to 5, 8 to 12, and 15 to 19, or days 1 to 4, 8 to 11, and 15 to 18	Total cohort: 5 DLTs were reported, including fatigue in 2 patients, thrombocytopenia in 2 patients, and elevated transaminases in 1 patient	RCC cohort: ORR (CR, PR): 27%	PFS: N/A OS: N/A
Bortezomib + bevacizumab	Phase 1	91	Bevacizumab administered at 2.5–15 mg/kg intravenously on day 1 of each 21 day cycle; bortezomib administered at 0.7–1.3 mg/m² intravenously on days 1, 4, 8, and 11 of each 21 day cycle	One patient with DLT from acute renal failure at highest dose level; 4 patients with partial response, 7 patients with stable disease at 6 months; toxicities included thrombocytopenia in 23% and fatigue in 19% of patients	CR: 0% PR: 4.4% SD: 42% PD: 47%	PFS: N/A OS: N/A
Sorafenib + bortezomib	Phase 2	17	Sorafenib administered orally twice daily in combination with bortezomib 1 mg/m² intravenously on days 1, 4, 8, and 11, then every 21 days	N/A	CR: 0% PR: 5.8% SD: 70% PD: 23%	mPFS: 13.7 weeks mOS: 110 weeks

CR: complete response; PR: partial response; SD: stable disease; PD: progressive disease; ORR: objective response rate; OS: overall survival; mOS: median overall survival; N/A: not available; HIF: hypoxia-inducible factor; RP2D: recommended phase II dose; RCC: renal cell carcinoma; PFS: Progression-free survival; mPFS: median progression free survival.

11. Conclusions

Insights into the molecular pathogenesis of ccRCC, especially the HIF pathway, have led to discovery of several therapeutic agents that have improved the treatment landscape. However, new strategies, which are durable and eventually a step closer to potential cure, are needed to further improve responses. HIF inhibition, either as monotherapy or in combination with other VEGF inhibitors, mTOR inhibitors, or immunotherapeutic agents, is promising. Numerous studies are underway evaluating these potentially synergistic combinations. (See Table 2). The preliminary results of SLM in early phase clinical trials of ccRCC are encouraging.

Table 2. Ongoing clinical trials of HIF inhibitors in metastatic renal cell carcinoma.

Clinicaltrials.gov NCT Identification Number	Phase	Title	N	Allocation/Treatment	Primary Objective/Outcome Measures	Status	Expected Completion
NCT03401788	Phase 2	A Phase 2 Study of PT2977 for the Treatment of Von Hippel-Lindau Disease-Associated Renal Cell Carcinoma	50	PT2977 (small molecule inhibitor of HIF2α)	Overall response rate	Recruiting	March 2023
NCT03592472	Phase 3	A Randomized, Double-blind, Placebo-controlled Study of Pazopanib with or without Abexinostat in Patients With Locally Advanced or Metastatic Renal Cell Carcinoma (RENAVIV)	413	Pazopanib + abexinostat vs. pazopanib + placebo	Progression-free survival; overall survival	Recruiting	January 2022
NCT02533533	Phase 1	A Therapeutic Trial for Safety and Preliminary Efficacy of the Combination of Axitinib and Selenomethionine (SLM) for Adult Patients with Advanced Metastatic Clear Cell Renal Cell Carcinoma	30	SLM administered orally twice daily for 14 days, followed by SLM once daily in combination with axitinib 5 mg twice daily	Safety	Recruiting	September 2020
NCT02974738	Phase 1	A Phase 1, Multiple-Dose, Dose-Escalation and Expansion Trial of PT2977, a HIF-2α Inhibitor, in Patients With Advanced Solid Tumors	125	PT2977	Safety	Recruiting	June 2019
NCT02293980	Phase 1	A Phase 1, Multiple-Dose, Dose-Escalation Trial of PT2385 Tablets, a HIF-2α Inhibitor, in Patients With Advanced Clear Cell Renal Cell Carcinoma	107	Part 1: PT2385 tablets; Part 2: PT2385 tablets in combination with nivolumab; Part 3: PT2385 tablets in combination with cabozantinib	Safety, DLT	Active, not recruiting	December 2018
NCT02619253	Phase 1/1b	A Phase 1/1b, Open Label, Dose Finding Study to Evaluate Safety, Pharmacodynamics and Efficacy of Pembrolizumab (MK-3475) in Combination with Vorinostat in Patients with Advanced Renal or Urothelial Cell Carcinoma	42	Pembrolizumab and vorinostat	Safety/DLT	Recruiting	May 2020
NCT03634540	Phase 2	A Phase 2 Trial of PT2977 in Combination with Cabozantinib in Patients with Advanced Clear Cell Renal Cell Carcinoma	118	PT2977 in combination with cabozantinib tablets	Overall response rate (CR, PR)	Not yet recruiting	September 2022

CR: complete response; PR: partial response; SD: stable disease; PD: progressive disease; ORR: objective response rate; OS: overall survival; N/A: not available; HIF: hypoxia inducible factor; RP2D: recommended phase II dose; RCC: renal cell carcinoma; DLT: dose-limiting toxicity.

Funding: Selenium clinical trial in renal cell cancer was supported by Pipeline Acceleration for Cancer Therapeutics (PACT) initiative by Holden Comprehensive Cancer Center and Rock 'n' Ride fundraiser, Washington, Iowa 52353. Additionally, the basic science research on selenium was supported by the NIH P42 ES013661, ISRP (Iowa Superfund Research Program) grant.

Conflicts of Interest: The authors declare no conflict of interest.

References

1. Siegel, R.L.; Miller, K.D.; Jemal, A. Cancer statistics, 2018. *CA Cancer J. Clin.* **2018**, *68*, 7–30. [CrossRef]
2. Gudbjartsson, T.; Thoroddsen, A.; Petursdottir, V.; Hardarson, S.; Magnusson, J.; Einarsson, G.V. Effect of incidental detection for survival of patients with renal cell carcinoma: Results of population-based study of 701 patients. *Urology* **2005**, *66*, 1186–1191. [CrossRef] [PubMed]
3. Novara, G.; Ficarra, V.; Antonelli, A.; Artibani, W.; Bertini, R.; Carini, M.; Cosciani Cunico, S.; Imbimbo, C.; Longo, N.; Martignoni, G.; et al. Validation of the 2009 TNM version in a large multi-institutional cohort of patients treated for renal cell carcinoma: Are further improvements needed? *Eur. Urol.* **2010**, *58*, 588–595. [CrossRef] [PubMed]
4. Chow, W.H.; Dong, L.M.; Devesa, S.S. Epidemiology and risk factors for kidney cancer. *Nat. Rev. Urol.* **2010**, *7*, 245–257. [CrossRef] [PubMed]
5. Oya, M. *Renal Cell Carcinoma: Molecular Features and Treatment Updates*; Springer: Berlin/Heidelberg, Germany; New York, NY, USA, 2017.
6. Eble, J.N.; Sauter, G.; Epstein, J.I.; Sesterhenn, I.A. *Pathology and Genetics of Tumours of the Urinary System and Male Genital Organs*; IARC Press: Lyon, France, 2004.
7. Patard, J.J.; Leray, E.; Rioux-Leclercq, N.; Cindolo, L.; Ficarra, V.; Zisman, A.; de la Taille, A.; Tostain, J.; Artibani, W.; Abbou, C.C.; et al. Prognostic value of histologic subtypes in renal cell carcinoma: A multicenter experience. *J. Clin. Oncol.* **2005**, *23*, 2763–2771. [CrossRef] [PubMed]
8. Linehan, W.M.; Walther, M.M.; Zbar, B. The genetic basis of cancer of the kidney. *J. Urol.* **2003**, *170 Pt 1*, 2163–2172. [CrossRef]
9. Motzer, R.J.; Jonasch, E.; Agarwal, N.; Bhayani, S.; Bro, W.P.; Chang, S.S.; Choueiri, T.K.; Costello, B.A.; Derweesh, I.H.; Fishman, M.; et al. Kidney Cancer, Version 2.2017, NCCN Clinical Practice Guidelines in Oncology. *J. Natl. Compr. Cancer Netw.* **2017**, *15*, 804–834. [CrossRef]
10. Méjean, A.; Ravaud, A.; Thezenas, S.; Colas, S.; Beauval, J.-B.; Bensalah, K.; Geoffrois, L.; Thiery-Vuillemin, A.; Cormier, L.; Lang, H.; et al. Sunitinib Alone or after Nephrectomy in Metastatic Renal-Cell Carcinoma. *N. Engl. J. Med.* **2018**, *379*, 417–427. [CrossRef] [PubMed]
11. Semenza, G.L. Hypoxia-inducible factors: Mediators of cancer progression and targets for cancer therapy. *Trends. Pharmacol. Sci.* **2012**, *33*, 207–214. [CrossRef]
12. Keith, B.; Johnson, R.S.; Simon, M.C. HIF1α and HIF2α: Sibling rivalry in hypoxic tumour growth and progression. *Nat. Rev. Cancer* **2011**, *12*, 9. [CrossRef]
13. Harris, A.L. Hypoxia—A key regulatory factor in tumour growth. *Nat. Rev. Cancer* **2002**, *2*, 38. [CrossRef] [PubMed]
14. Schodel, J.; Grampp, S.; Maher, E.R.; Moch, H.; Ratcliffe, P.J.; Russo, P.; Mole, D.R. Hypoxia, Hypoxia-inducible Transcription Factors, and Renal Cancer. *Eur. Urol.* **2016**, *69*, 646–657. [CrossRef] [PubMed]
15. Gossage, L.; Eisen, T.; Maher, E.R. VHL, the story of a tumour suppressor gene. *Nat. Rev. Cancer* **2015**, *15*, 55–64. [CrossRef] [PubMed]
16. Ohh, M.; Park, C.W.; Ivan, M.; Hoffman, M.A.; Kim, T.-Y.; Huang, L.E.; Pavletich, N.; Chau, V.; Kaelin, W.G. Ubiquitination of hypoxia-inducible factor requires direct binding to the β-domain of the von Hippel–Lindau protein. *Nat. Cell Biol.* **2000**, *2*, 423. [CrossRef] [PubMed]
17. Gnarra, J.R.; Tory, K.; Weng, Y.; Schmidt, L.; Wei, M.H.; Li, H.; Latif, F.; Liu, S.; Chen, F.; Duh, F.M.; et al. Mutations of the VHL tumour suppressor gene in renal carcinoma. *Nat. Genet.* **1994**, *7*, 85–90. [CrossRef] [PubMed]
18. Linehan, W.M.; Srinivasan, R.; Schmidt, L.S. The genetic basis of kidney cancer: A metabolic disease. *Nat. Rev. Urol.* **2010**, *7*, 277–285. [CrossRef] [PubMed]

19. Heng, D.Y.C.; Xie, W.; Regan, M.M.; Warren, M.A.; Golshayan, A.R.; Sahi, C.; Eigl, B.J.; Ruether, J.D.; Cheng, T.; North, S.; et al. Prognostic Factors for Overall Survival in Patients with Metastatic Renal Cell Carcinoma Treated with Vascular Endothelial Growth Factor–Targeted Agents: Results from a Large, Multicenter Study. *J. Clin. Oncol.* **2009**, *27*, 5794–5799. [CrossRef]

20. Motzer, R.J.; Hutson, T.E.; Tomczak, P.; Michaelson, M.D.; Bukowski, R.M.; Rixe, O.; Oudard, S.; Negrier, S.; Szczylik, C.; Kim, S.T.; et al. Sunitinib versus Interferon Alfa in Metastatic Renal-Cell Carcinoma. *N. Engl. J. Med.* **2007**, *356*, 115–124. [CrossRef]

21. Sternberg, C.N.; Davis, I.D.; Mardiak, J.; Szczylik, C.; Lee, E.; Wagstaff, J.; Barrios, C.H.; Salman, P.; Gladkov, O.A.; Kavina, A.; et al. Pazopanib in locally advanced or metastatic renal cell carcinoma: Results of a randomized phase III trial. *J. Clin. Oncol.* **2010**, *28*, 1061–1068. [CrossRef]

22. Motzer, R.J.; Hutson, T.E.; Cella, D.; Reeves, J.; Hawkins, R.; Guo, J.; Nathan, P.; Staehler, M.; de Souza, P.; Merchan, J.R.; et al. Pazopanib versus sunitinib in metastatic renal-cell carcinoma. *N. Engl. J. Med.* **2013**, *369*, 722–731. [CrossRef]

23. Motzer, R.J.; Hutson, T.E.; McCann, L.; Deen, K.; Choueiri, T.K. Overall survival in renal-cell carcinoma with pazopanib versus sunitinib. *N. Engl. J. Med.* **2014**, *370*, 1769–1770. [CrossRef] [PubMed]

24. Choueiri, T.K.; Hessel, C.; Halabi, S.; Sanford, B.; Michaelson, M.D.; Hahn, O.; Walsh, M.; Olencki, T.; Picus, J.; Small, E.J.; et al. Cabozantinib versus sunitinib as initial therapy for metastatic renal cell carcinoma of intermediate or poor risk (Alliance A031203 CABOSUN randomised trial): Progression-free survival by independent review and overall survival update. *Eur. J. Cancer* **2018**, *94*, 115–125. [CrossRef] [PubMed]

25. Choueiri, T.K.; Escudier, B.; Powles, T.; Mainwaring, P.N.; Rini, B.I.; Donskov, F.; Hammers, H.; Hutson, T.E.; Lee, J.L.; Peltola, K.; et al. Cabozantinib versus Everolimus in Advanced Renal-Cell Carcinoma. *N. Engl. J. Med.* **2015**, *373*, 1814–1823. [CrossRef] [PubMed]

26. Rini, B.I.; Escudier, B.; Tomczak, P.; Kaprin, A.; Szczylik, C.; Hutson, T.E.; Michaelson, M.D.; Gorbunova, V.A.; Gore, M.E.; Rusakov, I.G.; et al. Comparative effectiveness of axitinib versus sorafenib in advanced renal cell carcinoma (AXIS): A randomised phase 3 trial. *Lancet* **2011**, *378*, 1931–1939. [CrossRef]

27. Escudier, B.; Pluzanska, A.; Koralewski, P.; Ravaud, A.; Bracarda, S.; Szczylik, C.; Chevreau, C.; Filipek, M.; Melichar, B.; Bajetta, E.; et al. Bevacizumab plus interferon alfa-2a for treatment of metastatic renal cell carcinoma: A randomised, double-blind phase III trial. *Lancet* **2007**, *370*, 2103–2111. [CrossRef]

28. Abraham, R.T.; Gibbons, J.J. The Mammalian Target of Rapamycin Signaling Pathway: Twists and Turns in the Road to Cancer Therapy. *Clin. Cancer Res.* **2007**, *13*, 3109–3114. [CrossRef] [PubMed]

29. Motzer, R.J.; Bacik, J.; Murphy, B.A.; Russo, P.; Mazumdar, M. Interferon-alfa as a comparative treatment for clinical trials of new therapies against advanced renal cell carcinoma. *J. Clin. Oncol.* **2002**, *20*, 289–296. [CrossRef]

30. Hudes, G.; Carducci, M.; Tomczak, P.; Dutcher, J.; Figlin, R.; Kapoor, A.; Staroslawska, E.; Sosman, J.; McDermott, D.; Bodrogi, I.; et al. Temsirolimus, interferon alfa, or both for advanced renal-cell carcinoma. *N. Engl. J. Med.* **2007**, *356*, 2271–2281. [CrossRef]

31. Motzer, R.J.; Hutson, T.E.; Ren, M.; Dutcus, C.; Larkin, J. Independent assessment of lenvatinib plus everolimus in patients with metastatic renal cell carcinoma. *Lancet Oncol.* **2016**, *17*, e4–e5. [CrossRef]

32. Belldegrun, A.S.; Klatte, T.; Shuch, B.; LaRochelle, J.C.; Miller, D.C.; Said, J.W.; Riggs, S.B.; Zomorodian, N.; Kabbinavar, F.F.; Dekernion, J.B.; et al. Cancer-specific survival outcomes among patients treated during the cytokine era of kidney cancer (1989–2005): A benchmark for emerging targeted cancer therapies. *Cancer* **2008**, *113*, 2457–2463. [CrossRef]

33. Chow, L.Q.; Eckhardt, S.G. Sunitinib: From rational design to clinical efficacy. *J. Clin. Oncol.* **2007**, *25*, 884–896. [CrossRef] [PubMed]

34. Powles, T.; Albiges, L.; Staehler, M.; Bensalah, K.; Dabestani, S.; Giles, R.H.; Hofmann, F.; Hora, M.; Kuczyk, M.A.; Lam, T.B.; et al. Updated European Association of Urology Guidelines Recommendations for the Treatment of First-line Metastatic Clear Cell Renal Cancer. *Eur. Urol.* **2017**, *73*, 311–315. [CrossRef] [PubMed]

35. Motzer, R.J.; Escudier, B.; McDermott, D.F.; George, S.; Hammers, H.J.; Srinivas, S.; Tykodi, S.S.; Sosman, J.A.; Procopio, G.; Plimack, E.R.; et al. Nivolumab versus Everolimus in Advanced Renal-Cell Carcinoma. *N. Engl. J. Med.* **2015**, *373*, 1803–1813. [CrossRef] [PubMed]

36. Patel, P.H.; Chadalavada, R.S.; Ishill, N.M.; Patil, S.; Reuter, V.E.; Motzer, R.J.; Chaganti, R.S. Hypoxia-inducible factor (HIF) 1α and 2α levels in cell lines and human tumor predicts response to sunitinib in renal cell carcinoma (RCC). *J. Clin. Oncol.* **2008**, *26* (Suppl. 15), 5008. [CrossRef]

37. Papp, L.V.; Lu, J.; Holmgren, A.; Khanna, K.K. From selenium to selenoproteins: Synthesis, identity, and their role in human health. *Antioxid. Redox Signal* **2007**, *9*, 775–806. [CrossRef] [PubMed]

38. Micke, O.; Schomburg, L.; Buentzel, J.; Kisters, K.; Muecke, R. Selenium in oncology: From chemistry to clinics. *Molecules* **2009**, *14*, 3975–3988. [CrossRef] [PubMed]

39. Clark, L.C.; Combs, G.F., Jr.; Turnbull, B.W.; Slate, E.H.; Chalker, D.K.; Chow, J.; Davis, L.S.; Glover, R.A.; Graham, G.F.; Gross, E.G.; et al. Effects of selenium supplementation for cancer prevention in patients with carcinoma of the skin. A randomized controlled trial. *JAMA* **1996**, *276*, 1957–1963. [CrossRef]

40. Lippman, S.M.; Klein, E.A.; Goodman, P.J.; Lucia, M.S.; Thompson, I.M.; Ford, L.G.; Parnes, H.L.; Minasian, L.M.; Gaziano, J.M.; Hartline, J.A.; et al. Effect of selenium and vitamin E on risk of prostate cancer and other cancers: The Selenium and Vitamin E Cancer Prevention Trial (SELECT). *JAMA* **2009**, *301*, 39–51. [CrossRef]

41. Zakharia, Y.; Bhattacharya, A.; Rustum, Y.M. Selenium targets resistance biomarkers enhancing efficacy while reducing toxicity of anti-cancer drugs: Preclinical and clinical development. *Oncotarget* **2018**, *9*, 10765–10783. [CrossRef]

42. Saifo, M.S.; Rempinski, D.R., Jr.; Rustum, Y.M.; Azrak, R.G. Targeting the oncogenic protein beta-catenin to enhance chemotherapy outcome against solid human cancers. *Mol. Cancer* **2010**, *9*, 310. [CrossRef]

43. Cao, S.; Durrani, F.A.; Tóth, K.; Rustum, Y.M. Se-methylselenocysteine offers selective protection against toxicity and potentiates the antitumour activity of anticancer drugs in preclinical animal models. *Br. J. Cancer* **2014**, *110*, 1733–1743. [CrossRef] [PubMed]

44. Kong, W.; He, L.; Richards, E.J.; Challa, S.; Xu, C.X.; Permuth-Wey, J.; Lancaster, J.M.; Coppola, D.; Sellers, T.A.; Djeu, J.Y.; et al. Upregulation of miRNA-155 promotes tumour angiogenesis by targeting VHL and is associated with poor prognosis and triple-negative breast cancer. *Oncogene* **2014**, *33*, 679–689. [CrossRef] [PubMed]

45. Tang, K.; Xu, H. Prognostic value of meta-signature miRNAs in renal cell carcinoma: An integrated miRNA expression profiling analysis. *Sci. Rep.* **2015**, *5*, 10272. [CrossRef] [PubMed]

46. Fischer, J.L.; Mihelc, E.M.; Pollok, K.E.; Smith, M.L. Chemotherapeutic selectivity conferred by selenium: A role for p53-dependent DNA repair. *Mol. Cancer Ther.* **2007**, *6*, 355–361. [CrossRef] [PubMed]

47. Zakharia, Y.; Garje, R.; Brown, J.A.; Nepple, K.G.; Dahmoush, L.; Gibson-Corley, K.; Spitz, D.; Milhem, M.M.; Rustum, Y.M. Phase1 clinical trial of high doses of Seleno-L-methionine (SLM), in sequential combination with axitinib in previously treated and relapsed clear cell renal cell carcinoma (ccRCC) patients. *J. Clin. Oncol.* **2018**, *36* (Suppl. 6), 630. [CrossRef]

48. Courtney, K.D.; Infante, J.R.; Lam, E.T.; Figlin, R.A.; Rini, B.I.; Brugarolas, J.; Zojwalla, N.J.; Lowe, A.M.; Wang, K.; Wallace, E.M.; et al. Phase I Dose-Escalation Trial of PT2385, a First-in-Class Hypoxia-Inducible Factor-2α Antagonist in Patients with Previously Treated Advanced Clear Cell Renal Cell Carcinoma. *J. Clin. Oncol.* **2018**, *36*, 867–874. [CrossRef]

49. Pili, R.; Liu, G.; Chintala, S.; Verheul, H.; Rehman, S.; Attwood, K.; Lodge, M.A.; Wahl, R.; Martin, J.I.; Miles, K.M.; et al. Combination of the histone deacetylase inhibitor vorinostat with bevacizumab in patients with clear-cell renal cell carcinoma: A multicentre, single-arm phase I/II clinical trial. *Br. J. Cancer* **2017**, *116*, 874–883. [CrossRef]

50. Ellis, L.; Hammers, H.; Pili, R. Targeting tumor angiogenesis with histone deacetylase inhibitors. *Cancer Lett.* **2009**, *280*, 145–153. [CrossRef]

51. Zhang, C.; Yang, C.; Feldman, M.J.; Wang, H.; Pang, Y.; Maggio, D.M.; Zhu, D.; Nesvick, C.L.; Dmitriev, P.; Bullova, P.; et al. Vorinostat suppresses hypoxia signaling by modulating nuclear translocation of hypoxia inducible factor 1 alpha. *Oncotarget* **2017**, *8*, 56110–56125.

52. Aggarwal, R.; Thomas, S.; Pawlowska, N.; Bartelink, I.; Grabowsky, J.; Jahan, T.; Cripps, A.; Harb, A.; Leng, J.; Reinert, A.; et al. Inhibiting Histone Deacetylase as a Means to Reverse Resistance to Angiogenesis Inhibitors: Phase I Study of Abexinostat Plus Pazopanib in Advanced Solid Tumor Malignancies. *J. Clin. Oncol.* **2017**, *35*, 1231–1239. [CrossRef]

53. Shin, D.H.; Chun, Y.-S.; Lee, D.S.; Huang, L.E.; Park, J.-W. Bortezomib inhibits tumor adaptation to hypoxia by stimulating the FIH-mediated repression of hypoxia-inducible factor-1. *Blood* **2008**, *111*, 3131. [CrossRef] [PubMed]

54. Rao, A.; Lauer, R. Phase II study of sorafenib and bortezomib for first-line treatment of metastatic or unresectable renal cell carcinoma. *Oncologist* **2015**, *20*, 370–371. [CrossRef] [PubMed]

55. Falchook, G.S.; Wheler, J.J.; Naing, A.; Jackson, E.F.; Janku, F.; Hong, D.; Ng, C.S.; Tannir, N.M.; Lawhorn, K.N.; Huang, M.; et al. Targeting hypoxia-inducible factor-1α (HIF-1α) in combination with antiangiogenic therapy: A phase I trial of bortezomib plus bevacizumab. *Oncotarget* **2014**, *5*, 10280–10292. [CrossRef] [PubMed]

International Journal of
Molecular Sciences

MDPI

Review

Selenium Species: Current Status and Potentials in Cancer Prevention and Therapy

Heng Wee Tan †, Hai-Ying Mo †, Andy T. Y. Lau * and Yan-Ming Xu *

Laboratory of Cancer Biology and Epigenetics, Department of Cell Biology and Genetics, Shantou University Medical College, Shantou 515041, China; hwtan@stu.edu.cn (H.W.T.); 16hymo@stu.edu.cn (H.-Y.M.)
* Correspondence: andytylau@stu.edu.cn (A.T.Y.L.); amyymxu@stu.edu.cn (Y.-M.X.);
 Tel.: +86-754-8853-0052 (A.T.Y.L.); +86-754-8890-0437 (Y.-M.X.)
† These authors contributed equally to this work.

Received: 15 November 2018; Accepted: 20 December 2018; Published: 25 December 2018

Abstract: Selenium (Se) acts as an essential trace element in the human body due to its unique biological functions, particularly in the oxidation-reduction system. Although several clinical trials indicated no significant benefit of Se in preventing cancer, researchers reported that some Se species exhibit superior anticancer properties. Therefore, a reassessment of the status of Se and Se compounds is necessary in order to provide clearer insights into the potentiality of Se in cancer prevention and therapy. In this review, we organize relevant forms of Se species based on the three main categories of Se—inorganic, organic, and Se-containing nanoparticles (SeNPs)—and overview their potential functions and applications in oncology. Here, we specifically focus on the SeNPs as they have tremendous potential in oncology and other fields. In general, to make better use of Se compounds in cancer prevention and therapy, extensive further study is still required to understand the underlying mechanisms of the Se compounds.

Keywords: selenium species; Se-containing nanoparticles; anticancer; chemotherapeutics; epigenetics

1. Introduction

Selenium (Se) is an essential micronutrient for the human body that is mainly obtained through diet and/or nutritional supplement [1]. Trace amounts of Se are required for maintaining optimal health as Se is a component of the selenoproteins (mostly in the form of amino acid selenocysteine) that participate in a wide range of cellular physiological processes. These processes include, but are not limited to, thyroid hormone regulation [2], redox homeostasis [3–5], inflammatory and immunological responses [6–8], carbohydrate metabolism [9], cardiovascular [10] and reproductive [11,12] health, and brain function maintenance [13–15]. Se deficiency is associated with numerous human diseases with various degrees of illnesses [16,17]. For instance, the Keshan disease (fatal cardiomyopathy due to viral infection) [18] and Kashin-Beck disease (chronic osteochondropathy) [19] are a few typical examples of Se deficiency-related diseases, which often occur endemically in the population living in regions with Se-poor soil. Excessive Se can be toxic and may lead to selenosis [17,20]. Currently, the recommended dietary allowance of Se for adults is set at 55 μg (0.7 μmol)/day [21]. Individuals with daily Se intake less than ~15 μg appear to be at risk of Se deficiency-related diseases, whereas those who consume over 400 μg/day are prone to Se toxicity, although some studies have shown that safe levels of Se intake may be much lower than anticipated [16,21].

The relationship between Se and cancer, particularly in gastrointestinal and prostate cancer, was discovered in the middle of the 19th century, which then raised interest in the contribution of Se supplements to cancer prevention and therapy [22–24]. During half a century of exploration, many novel forms of Se compounds have been discovered and tested, and some have shown promising anticancer activity [25,26]. A double-blind, placebo-controlled, and randomized clinical trial carried out in the 1990s, the Nutritional Prevention of Cancer (NPC) trial, has provided us the early evidence supporting Se as a potential chemopreventive agent [27–29]. However, to date, none of the Se compounds have been clinically recognized as anticancer drugs, partly because, over the years, researchers have obtained conflicting results within and between epidemiological, clinical, and laboratory studies [30,31]. Notably, in contrast with the NPC trial, subsequent clinical trials such as the Selenium and Vitamin E Cancer Prevention Trial (SELECT) failed to demonstrate the anticancer effects of Se [32–34]. Some of these conflicting results showed that Se compounds not only failed to exert their cancer prevention or anticancer ability as anticipated but, in some cases, may even promote cancer [35]. Recent epidemiologic evidence suggests that chronic exposure to inorganic Se may increase cancer risk [36]. As a result, the dual role of Se compounds in carcinogenesis, especially in relation to the aspects of oxidative stress and angiogenesis, has been proposed and recently summarized [31]. So far, there is no clear conclusion on the circumstances under which a particular Se compound prevents or enhances carcinogenesis, perhaps due to the wide variety of Se speciations and their diverse effects at different concentrations on different metabolic pathways of cells and tissues [37,38]. Se or Se-containing compounds can be grouped into three main categories: inorganic, organic (also known as the organoselenium compounds), and Se-containing nanoparticles (SeNPs). In order to better utilize the anticancer properties of Se species, it is necessary to thoroughly evaluate the current status of Se species. Here, we systematically organize the relevant forms of Se species, with slightly more emphasize on SeNPs, and review their recent developments and potential in cancer prevention and therapy.

2. Anti- or Pro-Cancer?

Several Se compounds derived from all the three groups of Se (inorganic compounds, organoselenium compounds, and SeNPs) have shown possible anticancer ability. It is generally accepted that Se compounds exert their anticancer ability mainly through their direct or indirect antioxidant properties that intracellularly maintain the redox status and protect healthy cells from reactive oxygen species (ROS)-induced oxidative damage [39]. ROS are free radicals with unpaired electrons generated during normal biophysiological function. The evidence is strong that excessive ROS promotes carcinogenesis via elevated oxidative stress and increased DNA mutation [40]. Cancer cells are often characterized by their ability to produce and cope with an increased amount of ROS [41]. In other words, increased dependence on an antioxidant defense system is one of the principal characteristics of cancer cells. Despite the links between ROS and cancer formation, however, optimal (usually low) levels of ROS are actually beneficial as they play important roles in regulating many biological functions. Some enzymes and cells (e.g., white blood cells) can deliberately produce a range of superoxide radicals to kill invading pathogens [39,42]. ROS can also destroy damaged cells by promoting cellular senescence and apoptosis and thus eliminate the formation of cancer [41,42]. Such a dual role of ROS may explain why conflicting results for Se species, as "antioxidants", are often observed in cancer research. To further complicate the situation, other research found that some selenoproteins could actually behave as prooxidants instead of antioxidants, demonstrating both cancer-inhibiting and -promoting features in a cell type-, genotype-, and dosage-dependent manner [43–47]. For example, thioredoxin reductase 1, an essential redox regulating selenoprotein, can change from an anti- to a pro-oxidant and can both inhibit or promote carcinogenesis [44,46].

In addition to oxidative stress regulation, the duality of Se compounds on angiogenesis has been discovered [31,38]. Angiogenesis refers to the physiological process responsible for the formation and growth of micro-blood vessels from pre-existing vasculature, which is one of the most important mechanisms for cells to obtain oxygen and nutrients. The roles of angiogenesis in relation to cancer development and metastasis have been studied extensively, and therapy explicitly targeting angiogenesis has become a promising approach for cancer treatment [48]. In vitro and in vivo studies showed that some Se compounds, such as the monomethylated Se amino acid methylselenocysteine (MSC), could inhibit cancer growth through its antiangiogenic properties [49–51]. MSC might also normalize the blood vessels and thus enhance delivery of a range of chemotherapeutic drugs and simultaneously reduce their toxicity [52–56]. Conversely, opposite results in which pro-angiogenic responses of Se-selenoproteins/compounds in normal or cancer cells have also been reported [57,58]. Thus, the dual role and narrow window between the beneficial and toxic effects of Se compounds often limit their potential for clinical application.

The dual effect of Se mentioned in this section is often restricted to the inorganic and organic Se compounds, and so far, research on anticancer activity of SeNPs, the emerging special form of Se species, appears to be positive. The use of SeNPs has had a revolutionary impact on cancer therapy, and they have shown tremendous potential compared to "ordinary" inorganic and organic Se compounds [59–62]. However, knowledge regarding the cytotoxicity and other possible adverse effects of these SeNPs in humans is still lacking, and further extensive research is required [63,64]. So far, all Se compounds are considered non-carcinogenic with the exception of selenium sulfide, which is categorized as a probable human carcinogen. Overall, in order to determine the potential of Se compounds in cancer prevention and therapy, multiple factors (e.g., speciation, concentration, targeting cell type, and cell state/condition) must be considered. In the following sections, we organize relevant forms of Se species and discuss their potential roles in cancer treatment based on recently published data.

3. Se-Containing Compounds and Their Usage in Oncology

All three main categories of Se (inorganic, organic, and SeNPs) contain compounds with potential anticancer properties. For inorganic and organic Se compounds, research has found that they are both metabolized differently and have varied mechanisms of action in diverse bio-physiological processes, including their roles in cancer [65]. Both forms of Se compounds can be readily absorbed by the human body, but only organic Se compounds, usually in the forms of amino acids (e.g., selenomethionine (SeMet) and selenocysteine), can be better retained and used [65]. The cancer prevention ability of a range of inorganic and organic Se compounds has been supported by a large number of publications from a wide range of studies under different settings, including biochemical, epidemiological, clinical, and animal studies [38,44,46,66–68]. However, toxicity risks accompanied by the use of these Se compounds have also been recorded. Although organic forms of Se may have lesser toxic effects than inorganic Se compounds [69], in reality, the toxic effects of Se are determined by multiple factors, with the forms of Se and dosage exposure being two of the most important parameters [26]. Despite the greater toxic effects, inorganic Se compounds may have an advantage in certain aspects of cancer therapy as described below.

3.1. Inorganic Se Compounds

Se exists in four natural valence states: elemental Se (0), selenide (-2; Se^{2-}), selenite ($+4$; SeO_3^{2-}), and selenate ($+6$; SeO_4^{2-}). In Figure 1, we display the chemical structures of some of the representative Se compounds. A more detailed list of classification of Se compounds based on their structural features is summarized by Sanmartín et al. [70]. The functional and toxic effects of inorganic Se compounds differ according to their valence states. In Choi et al. [71], various concentrations of sodium selenate (Na_2SeO_4) (5, 10, 30, and 50 µM for 48 h) and sodium selenite (Na_2SeO_3) (0.1, 0.25, and 0.5 µM for 48 h) along with three other organic Se compounds (SeMet, MSC, and methylseleninic acid (MSA)) were tested for their ability to sensitize human oral squamous carcinoma (KB) cells resistant to chemotherapeutic drug vincristine (KBV20C). They found that although all five Se compounds appeared to be able to sensitize KBV20C to the same extent as the sensitive parent KB cells, only selenate produced a higher sensitizing effect on the KBV20C cells by arresting the cell cycle at G2-phase and activating apoptotic pathways. However, opposite results were obtained by Takahashi et al. [72], where they showed that human oral squamous carcinoma (HSC-3) cells were more sensitive to Na_2SeO_3 and selenium dioxide (SeO_2), but not sodium selenate at concentrations ranging from 1 to 1000 µM (72 h). Pronounced anti-proliferative effect of selenite (5–100 µM for 2–5 days) against three oral cancer cell lines (HSC-3, HSC-4, and SAS) was reported [73]. This study also suggested that selenite had a better anticancer effect than the two other organoselenium compounds (SeMet and MSC) tested.

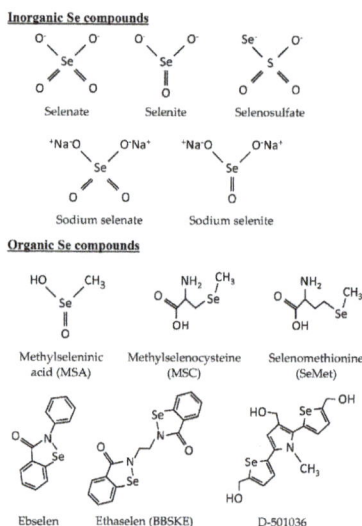

Figure 1. Chemical structures of selected representative inorganic and organic Se compounds discussed in this review.

Selenite is the most studied form of inorganic Se compounds as it exhibits excellent chemopreventive and anticancer features [74]. Selenite could effectively inhibit cell proliferation of various types of cancer cells, including lung cancer, which is the most common and deadliest cancer worldwide [75]. Among different human cancer cell lines tested, lung cancer cells, in general, appeared to be especially sensitive to selenite [76–79]. Olm et al. [80] indicated that selenite cytotoxicity (5 µM for 5 h) was correlated with Se uptake of three lung cancer cell lines (H157, H611, and U2020) and that high concentrations (>1 mM) of selenate were non-toxic for these cell lines. Selenite was suggested to play a role in natural killer (NK) cell-based anticancer immunotherapy where it could increase the susceptibility of cancer cells to CD94/NK group 2A-positive NK cells, and has possible clinical applications in lung cancer patients [81]. The synergistic effect of selenite and thioredoxin reductase inhibitors (e.g., ethaselen and auranofin) was detected in human ovarian and lung cancer cell lines [82,83]. These results demonstrated the potential of Se compounds to enhance the activity and reduce the toxicity of anticancer drugs including those commonly used in chemotherapy (e.g., cisplatin, docetaxel, 5-FU, oxaliplatin, and irinotecan) [84–86]. Se compounds appear to be more effective in inhibiting the growth of anticancer drug-resistant cancer cells compared with drug-sensitive cancer cells via deactivating various resistance mechanisms used by the cancer cells. Chemotherapeutic drug-resistant lung cancer cells were revealed to be generally more sensitive to selenite (ranging from 0.1 to 100 µM for 48 h and up to 4 days) compared to drug-sensitive cancer cells [82,87,88]. In addition to the above-mentioned in vitro studies, a phase I clinical trial published demonstrated the beneficial effects of selenite in cancer patients, especially in lung cancer patients who were resistant to cytostatic drugs [76].

Notably, the above results, however, do not necessarily mean that selenite is superior to other inorganic Se compounds in terms of cancer prevention and therapy. For instance, selenate might be more effective than selenite under certain circumstances as described earlier, even in the same type of cancer albeit different cell lines [71,72]. Inorganic Se such as selenosulfate ($SeSO_3^-$) was reported to have generally greater cytotoxic effects on cancer cells but was less toxic in healthy cells than selenite, depending on the cell types and the presence or absence of supplement amino acids that may affect the uptake of selenite [89]. Both $SeSO_3^-$ and selenite might have the potential to work as a remedy against chemotherapy toxicity [76,84]. It was shown that hydrogen selenide (H_2Se), a common intermediate of dietary Se metabolism produced by reduced selenite, could trigger the apoptosis of cancer cells (HepG2, HeLa, and MCF-7 cells) via accumulation in mitochondria induced by selenite, which would subsequently damage mitochondrial function and structure and lead to cell death [90].

3.2. Organic Se Compounds

Organic Se compounds attract considerable attention in the field of cancer research mainly due to their lower toxicity risk and ability to deliver significant anticancer activity, comparable or sometimes even better than inorganic Se compounds [69]. Organic Se compounds can be classified into different families based on their functional chemical structures: Selenides/diselenides, selenocyanates, selenoaminoacid derivatives (e.g., SeMet and MSC), methylseleninic acid (MSA; CH_3SeO_2H), Se-heterocyclic compounds, and other miscellaneous Se-containing compounds (Figure 1). These organoselenium compounds exhibit anticancer and chemopreventive activity through diverse mechanisms of action, including reduction of oxidative stress [91], induction of apoptotic events [92–94], and enhancement of chemotherapeutic drug activity [95–97]. To date, many scientific studies on organoselenium compounds is available, many of which investigated their roles in cancer prevention and therapy. Several excellent reviews have listed a range of organoselenium compounds based on different classifications and summarized their functions in oncology [31,38,70,98]. Additionally, the potential anticancer and chemopreventive activity of selenides/diselenides [99] and selenocyanates [100] have been extensively reviewed.

Organoselenium compounds have the potential to be used as anti-neoplastic agents against solid tumors. Since necrosis of cancer cells is linked to host inflammatory response and may lead to treatment complications, the anti-necrotic and pro-apoptotic feature of some organic Se compounds is therefore largely preferred in cancer therapy [101]. Selenoaminoacid derivatives, such as SeMet and MSC, at low concentrations (as low as 0.113 µM), were shown to be able to promote apoptosis in solid tumors of various types of human cancer, whereas the control non-tumorigenic mammary epithelial cells (MCF-10A) required substantially high concentrations (up to 87.9 µM for 72 h) of organic Se to display sensitivity to apoptosis [92]. Owing to the ability of organoselenium compounds to induce apoptosis, their synergistic effects on chemotherapeutic drugs against cancer were observed [53,95–97]. MSC may also provide additional protection from the toxicity of anticancer drugs [54,56].

MSA is an oxidized form of methylselenol (CH_3Se^-) converted from selenoaminoacids (e.g., SeMet and MSC) [43]. In vivo and in vitro studies have indicated that MSA is an excellent anticancer agent comparable to SeMet or selenite against a range of cancer models, including lung [102], breast [103,104], melanoma [105], and prostate cancer in particular [106–108]. In a recent study, MSA showed significant cytotoxic effects toward monocytic leukemia cells (THP1) compared with the healthy peripheral blood mononuclear (PBM) cells [109]. MSA also enhanced the anticancer activity of radiation and chemotherapeutic drugs (cytosine arabinoside and doxorubicin) in the malignant THP1 cells in a dose-dependent manner (2.5, 5, and 15 µM for 48 h) [109]. At lower concentrations, MSA protected normal PBM cells from radiation and chemotherapeutic drugs, whereas at higher concentrations, MSA was considered toxic and could increase the cytotoxicity of radiation but not chemotherapy [109]. In another study, MSA was able to inhibit the proliferation, migration, and adhesion of HeLa cells more effectively than SeMet and MSC, and it showed synergistic anticancer activity with S-adenosyl-methionine—a universal methyl group co-substrate involved in multiple intermediary metabolites [110]. MSA was shown to be able to reverse the tamoxifen resistance of breast cancer cells when used in combination with tamoxifen through the activation of caspase-9 and then caspase-8, resulting in the induction of the intrinsic, mitochondrial apoptotic pathway [111]. A novel programmed cell death mechanism (entosis) induced by MSA and methylselenoesters was identified in pancreatic cancer Panc-1 cells [112]. Entosis is characterized by the invasion of a living cell to another cell's cytoplasm resulting in endophagocytosis and the formation of cell-in-cell structures.

Other groups of selenocompounds, such as the selenocynates and Se-containing heterocycles, also contained several Se compounds with promising chemopreventive or anticancer properties, ranging from the well-studied *p*-xylene selenocyanate and benzyl selenocyanate to the recently reported novel active compounds that combine the selenocyanate moiety with different heterocycles, quinones, or steroids [100]. Heterocyclic organoselenium compounds, such as ebselen and ethaselen (also known as BBSKE) are small molecules that have potential in cancer therapy [113,114]. A list of these heterocyclic compounds and their anticancer ability is summarized in Fernandes and Gandin [38] and Sanmartíin et al. [70]. The selenophene-based triheterocyclic derivative 2,5-bis(5-hydroxymethyl-2-selenienyl)-3-hydroxymethyl-*N*-methylpyrrole (D-501036) has received increasing attention due to its broad-spectrum anticancer activity against several human cancer cells in a dose- and time-dependent manner [115–117]. D-501036 selectively induces apoptosis and double-strand DNA breaks in cancer cells, and is especially effective against chemotherapeutic drug-resistant cancer cells with overexpression of P-glycoprotein/multidrug-resistant protein [116]. Further study has suggested that enhanced non-homologous end-joining DNA repair activity was involved in the development of D-501036-resistance in cancer cells [117]. Previously in a cisplatin-resistant prostate cancer model, the degree of drug-resistance was found to be associated with the oxidative system in the cells [118]. Thus, it was likely that cancer cells, and in particularly drug-resistant cancer cells, have higher Se uptake compared with benign cells due to the redox state of Se and the oxidative system of the cancer cells [118,119].

Previous clinical trials (e.g., NPC) have shown that dietary Se supplements (Se-enriched yeast) could reduce the risk of multiple cancers [27–29]. However, subsequent trials (e.g., SELECT and Southwest Oncology Group (SWOG) S9917), tested using SeMet, showed no such beneficial effects [32,33]. In a clinical trial, it was indicated that SeMet did not improve quality of life or survival outcomes of patients with head and neck squamous cell cancer undergoing concurrent chemoradiation [35]. These conflicting results may be partially explained by the selection of the Se compound administered: 200 µg/day of Se-enriched yeast in the NPC trial and 200 µg/day of SeMet in SELECT and SWOG S9917. Although SeMet is the major component of Se-enriched yeast, it is possible that other components in Se-enriched yeast (e.g., MSC) may be contributing to the overall chemopreventive effect of Se observed in the NPC trial. This speculation, however, cannot be verified since there were substantial batch-to-batch variations in specific organoselenium compounds in the samples of NPC yeast [32]. Additionally, no significant benefit in the prevention of second primary tumors was detected in a phase III clinical trial (ECOG 5597) that administrated 200 µg/day of Se-enriched yeast to patients with completely resected stage I non-small-cell lung cancer [34]. Despite these negative findings, most of these trials suggested that Se supplement was safe to consume. Thus, organic Se may be useful in cancer therapy but appears to confer no significant benefit in the prevention of cancer.

3.3. SeNPs

Since the late 1990s, application of nanotechnology in the bio-medical field has received extensive attention. Owing to their biocompatibility, biodegradability, and designability, nanomaterials are increasingly utilized in cancer therapy and diagnosis as pharmaceutical products, drug carriers, imaging agents, and diagnostic reagents to overcome some limitations of the traditional materials [120]. SeNPs, as emerging Se species, are considered to be promising medical materials, according to their reported chemotherapeutical properties [121,122], nutritional effects [123], and relatively low toxicity compared with some other Se compounds [3,124]. The chemotherapeutical potency of SeNPs and their proposed anticancer mechanisms have been reviewed by Menon et al. [125]. Here, we systematically tabulated a list of SeNPs based on data published in 2017 and the first half of 2018 and arranged them according to their reported functions: chemotherapy (Table 1), drug delivery (Table 2), diagnosis (Table 3), and SeNPs with multiple functions (Table 4). The potential of these SeNPs in cancer prevention and treatment, along with their synthesis methods, are discussed below.

Table 1. Summary of recent work on Se-containing nanoparticles (SeNPs) with potential in cancer chemotherapy.

SeNP	Material	Shape and Size (nm)	Effects	Dosage	Pathway	Model	Reference
Acinetobacter sp. SW30 SeNPs	Acinetobacter sp. SW30	Amorphous nanospheres, 78 nm; Polygonal-shaped, 79 nm	Selectively against breast cancer cells and non-toxic to normal cells	-	-	Breast cancer cells (4T1, MCF-7) and noncancer cells (NIH/3T3, HEK293)	[126]
Bacillus licheniformis JS2 derived biogenic SeNPs	Bacillus licheniformis JS2, aerobic condition in 1.8 mM Na$_2$SeO$_3$ stress	Spherical, 110 nm	Stimulated ROS production and caused damage to the mitochondria without affecting the cell membrane integrity. Induced overexpression of necroptotic genes and promoted RIP3-independent necroptosis	Concentration of 2 μg Se/mL	-	Human prostate adenocarcinoma cell line (PC-3)	[127]
Blg stabilized SeNPs	Ascorbic acid, Blg, Na$_2$SeO$_3$	Spherical, mean particle size of 36.8 ± 4.1 nm	Lower cytotoxicity than Na$_2$SeO$_3$. Similar cell growth inhibition on both colon cancer cell and corresponding normal cell	-	-	Human colon adenocarcinoma cells (HCT116) and colon normal cell (CCD112)	[124]
Ferulic acid-modified SeNPs	Na$_2$SeO$_3$, ascorbic acid, ferulic acid solution	Amorphous, average diameter of 109 nm	Induced intracellular ROS generation and MMP disruption	>100 μg/mL	Caspase-3/9, mitochondrial pathway	HepG2	[128]
Folic acid surface-coated Se nanoparticles (FA@SeNPs)	Folic acid, SeO$_2$ solution, ascorbic acid solution	Rod-shaped (400 × 100 nm)	Showed antiproliferative effect against 4T1 cells. Significantly increased the lifespan and reduced the tumor size of cancerous animals. Had better absorption toward cancer cells. Exhibited a better in vivo anticancer effect compared to SeNPs	200 μg/mL in vitro and 300 mg/week in vivo	-	4T1 breast cancer cell line and inbred Balb/c mice	[129]
Folic acid-conjugated SeNPs (FA@SeNPs)	Folic acid, CTS, Na$_2$SeO$_3$, ascorbic acid	Spherical, ~192 nm	Able to synergistically enhance the anticancer efficacy and colony formation inhibition ability of radioactive ^{125}I seeds. Increased ROS overproduction. Induced DNA damage and activated the mitogen-activated protein kinase and TP53 signaling pathways	5 mg/kg of FA@SeNPs with an intratumor injection strategy every other day and/or implanted with radioactive ^{125}I seeds	DNA damage-mediated p53 and MAPK signaling pathways	Michigan Cancer Foundation-7 cell (MCF-7) and female nude mice	[130]

Table 1. *Cont.*

SeNP	Material	Shape and Size (nm)	Effects	Dosage	Pathway	Model	Reference
PEC-decorated Se nanoparticles	Selenite and ascorbic acid, PEC	Spherical, average size of ~41 nm	PEC as a surface decorator could be effectively used to improve the stability and antioxidant capacity of SeNPs	-	-	Cancer cells (SPCA-1 and HeLa) and normal cells (RWPE-1)	[131]
Pleurotus tuber-regium (PTR)-conjugated SeNPs (PTR-SeNPs)	Sclerotia of tiger milk mushrooms, Na_2SeO_3, ascorbic acid	Se concentration: 1.35 ± 0.12 μM; particle size: 80.0 ± 12.3 nm	Triggered intracellular G2/M phase arrest and apoptosis. Activated autophagy to promote the death of cancer cells	-	Beclin 1-related signaling pathways	Human colon cancer cells (HCT 116)	[132]
SeNPs	Se powder, Na_2SO_4, acetic acid	Spherical, 12–30 nm	Induced TAMs isolated from DL-bearing mice. Induced ROS generation, macrophage polykaryon formation, and adhesion molecules (CD54 or ICAM-1), and fusion receptors (CD47 and CD172α) expression on TAMs. Decreased tumor cell proliferation	20–50 ng for 10^6 cells	-	Daltons lymphoma cells and DL-bearing BALB/c (H2d) strain of mice	[133]
SeNPs	SeO_2	-	Combination of AET and SeNP supplementation effects anti-tumor immune responses in splenocytes	6 weeks of AET and SeNP administration (100 mg three times/week). Oral administration in doses of 100 and 200 mL per mouse	-	Mice bearing the 4T1 mammary carcinoma	[122]
SeNPs	Na_2SeO_3, GSH, BSA	20–70 nm, average size of 40 nm	Able to rapidly, massively, and selectively accumulate in cancer cells. Showed stronger pro-oxidant property than selenite	-	-	Male Kunming mice and Murine H22 hepatocarcinoma cells	[134]

Abbreviations, AET: aerobic exercise training; Blg: beta-lactoglobulin; BSA: bovine serum albumin; CTS: chitosan; DL: dalton's lymphoma; DL: dalton's lymphoma; MMP: mitochondrial membrane potential; PEC: pectin; ROS: reactive oxygen species; SeNPs: Se-containing nanoparticles; SeNPs: Se-containing nanoparticles; TAMs: tumor-associated macrophages; TrxR: thioredoxin reductase.

Table 2. Summary of recent work on SeNPs with potential for anti-cancer drug delivery.

SeNP	Material	Shape and Size (nm)	Effects	Dosage	Pathway	Model	Reference
CPT and DOX-loaded PEG-b-PBSe core crosslinked micelles (CPT/DOX-CCM)	Diselenide diols precursors, PEG45-based RAFT agent, CPT, DOX	Spherical, ~129 nm	Features include high drug loading, visible light-induced in situ crosslinking, improved physiological stability, optimized pharmacokinetics, and tumor-specific combined drug release	1 mg CPT or DOX equivalent per kg every four days in 24 days	–	Human breast cancer cell line (MCF-7) and mouse mammary tumor cell line (EMT-6)	[135]
CTS-modified Se nanoparticle	Na$_2$SeO$_3$, ascorbic acid solution (vitamin C), CTS	400–4000 cm^{-1}	Slow-release carrier conjugated to the TNF-α-derived peptide P16, G0/G1 cell-cycle arrest, and apoptosis	–	p38MAPK/JNK pathway	Prostate cancer cells (DU145) and normal human prostate epithelial cells (RWPE-1)	[136]
DOX-SeNPs@TMC-FA (pH-sensitive)	Selenite, ascorbic acid, folic acid-N trimethyl CTS (TMC-FA)	An average diameter of 50 nm	Enhanced the activity of DOX by approximately 10-fold for a reduced IC$_{50}$ value compared to free DOX	–	Apoptosis pathway involved caspase-3 and PARP proteins	Ovarian cancer DOX sensitive (OVCAR8) and resistant (NCI/ADR-RES) cells	[137]
EPI-loaded-NAS-24-functionalized PEIPEG-5TR1 aptamer coated SeNPs (ENPPASe complex)	Na$_2$SeO$_3$, EPI-loaded-NAS-24-functionalized PPA complex	68.2 ± 6 nm	Able to provide high loading of EPI and NAS-24. Reduced the toxicity in non-target cells. Reduced the cell viability in the target cancer cells. Reduced the tumor growth in cancer-bearing mice compared to EPI treatment alone	–	–	Human breast carcinoma cell (MCF7), murine colon carcinoma cell (C26) and human hepatocellular carcinoma cell (HepG2)	[138]
FA-CP/SeNPs	Na$_2$SeO$_3$, folic acid decorated cationic pullulan (FA-CP)	Flower-like structure, approximately 50 nm	Higher loading capacity of DOX. Less toxicity against normal cells	–	–	KB cancer cells line and normal cell line (L292)	[139]
Hyaluronic acid Se-PEI nanoparticle	Na$_2$SeO$_3$, ascorbic acid, hyaluronic acid, PEI	70–180 nm	Showed higher transfection efficiency, greater gene silencing ability, and stronger cytotoxicity	–	–	HepG2 cell, Lo2 cell and xenograft mouse model	[140]
Nano-Se + Nano-fluorouracil	Na$_2$SeO$_3$, GSH, BSA	Spherical, ranged from 66.43 nm to 98.9 nm	Induced chemo-sensitivity of 5-fluorouracil-encapsulated poly (D, L-lactide-co-glycolide) nanoparticles (nano-FU) in cancer cells	0, 2, 4, 6, 8, 10, 30, and 50 μM	Glucose uptake slight blockage, interaction with Zn	Human breast cancer (MCF7) and human colorectal cancer Cell (Caco-2)	[141]

Table 2. *Cont.*

SeNP	Material	Shape and Size (nm)	Effects	Dosage	Pathway	Model	Reference
Oil-soluble CdSe QD	CdO, mineral oil, oleic acid, Se	DG-PEG-OC-9R, near spherical, 112.0 ± 1.63 nm; FA-PEG-OC-9R, near spherical, 115.2 ± 1.94 nm	Could be used to evaluate the hypoxic tumor cell-targeting properties of the wrapped CTS-based micelles	-	-	Normoxic/hypoxic HepG2 and HeLa cells	[142]
Oridonin-loaded and GE11 peptide conjugated SeNPs (GE11-Ori-SeNPs)	Na$_2$SeO$_3$, oridonin, ascorbic acid, GE11 polypeptide	Near-spherical, average diameter of 70 nm	The GE11 surface modification provides targeting towards cancer cells: oridonin releasing induced cancer cell apoptosis. Inhibited tumor growth via inhibition of tumor angiogenesis by reducing the angiogenesis-marker CD31 and activation of the immune system by enhancing IL-2 and TNF-a production	2.5, 5, and 7.5 mg/kg/day for 15 days through tail intravenous injection	EGFR-mediated PI3K/AKT and Ras/Raf/MEK/ERK pathway, mitochondria-dependent pathway	Human esophageal cancer cell lines (KYSE-150 and EC9706) and KYSE-150 xenograft mice model	[143]
Se@MIL-101-(P + V) siRNA	MIL-101(Fe), cysteine, Na$_2$SeO$_3$, siRNA	Spherical, particle size 160 nm, pore diameter 2.19 nm	Enhanced protection of siRNAs against nuclease degradation. Increased siRNA cellular uptake and promoted siRNA escape from endosomes/lysosome to silence MDR genes in MCF-7/T (Taxol-resistance) cells. Enhanced cancer therapeutic efficacy and decreased systemic toxicity in vivo	10 mg/kg by intravenous injection for 15 d (12 μg of siRNA per mouse)	p53, MAPK, and PI3K/Akt	MCF-7/T cells, paclitaxel resistance MCF-7/T cells and nude mice	[144]

BSA: bovine serum albumin; CPT: camptothecin; CTS: chitosan; DOX: doxorubicin; EPI: epirubicin; GSH: glutathione; MDR: multidrug resistance; PPa: pyropheophorbide a; QD: quantum dots; RIS: risedronate sodium; SeNPs: Se-containing nanoparticles.

Table 3. Summary of recent work on SeNPs with potential in cancer diagnosis.

SeNP	Material	Shape and Size (nm)	Effects	Dosage	Model	Reference
Anti-HE4 IgG-HE4-anti-HE4 CdSe/ZnS immunocomplex	Anti-HE4 IgG antibodies, CdSe/ZnS QD	-	Electrochemical immunosensor for HE4 protein detection using QD as electrochemically active labels of specific antibodies. Contributed significantly to the analytical performance of tumor marker detection and met the exacting requirements for HE4 protein clinical monitoring	-	HE4 in human serum	[145]
Aptamer-modified SeNPs (Apt-SeNPs)	-	Spherical structures, 88 ± 30 nm	Good chemical stability, water solubility, and biocompatibility. Strong green scattering light with a characterized scattering peak at 570 nm. Precisely and specifically target and image nucleolin overexpressed cancer cells after being modified with aptamers	-	Human epidermoid cancer (Hep-2) cells	[146]
CdTe/ZnSe core/shell QDs	CdTe, Na₂TeO₃, Zn(CH₃CO₂)₂, Na₂SeO₃	QD (10 ± 2 nm), QD+T (13 ± 2 nm), QD+C (18 ± 2 nm), QD+A (71 ± 2 nm), QD+G (95 ± 2 nm)	Able to detect DNAs (directly from cell extracts), damages to the DNA, and mutations	-	Prostate cancer (PC3) and normal human cells (PNT1A)	[147]
IL13 conjugated QD (IL13QD)	CdSe-based QD, 1-ethyl-3-(3-dimethylaminopropyl) carbodiimide (EDC), interleukin-13 (IL13)	Core-shell structure, a size range of 15–20 nm	IL13Rα2 can be detected in cerebrospinal fluid by IL13QD. A higher force of binding interaction between the IL13QD and IL13Rα2 expressing glioma cells and exosomes secreted by glioma stem cells was observed	-	U251 human glioma cells and CD133 positive glioma initiating cells (T3691)	[148]
SeNP loaded imprinted core-shell microcomposites (SIMs)	CTS, zeolite, TiO₂, Na₂SeO₄, Na₂SeO₃	Spherical, average size of 80 nm	Could be used for dot-blot immunoassays for rapid serodiagnosis of human lung cancer. The detection time of the colloidal Se dot test for the progastrin-releasing peptide (as a tumor marker for small cell lung cancers) was only 5 min	Linear with the concentration of antigen within the concentration range of 0–105 pg/mL. The lowest concentration to distinguish significant positive results was observed to be 75 pg/mL	Human progastrin releasing-peptide	[149]

CTS: chitosan; HE4: human epididymis protein 4; QD: quantum dots; SeNPs: Se-containing nanoparticles.

Table 4. Summary of recent work on multi-functional SeNPs with potential in cancer-related research.

SeNPs	Material	Shape and Size (nm)	Function [1]	Effects	Dosage	Pathway	Model	Reference
Ag$_2$Se-cetuximab nanoprobes	Bis(trimethylsilyl) selenide, silver acetate, cetuximab	Spherical, diameter of 2.8 ± 0.5 nm	C and E	Displayed faster and more enrichment at the site of cancer. Inhibited the tumor growth and improved the survival rate of the cancer-bearing nude mice model. Combined targeted imaging and therapy	-	-	Human tongue squamous cell carcinoma cells (CAL-27) and human immortalized noncancerous keratinocytes cells (HaCaT) and Balb/c mice	[150]
DOX-loaded selenopolymeric nanocarriers (Se@CMHA-DOX NPs)	Na$_2$SeO$_3$, ascorbic acid, poly (ethylene glycol) (PEG), cetyl-modified hyaluronic acid, DOX	Spherical, 244 ± 6.8 nm	A and C	Inhibited TrxR activity and augmented the anticancer efficacy of DOX. Induced G2/M cell cycle arrest and TP53-mediated caspase-independent apoptosis. Reduced tumor activity in a three-dimensional tumor sphere model	5 µg/mL for 48 h	Apoptotic pathway	MCF7 breast adenocarcinoma cells and MCF7 tumor sphere model	[151]
HSAMSe@DOX	Na$_2$SeO$_3$, L-ascorbic acid, Human serum albumin, DOX	Homogeneous spherical, ~80 nm	A and C	Synergistically enhanced the antitumor activity of DOX and decreased the side effects associated with DOX. Increased tumor-targeting effects and enhanced cellular uptake through nanoparticle interact with SPARC protein	10 mg/mL, 100 µL into the veins of the tails	-	MCF-7, MCF-10A, MDA-MB-231, SKBR3 and female BALB/C nude mice	[152]
MoSe$_2$(Gd^{3+})-PEG nanosheets	NaMoO$_4$·2H$_2$O, Se, NaBH$_4$, gadolinium(III) chloride hexahydrate	Lamellar, 100–150 nm	B, F and G	Able to provide a strong contrast for T1 weighted magnetic resonance imaging. Could be used as contrast agent for photoacoustic imaging (PAI). Increased the temperature to help kill cancer cells under laser irradiation. Enhanced permeation and retention effect in the tumor using magnetic resonance photoacoustic bimodal imaging in vivo. Suppressed tumors in mice by injection with laser irradiation	-	-	Hep G2 human hepatoma carcinoma cells, BALB/c nude mice	[153]
Paclitaxel-loaded SeNPs	SeO$_2$, paclitaxel, ascorbic acid, pluronic F-127	Hydrodynamic diameter, 87 nm, spherical	A and B	Significant antiproliferative activity against cancer cells. G2/M phase arrest in a dose-dependent manner leading to apoptosis. Disruption of mitochondrial membrane potential orchestrated with the induction of reactive oxygen species leading to the activation of caspases	-	MMP, caspases	Lung cells (L-132), cervical cancer cells (HeLa), breast cancer cells, (MCF7), non-small lung carcinoma cells (A549) and colorectal adenocarcinoma cells (HT29)	[154]
PPa@CTX-Se-OA/ DSPE-PEG2k	Cabazitaxel, PPa, oleic acid, Se powder, DSPE-PEG2k	Spherical, average diameter of 104.1 ± 3.1 nm	C and D	Light irradiation disassembles the structure of ROS-responsive prodrug nanosystems by cleaving the ROS-responsive linkers to accelerate the release of the parent drug	200 ng/mL for 4 h or 24 h	-	4T1 murine breast cancer cells	[19]

Table 4. *Cont.*

SeNPs	Material	Shape and Size (nm)	Function [1]	Effects	Dosage	Pathway	Model	Reference
Se/iron oxide nanoparticles (Se:IONP)	Na$_2$SeO$_3$, acetic acid, CTS, hydrophobic IONP	Spherical in a transmission electron microscope, irregular in a scanning electron microscope, 5–9 nm	A and H	An iron oxide core produced by thermal decomposition, followed by a silane ligand exchange, a CTS coating, and Se decoration. Reduced cancer cell viability	-	-	MB-231 breast cancer cells	[155]
Se-containing hydroxyapatite/alginate (SeHA/ALG) composite granules	(NH$_4$)$_2$HPO$_4$, Na$_2$SeO$_3$·5H$_2$O, Ca(NO$_3$)$_2$·4H$_2$O, hydroxyapatite, alginate sodium, RIS	Spherical, 1.1–1.5 mm	A and C	Biphasic process of releasing sufficient Se and RIS against osteosarcoma cells	-	-	Human osteoblast-like cell line (Saos-2) and normal human fetal osteoblasts (hFOB 1.19)	[156]
SeNPs-DOX-ICG-RP	L-ascorbic acid, Na$_2$SeO$_3$, dual-target (RC-12 and PG-6 peptides), loaded with both DOX and ICG	Sphere-like morphology with an average size of 110 nm	B and C	NIR-laser irradiation that raised the temperature of the nanosystem and allowed nanoparticles to decompose and release drugs accurately in the tumor site. Reduced the damage of chemotherapy drugs to normal tissue	-	-	HepG2 and normal L02 cells	[157]
Ultra-small Nano-Se	Na$_2$SeO$_3$, GSH, BSA	Monodisperse spherical shape with a diameter about 27.5 ± 4.3 nm	A and F	Reinforced the toxic effects of irradiation, leading to a higher mortality rate than either treatment used alone. Induced cell cycle arrest at the G2/M phase and the activation of autophagy. Increased both endogenous and irradiation-induced ROS formation. Improved cancer cell sensitivity to the toxic effects of irradiation	0.15 and 0.3 µg/mL, X-rays (6-MeV, 200 cGy/min)	-	MCF-7 breast carcinoma cells	[158]

[1] Function, A: chemotherapy; B: photothermal therapy; C: drug delivery; D: photodynamic therapy; E: diagnosis; F: radiosensitizer; G: imaging; H: magnetically-targeted. Abbreviations, BSA: bovine serum albumin; CTS: chitosan; DOX: doxorubicin; GSH: glutathione; ICG: indocyanine green; MMP: mitochondrial membrane potential; PPa: pyropheophorbide a; RIS: risedronate sodium; ROS: reactive oxygen species; SeNPs: Se-containing nanoparticles; TrxR: thioredoxin reductase.

Many methods have been reported for SeNPs preparation, which are generally classified into three broad categories based on different producing principles: chemosynthesis, biosynthesis, and physical synthesis [159]. Among them, chemosynthesis is considered the most common method to prepare SeNPs. In chemosynthesis, Se in the +4-valence state, such as selenite, selenious acid, or SeO_2, is frequently employed as precursors, whereas reducing agents (e.g., ascorbic acid and glutathione [GSH]) and stabilizing agents (e.g., chitosan and pectin) are used for SeNPs formation and maintenance [131,141,159,160]. SeNPs synthesized in the Na_2SeO_3-GSH redox system tended to gather in cancer cells and presented stronger pro-oxidant activity comparing to selenite [134]. However, to optimize their function in cancer therapy and prevention, chemosynthetic SeNPs are usually decorated with specific molecules, which endow them with ideal features to meet the demands of practical applications. For example, decorating SeNPs with other bioactive molecules can enhance the therapeutic effects upon certain types of cancers when compared with non-decorated SeNPs [128,129,161]. Aside from direct therapeutic effects, chemically modified SeNPs can function as vehicles that endow the carried objects with favorable properties like tumor targeting [150,157], high efficacy [137,144], and low toxicity [139]. Notably, chemosynthetic SeNPs were also examined as diagnostic agents [145,147,149], imaging agents [146,153], and radiosensitizers [130,158]. Overall, chemosynthesis is the most common method used to obtain and modify SeNPs, because the process is easy to implement and control. Nevertheless, environmental pollution and accumulation of chemosynthetic materials in the body should also be considered.

In contrast to chemosynthetic SeNPs, biosynthetic SeNPs appear to be environmentally friendlier and biologically safer. Thus, there has been increased focus upon biosynthetic SeNPs, partly also due to their extraordinary biocompatibility, sustainability, and economy [162]. These organismal materials mediating SeNPs are extracellularly or intracellularly manufactured with selective plants [64], bacteria [4], fungi [163], and other organisms [163,164]. Taking biological extracts as ingredients, researchers have successfully synthesized some chemical pollution-free SeNPs that display diverse biological effects, such as UVB-induced DNA damage prevention [165] and cancer cells proliferating inhibition [126]. So far, bacteria are the most important source of biosynthesized SeNPs. Bacteria such as *Bacillus licheniformis* JS2 [127], *Ochrobactrum* sp. MPV1 [166], *Streptomyces minutiscleroticus* M10A62 [167], and *Acinetobacter* sp. SW30 [126] have been employed to fabricating SeNPs. These bacteria-based SeNPs are synthesized by culturing bacterial strain with sodium selenite (0.5–2 mM). Zonaro et al. [166] demonstrated that, under the stress of Na_2SeO_3 (0.5 or 2 mM), *Ochrobactrum* sp. MPV1 was capable of converting selenite to elemental Se and synthesizing SeNPs intracellularly; however, medical applications of these SeNPs remain underexploited. The synthesizing protocol of bacteria-based SeNPs is now gradually developing; however, SeNPs should always be purified to avoid the toxicity caused by bacteria. Similar to bacteria, fungi with properties such as large output, easy accessibility, and economic feasibility could be used as candidates for biosynthesizing SeNPs [168]. However, the investigation into the practicability of fungi synthesized SeNPs in medical fields is relatively deficient. In Vetchinkina et al. [163], fungus-based spherical SeNPs were obtained from the medicinal basidiomycete *Lentinula edodes* cultured with inorganic selenium (10^{-2} mol) or organoselenium (10^{-7} to 10^{-3} mol). The reduction in selenium (Se^{IV} to Se^0) was observed, and the synthesis mechanism of *L. edodes* based SeNPs was revealed by transmission electron microscopy, electron energy loss spectroscopy, and X-ray fluorescence [163]. Overall, bacteria and fungi are expected to be competent media for SeNPs assembling, given their capability of producing a hypotoxic form of Se by stepwise transforming Se oxyanions to elemental Se [169,170]. Intracellular synthesis of SeNPs could be achieved under ideal experimental conditions that usually require Se-involved culture [171], so that Se oxyanions would be transported into cells for downstream reduction and reassembly. In a study carried out by Sonkusre et al. [127], SeNPs synthesized in *Bacillus licheniformis* strain JS2 were able to initiate necroptosis in PC-3 cell by the ROS-mediated activation regulated by a RIP1 kinase, thus taken as undeveloped anti-cancer substances with high producing efficiency. Plant-based SeNPs are another interesting aspect of biosynthetic SeNPs [168]. For instance, Sharma et al. [64]

synthesized selenium nanoballs with uniform size (3–18 nm) and shape using dried *Vitis vinifera* (raisin) extracts and selenous acid, with a simple refluxing method. Although the extract mechanism of plant-based SeNPs and the pharmaceutical applications remain largely underexplored, SeNPs synthesized with/in plants are likely to have vast potential for future development due to the variety of plants and their cleanliness.

As for the physically synthesized functional SeNPs, approaches such as pulsed laser ablation [172] and γ-radiation [173] have been used for their generation. For instance, Guisbiers et al. [174] obtained pure SeNPs by pulsed laser ablation in liquids and showed that these SeNPs were able to disturb the biofilm formation of a human pathogen *Candida albicans*, highlighting the potential for medical application of physical synthesis SeNPs. However, the application of physical-synthetic SeNPs in cancer treatment and prevention is still immature, which may be due to the equipment requirements.

From the above and Tables 1–4, SeNPs have great potential not only in cancer treatment but also as diagnostic/imaging agents and more. Among all, the chemotherapeutic effects of SeNPs could be easily considered the most promising application of SeNPs. So far, results from laboratories regarding the anticancer property of SeNPs have mainly been positive. SeNPs showed anticancer effects in a range of cancers, including hepatocarcinoma [128,140], breast cancer [122,129,141], colon adenocarcinoma [124,132], lymphoma [133], esophageal cancer [143], prostate cancer [136,147], ovarian cancer [137], and glioma [148]. Further studies and clinical trials are needed to elucidate the possible applications of SeNPs in oncology. Also, concerns such as toxicity of nanoparticles accumulation in the human body and the environment should be cautiously addressed.

4. Se and Epigenetics: Possible Roles in Cancer Prevention and Therapy

Epigenetic refers to the study of heritable changes in gene expression that do not involve changes to the underlying DNA sequence [1,75]. These changes are controlled by epigenetic factors such as epimodifications of DNA, post-translational modification (PTM) of histone, and expression of non-coding RNA (ncRNA) [1]. Epigenetics plays a vital role in cancer development, and currently, therapy targeting epigenetic changes is considered one of the most promising approaches in cancer treatment [175,176]. The epigenetic effects of Se and their implications for human health, including cancer, have been reviewed [1,177,178]. Studies have revealed that Se and Se compounds could affect the epigenome of a cell through all three major epigenetic controls: DNA methylation, histone modifications, and ncRNA expression [1]. In regards to cancer therapy, inorganic and organic Se compounds, such as MSA, SeMet, MSC, Se-allylselenocysteine, and selenite, could effectively inhibit the activities of histone deacetylases and DNA methyltransferases, which expressions are usually up-regulated in various types of cancer cells [179–182]. Notably, the epigenetic inhibition mechanisms of Se compounds appear to be quite distinctive depending on their chemical forms [180,183]. A genome-wide epigenetic analysis indicated that both inorganic (selenite) and organic (MSA) Se could epigenetically affect distinct gene sets in human chronic myeloid leukemia K562 cells: selenite affected genes involved in response to oxygen and hypoxia, whereas MSA affected genes associated with cell adhesion and glucocorticoid receptors [184]. So far, the epigenetic effects of SeNPs and mechanism of their action on gene expression remain largely unknown.

5. Other Potential Applications of Se Compounds

Se is an essential trace element required for the maintenance of human health. In addition to its potential roles and use in cancer prevention and therapy, Se compounds have many other useful non-cancer-related features. Among them, one of the most interesting features of Se is the antimicrobial and antiviral activity observed in specific Se compounds. Previous studies demonstrated the relationship between selenocompounds and the immune system and showed that Se supplementation could enhance the activity and cytotoxic response of NK cells [6–8,38,81]. Specifically, sufficient intake of dietary Se is crucial for handling viral infections and for preventing Se deficiency-related diseases caused by bacteria and viruses (e.g., Keshan disease) [6,18]. Researchers have also discovered the antibacterial, antifungal, and antiviral properties of SeNPs, and have demonstrated how SeNPs can be potentially used in various settings [168,185,186]. For example, Nguyen et al. [63] examined the antimicrobial activity of SeNPs against foodborne pathogens and indicated the possible use of SeNPs for food safety applications. Overall, it is anticipated that new and novel applications of Se compounds in various fields of life sciences will be extensively investigated and explored due to rapid advancement in nanotechnology and further understanding of the mechanisms underlying Se compounds at the nano-scale level.

6. Conclusions

Although several clinical trials indicated no significant benefit of Se in preventing cancer, overwhelming evidence has demonstrated that Se and many Se compounds, under certain circumstances, are potent anticancer agents. In vivo and in vitro studies have hinted that the Se compounds exert their anticancer ability through multiple mechanisms. However, further research and clinical trials are still required before these Se compounds can be clinically recognized as anticancer drugs. In addition to cancer therapy, Se compounds have been proven to be quite useful in other cancer-related fields such as chemoprevention, diagnosis, and imaging, as well as in non-cancer-related fields as described in this review and summarized in Tables 1–4. Among the Se compounds, SeNPs, as the emerging form of Se species, have attracted considerable attention. Judging from the current positive results of SeNPs against a range of cancers, SeNPs will play more critical roles in cancer prevention and therapy in the near future, especially in the era of precision medicine, where patients are provided with personalized and tailored treatment based on individual conditions and needs [75]. To conclude, in order to fully determine the potential of Se compounds, in particular the SeNPs, in cancer prevention and therapy, extensive further studies are required to better understand the underlying mechanisms behind the biophysiological effects of Se.

Author Contributions: Writing—Original Draft Preparation, H.W.T., H.-Y.M., A.T.Y.L., Y.-M.X.; Writing—Review and Editing, H.W.T., H.-Y.M., A.T.Y.L., Y.-M.X.; Supervision, A.T.Y.L., Y.-M.X.; Funding Acquisition, A.T.Y.L., Y.-M.X.

Funding: This work was supported by the grants from the National Natural Science Foundation of China (No. 31271445 and 31771582), the Science and Technology Planning Project of Guangdong Province of China (No. 2016A020215144), the Guangdong Natural Science Foundation of China (No. 2017A030313131), the "Thousand, Hundred, and Ten" project of the Department of Education of Guangdong Province of China (No. 124), the Basic and Applied Research Major Projects of Guangdong Province of China (2017KZDXM035), the Colleges and Universities Innovation Project of Guangdong Province of China (No. 2016KTSCX041 and 2016KTSCX042), and the "Yang Fan" Project of Guangdong Province of China (A.T.Y.L.-2016; Y.-M.X.-2015).

Acknowledgments: We would like to thank members of the Lau and Xu laboratory for critical review of this manuscript.

Conflicts of Interest: The authors declare no conflict of interest.

Abbreviations

D-501036	2,5-bis(5-hydroxymethyl-2-selenienyl)-3-hydroxymethyl-*N*-methylpyrrole
MSA	Methylseleninic acid
MSC	Methylselenocysteine
ncRNA	non-coding RNA
NK	Natural killer
NPC	Nutritional prevention of cancer
PBM	Peripheral blood mononuclear
PTM	Post-translational modification
ROS	Reactive oxygen species
Se	Selenium
SELECT	Selenium and vitamin E cancer prevention trial
SeMet	Selenomethionine
SeNPs	Se-containing nanoparticles
SWOG	Southwest Oncology Group

References

1. Lau, A.T.Y.; Tan, H.W.; Xu, Y.M. Epigenetic effects of dietary trace elements. *Curr. Pharmacol. Rep.* **2017**, *3*, 232–241. [CrossRef]

2. Wu, Q.; Rayman, M.P.; Lv, H.; Schomburg, L.; Cui, B.; Gao, C.; Chen, P.; Zhuang, G.; Zhang, Z.; Peng, X.; et al. Low population selenium status is associated with increased prevalence of thyroid disease. *J. Clin. Endocrinol. Metab.* **2015**, *100*, 4037–4047. [CrossRef] [PubMed]

3. Li, Y.; Li, X.; Wong, Y.S.; Chen, T.; Zhang, H.; Liu, C.; Zheng, W. The reversal of cisplatin-induced nephrotoxicity by selenium nanoparticles functionalized with 11-mercapto-1-undecanol by inhibition of ROS-mediated apoptosis. *Biomaterials* **2011**, *32*, 9068–9076. [CrossRef] [PubMed]

4. Forootanfar, H.; Adeli-Sardou, M.; Nikkhoo, M.; Mehrabani, M.; Amir-Heidari, B.; Shahverdi, A.R.; Shakibaie, M. Antioxidant and cytotoxic effect of biologically synthesized selenium nanoparticles in comparison to selenium dioxide. *J. Trace Elem. Med. Biol.* **2014**, *28*, 75–79. [CrossRef] [PubMed]

5. Xiao, Y.; Huang, Q.; Zheng, Z.; Guan, H.; Liu, S. Construction of a *Cordyceps sinensis* exopolysaccharide-conjugated selenium nanoparticles and enhancement of their antioxidant activities. *Int. J. Biol. Macromol.* **2017**, *99*, 483–491. [CrossRef]

6. Broome, C.S.; McArdle, F.; Kyle, J.A.; Andrews, F.; Lowe, N.M.; Hart, C.A.; Arthur, J.R.; Jackson, M.J. An increase in selenium intake improves immune function and poliovirus handling in adults with marginal selenium status. *Am. J. Clin. Nutr.* **2004**, *80*, 154–162. [CrossRef] [PubMed]

7. Narayan, V.; Ravindra, K.C.; Liao, C.; Kaushal, N.; Carlson, B.A.; Prabhu, K.S. Epigenetic regulation of inflammatory gene expression in macrophages by selenium. *J. Nutr. Biochem.* **2015**, *26*, 138–145. [CrossRef]

8. Wu, F.; Cao, W.; Xu, H.; Zhu, M.; Wang, J.; Ke, X. Treatment with a selenium-platinum compound induced T-cell acute lymphoblastic leukemia/lymphoma cells apoptosis through the mitochondrial signaling pathway. *Oncol. Lett.* **2017**, *13*, 1702–1710. [CrossRef]

9. Kohler, L.N.; Florea, A.; Kelley, C.P.; Chow, S.; Hsu, P.; Batai, K.; Saboda, K.; Lance, P.; Jacobs, E.T. Higher plasma selenium concentrations are associated with increased odds of prevalent type 2 diabetes. *J. Nutr.* **2018**, *148*, 1333–1340. [CrossRef]

10. Casaril, A.M.; Ignasiak, M.T.; Chuang, C.Y.; Vieira, B.; Padilha, N.B.; Carroll, L.; Lenardao, E.J.; Savegnago, L.; Davies, M.J. Selenium-containing indolyl compounds: Kinetics of reaction with inflammation-associated oxidants and protective effect against oxidation of extracellular matrix proteins. *Free Radic. Biol. Med.* **2017**, *113*, 395–405. [CrossRef]

11. Mistry, H.D.; Broughton Pipkin, F.; Redman, C.W.; Poston, L. Selenium in reproductive health. *Am. J. Obstet. Gynecol.* **2012**, *206*, 21–30. [CrossRef] [PubMed]

12. Riaz, M.; Mahmood, Z.; Shahid, M.; Saeed, M.U.; Tahir, I.M.; Shah, S.A.; Munir, N.; El-Ghorab, A. Impact of reactive oxygen species on antioxidant capacity of male reproductive system. *Int. J. Immunopathol. Pharmacol.* **2016**, *29*, 421–425. [CrossRef] [PubMed]

13. Gao, S.; Jin, Y.; Hall, K.S.; Liang, C.; Unverzagt, F.W.; Ji, R.; Murrell, J.R.; Cao, J.; Shen, J.; Ma, F.; et al. Selenium level and cognitive function in rural elderly Chinese. *Am. J. Epidemiol.* **2007**, *165*, 955–965. [CrossRef] [PubMed]

14. Jin, N.; Zhu, H.; Liang, X.; Huang, W.; Xie, Q.; Xiao, P.; Ni, J.; Liu, Q. Sodium selenate activated Wnt/β-catenin signaling and repressed amyloid-β formation in a triple transgenic mouse model of Alzheimer's disease. *Exp. Neurol.* **2017**, *297*, 36–49. [CrossRef] [PubMed]

15. Skröder, H.; Kippler, M.; Tofail, F.; Vahter, M. Early-life selenium status and cognitive function at 5 and 10 years of age in Bangladeshi children. *Environ Health Perspect* **2017**, *125*, 117003. [CrossRef] [PubMed]

16. Vinceti, M.; Filippini, T.; Cilloni, S.; Bargellini, A.; Vergoni, A.V.; Tsatsakis, A.; Ferrante, M. Health risk assessment of environmental selenium: Emerging evidence and challenges. *Mol. Med. Rep.* **2017**, *15*, 3323–3335. [CrossRef] [PubMed]

17. Vinceti, M.; Wei, E.T.; Malagoli, C.; Bergomi, M.; Vivoli, G. Adverse health effects of selenium in humans. *Rev. Environ. Health* **2001**, *16*, 233–251. [CrossRef] [PubMed]

18. Loscalzo, J. Keshan disease, selenium deficiency, and the selenoproteome. *N. Engl. J. Med.* **2014**, *370*, 1756–1760. [CrossRef]

19. Yang, B.; Wang, K.; Zhang, D.; Sun, B.; Ji, B.; Wei, L.; Li, Z.; Wang, M.; Zhang, X.; Zhang, H.; et al. Light-activatable dual-source ROS-responsive prodrug nanoplatform for synergistic chemo-photodynamic therapy. *Biomater. Sci.* **2018**, *6*, 2965–2975. [CrossRef]

20. Sutter, M.E.; Thomas, J.D.; Brown, J.; Morgan, B. Selenium toxicity: A case of selenosis caused by a nutritional supplement. *Ann. Intern. Med.* **2008**, *148*, 970–971. [CrossRef]

21. Institute of Medicine (US) Panel on Dietary Antioxidants and Related Compounds. *Dietary Reference Intakes for Vitamin C, Vitamin E, Selenium, and Carotenoids*; National Academy Press: Washington, DC, USA, 2000; pp. 284–324.

22. Weisberger, A.S.; Suhrland, L.G. Studies on analogues of L-cysteine and L-cystine. II. The effect of selenium cystine on Murphy lymphosarcoma tumor cells in the rat. *Blood* **1956**, *11*, 11–18. [PubMed]

23. Weisberger, A.S.; Suhrland, L.G. Studies on analogues of L-cysteine and L-cystine. III. The effect of selenium cystine on leukemia. *Blood* **1956**, *11*, 19–30. [PubMed]

24. Mautner, H.G.; Jaffe, J.J. The activity of 6-selenopurine and related compounds against some experimental mouse tumors. *Cancer Res.* **1958**, *18*, 294–298. [PubMed]

25. Weekley, C.M.; Harris, H.H. Which form is that? The importance of selenium speciation and metabolism in the prevention and treatment of disease. *Chem. Soc. Rev.* **2013**, *42*, 8870–8894. [CrossRef] [PubMed]

26. Misra, S.; Boylan, M.; Selvam, A.; Spallholz, J.E.; Bjornstedt, M. Redox-active selenium compounds–from toxicity and cell death to cancer treatment. *Nutrients* **2015**, *7*, 3536–3556. [CrossRef] [PubMed]

27. Clark, L.C.; Combs, G.F., Jr.; Turnbull, B.W.; Slate, E.H.; Chalker, D.K.; Chow, J.; Davis, L.S.; Glover, R.A.; Graham, G.F.; Gross, E.G.; et al. Effects of selenium supplementation for cancer prevention in patients with carcinoma of the skin. A randomized controlled trial. Nutritional Prevention of Cancer Study Group. *JAMA* **1996**, *276*, 1957–1963. [CrossRef] [PubMed]

28. Clark, L.C.; Dalkin, B.; Krongrad, A.; Combs, G.F., Jr.; Turnbull, B.W.; Slate, E.H.; Witherington, R.; Herlong, J.H.; Janosko, E.; Carpenter, D.; et al. Decreased incidence of prostate cancer with selenium supplementation: Results of a double-blind cancer prevention trial. *Br. J. Urol.* **1998**, *81*, 730–734. [CrossRef] [PubMed]

29. Duffield-Lillico, A.J.; Reid, M.E.; Turnbull, B.W.; Combs, G.F., Jr.; Slate, E.H.; Fischbach, L.A.; Marshall, J.R.; Clark, L.C. Baseline characteristics and the effect of selenium supplementation on cancer incidence in a randomized clinical trial: A summary report of the Nutritional Prevention of Cancer Trial. *Cancer Epidemiol. Biomarkers Prev.* **2002**, *11*, 630–639. [PubMed]

30. Vinceti, M.; Filippini, T.; Cilloni, S.; Crespi, C.M. The epidemiology of selenium and human cancer. *Adv. Cancer Res.* **2017**, *136*, 1–48.

31. Collery, P. Strategies for the development of selenium-based anticancer drugs. *J. Trace Elem. Med. Biol.* **2018**, *50*, 498–507. [CrossRef]

32. Lippman, S.M.; Klein, E.A.; Goodman, P.J.; Lucia, M.S.; Thompson, I.M.; Ford, L.G.; Parnes, H.L.; Minasian, L.M.; Gaziano, J.M.; Hartline, J.A.; et al. Effect of selenium and vitamin E on risk of prostate cancer and other cancers: The Selenium and Vitamin E Cancer Prevention Trial (SELECT). *JAMA* **2009**, *301*, 39–51. [CrossRef] [PubMed]

33. Marshall, J.R.; Tangen, C.M.; Sakr, W.A.; Wood, D.P., Jr.; Berry, D.L.; Klein, E.A.; Lippman, S.M.; Parnes, H.L.; Alberts, D.S.; Jarrard, D.F.; et al. Phase III trial of selenium to prevent prostate cancer in men with high-grade prostatic intraepithelial neoplasia: SWOG S9917. *Cancer Prev. Res.* **2011**, *4*, 1761–1769. [CrossRef] [PubMed]

34. Karp, D.D.; Lee, S.J.; Keller, S.M.; Wright, G.S.; Aisner, S.; Belinsky, S.A.; Johnson, D.H.; Johnston, M.R.; Goodman, G.; Clamon, G.; et al. Randomized, double-blind, placebo-controlled, phase III chemoprevention trial of selenium supplementation in patients with resected stage I non-small-cell lung cancer: ECOG 5597. *J. Clin. Oncol.* **2013**, *31*, 4179–4187. [CrossRef] [PubMed]

35. Mix, M.; Singh, A.K.; Tills, M.; Dibaj, S.; Groman, A.; Jaggernauth, W.; Rustum, R.; Jameson, M.B. Randomized phase II trial of selenomethionine as a modulator of efficacy and toxicity of chemoradiation in squamous cell carcinoma of the head and neck. *World J. Clin. Oncol.* **2015**, *6*, 166. [CrossRef] [PubMed]

36. Vinceti, M.; Vicentini, M.; Wise, L.A.; Sacchettini, C.; Malagoli, C.; Ballotari, P.; Filippini, T.; Malavolti, M.; Rossi, P.G. Cancer incidence following long-term consumption of drinking water with high inorganic selenium content. *Sci. Total Environ.* **2018**, *635*, 390–396. [CrossRef] [PubMed]

37. Akbaraly, N.T.; Arnaud, J.; Hininger-Favier, I.; Gourlet, V.; Roussel, A.M.; Berr, C. Selenium and mortality in the elderly: Results from the EVA study. *Clin. Chem.* **2005**, *51*, 2117–2123. [CrossRef] [PubMed]

38. Fernandes, A.P.; Gandin, V. Selenium compounds as therapeutic agents in cancer. *Biochim. Biophys. Acta* **2015**, *1850*, 1642–1660. [CrossRef]
39. Rahmanto, A.S.; Davies, M.J. Selenium-containing amino acids as direct and indirect antioxidants. *IUBMB Life* **2012**, *64*, 863–871. [CrossRef]
40. Prasad, S.; Gupta, S.C.; Tyagi, A.K. Reactive oxygen species (ROS) and cancer: Role of antioxidative nutraceuticals. *Cancer Lett.* **2017**, *387*, 95–105. [CrossRef]
41. Georgieva, E.; Ivanova, D.; Zhelev, Z.; Bakalova, R.; Gulubova, M.; Aoki, I. Mitochondrial dysfunction and redox imbalance as a diagnostic marker of "free radical diseases". *Anticancer Res.* **2017**, *37*, 5373–5381.
42. Valko, M.; Rhodes, C.J.; Moncol, J.; Izakovic, M.; Mazur, M. Free radicals, metals and antioxidants in oxidative stress-induced cancer. *Chem. Biol. Interact.* **2006**, *160*, 1–40. [CrossRef] [PubMed]
43. Drake, E.N. Cancer chemoprevention: Selenium as a prooxidant, not an antioxidant. *Med. Hypotheses* **2006**, *67*, 318–322. [CrossRef] [PubMed]
44. Hatfield, D.L.; Yoo, M.H.; Carlson, B.A.; Gladyshev, V.N. Selenoproteins that function in cancer prevention and promotion. *Biochim. Biophys. Acta* **2009**, *1790*, 1541–1545. [CrossRef]
45. Brozmanová, J.; Mániková, D.; Vlčková, V.; Chovanec, M. Selenium: A double-edged sword for defense and offence in cancer. *Arch. Toxicol.* **2010**, *84*, 919–938. [CrossRef] [PubMed]
46. Brigelius-Flohé, R.; Flohé, L. Selenium and redox signaling. *Arch. Biochem. Biophys.* **2017**, *617*, 48–59. [CrossRef] [PubMed]
47. Varlamova, E.G.; Cheremushkina, I.V. Contribution of mammalian selenocysteine-containing proteins to carcinogenesis. *J. Trace Elem. Med. Biol.* **2017**, *39*, 76–85. [CrossRef] [PubMed]
48. Li, T.; Kang, G.; Wang, T.; Huang, H. Tumor angiogenesis and anti-angiogenic gene therapy for cancer. *Oncol. Lett.* **2018**, *16*, 687–702. [CrossRef] [PubMed]
49. Unni, E.; Koul, D.; Yung, W.K.; Sinha, R. Se-methylselenocysteine inhibits phosphatidylinositol 3-kinase activity of mouse mammary epithelial tumor cells in vitro. *Breast Cancer Res.* **2005**, *7*, R699–R707. [CrossRef] [PubMed]
50. Schroterova, L.; Kralova, V.; Voracova, A.; Haskova, P.; Rudolf, E.; Cervinka, M. Antiproliferative effects of selenium compounds in colon cancer cells: Comparison of different cytotoxicity assays. *Toxicol. In Vitro* **2009**, *23*, 1406–1411. [CrossRef] [PubMed]
51. Suzuki, M.; Endo, M.; Shinohara, F.; Echigo, S.; Rikiishi, H. Differential apoptotic response of human cancer cells to organoselenium compounds. *Cancer Chemother. Pharmacol.* **2010**, *66*, 475–484. [CrossRef]
52. Cao, S.; Durrani, F.A.; Rustum, Y.M. Selective modulation of the therapeutic efficacy of anticancer drugs by selenium containing compounds against human tumor xenografts. *Clin. Cancer Res.* **2004**, *10*, 2561–2569. [CrossRef] [PubMed]
53. Li, Z.; Carrier, L.; Belame, A.; Thiyagarajah, A.; Salvo, V.A.; Burow, M.E.; Rowan, B.G. Combination of methylselenocysteine with tamoxifen inhibits MCF-7 breast cancer xenografts in nude mice through elevated apoptosis and reduced angiogenesis. *Breast Cancer Res. Treat.* **2009**, *118*, 33–43. [CrossRef] [PubMed]
54. Chintala, S.; Toth, K.; Cao, S.; Durrani, F.A.; Vaughan, M.M.; Jensen, R.L.; Rustum, Y.M. Se-methylselenocysteine sensitizes hypoxic tumor cells to irinotecan by targeting hypoxia-inducible factor 1α. *Cancer Chemother. Pharmacol.* **2010**, *66*, 899–911. [CrossRef] [PubMed]
55. Bhattacharya, A. Methylselenocysteine: A promising antiangiogenic agent for overcoming drug delivery barriers in solid malignancies for therapeutic synergy with anticancer drugs. *Expert. Opin. Drug Deliv.* **2011**, *8*, 749–763. [CrossRef] [PubMed]
56. Cao, S.; Durrani, F.A.; Toth, K.; Rustum, Y.M. Se-methylselenocysteine offers selective protection against toxicity and potentiates the antitumour activity of anticancer drugs in preclinical animal models. *Br. J. Cancer* **2014**, *110*, 1733–1743. [CrossRef] [PubMed]
57. McAuslan, B.R.; Reilly, W. Selenium-induced cell migration and proliferation: Relevance to angiogenesis and microangiopathy. *Microvasc. Res.* **1986**, *32*, 112–120. [CrossRef]
58. Streicher, K.L.; Sylte, M.J.; Johnson, S.E.; Sordillo, L.M. Thioredoxin reductase regulates angiogenesis by increasing endothelial cell-derived vascular endothelial growth factor. *Nutr. Cancer* **2004**, *50*, 221–231. [CrossRef]
59. Zhang, J.; Wang, H.; Yan, X.; Zhang, L. Comparison of short-term toxicity between Nano-Se and selenite in mice. *Life Sci.* **2005**, *76*, 1099–1109. [CrossRef]

60. Zhang, J.; Wang, X.; Xu, T. Elemental selenium at nano size (Nano-Se) as a potential chemopreventive agent with reduced risk of selenium toxicity: Comparison with Se-methylselenocysteine in mice. *Toxicol. Sci.* **2008**, *101*, 22–31. [CrossRef]

61. Chaudhary, S.; Umar, A.; Mehta, S.K. Surface functionalized selenium nanoparticles for biomedical applications. *J. Biomed. Nanotechnol.* **2014**, *10*, 3004–3042. [CrossRef]

62. Bai, K.; Hong, B.; He, J.; Hong, Z.; Tan, R. Preparation and antioxidant properties of selenium nanoparticles-loaded chitosan microspheres. *Int. J. Nanomed.* **2017**, *12*, 4527–4539. [CrossRef] [PubMed]

63. Nguyen, T.H.D.; Vardhanabhuti, B.; Lin, M.J.; Mustapha, A. Antibacterial properties of selenium nanoparticles and their toxicity to Caco-2 cells. *Food Control* **2017**, *77*, 17–24. [CrossRef]

64. Sharma, G.; Sharma, A.R.; Bhavesh, R.; Park, J.; Ganbold, B.; Nam, J.S.; Lee, S.S. Biomolecule-mediated synthesis of selenium nanoparticles using dried *Vitis vinifera* (raisin) extract. *Molecules* **2014**, *19*, 2761–2770. [CrossRef] [PubMed]

65. Fairweather-Tait, S.J.; Collings, R.; Hurst, R. Selenium bioavailability: Current knowledge and future research requirements. *Am. J. Clin. Nutr.* **2010**, *91*, 1484S–1491S. [CrossRef] [PubMed]

66. Rayman, M.P. Selenium in cancer prevention: A review of the evidence and mechanism of action. *Proc. Nutr. Soc.* **2005**, *64*, 527–542. [CrossRef] [PubMed]

67. Papp, L.V.; Lu, J.; Holmgren, A.; Khanna, K.K. From selenium to selenoproteins: Synthesis, identity, and their role in human health. *Antioxid. Redox Signal.* **2007**, *9*, 775–806. [CrossRef] [PubMed]

68. Jackson, M.I.; Combs, G.F., Jr. Selenium and anticarcinogenesis: Underlying mechanisms. *Curr. Opin. Clin. Nutr. Metab. Care* **2008**, *11*, 718–726. [CrossRef]

69. Ronai, Z.; Tillotson, J.K.; Traganos, F.; Darzynkiewicz, Z.; Conaway, C.C.; Upadhyaya, P.; el-Bayoumy, K. Effects of organic and inorganic selenium compounds on rat mammary tumor cells. *Int. J. Cancer* **1995**, *63*, 428–434. [CrossRef]

70. Sanmartín, C.; Plano, D.; Sharma, A.K.; Palop, J.A. Selenium compounds, apoptosis and other types of cell death: An overview for cancer therapy. *Int. J. Mol. Sci.* **2012**, *13*, 9649–9672. [CrossRef]

71. Choi, A.; Jo, M.J.; Jung, M.J.; Kim, H.S.; Yoon, S. Selenate specifically sensitizes drug-resistant cancer cells by increasing apoptosis via G2 phase cell cycle arrest without P-GP inhibition. *Eur. J. Pharmacol.* **2015**, *764*, 63–69. [CrossRef]

72. Takahashi, M.; Sato, T.; Shinohara, F.; Echigo, S.; Rikiishi, H. Possible role of glutathione in mitochondrial apoptosis of human oral squamous cell carcinoma caused by inorganic selenium compounds. *Int. J. Oncol.* **2005**, *27*, 489–495. [CrossRef] [PubMed]

73. Endo, M.; Hasegawa, H.; Kaneko, T.; Kanno, C.; Monma, T.; Kano, M.; Shinohara, F.; Takahashi, T. Antitumor activity of selenium compounds and its underlying mechanism in human oral squamous cell carcinoma cells: A preliminary study. *J. Oral Maxillofac. Surg. Med. Pathol.* **2017**, *29*, 17–23. [CrossRef]

74. Kieliszek, M.; Lipinski, B.; Blazejak, S. Application of sodium selenite in the prevention and treatment of cancers. *Cells* **2017**, *6*, 39. [CrossRef]

75. Tan, H.W.; Xu, Y.M.; Wu, D.D.; Lau, A.T.Y. Recent insights into human bronchial proteomics – how are we progressing and what is next? *Expert. Rev. Proteomics* **2018**, *15*, 113–130. [CrossRef] [PubMed]

76. Brodin, O.; Eksborg, S.; Wallenberg, M.; Asker-Hagelberg, C.; Larsen, E.H.; Mohlkert, D.; Lenneby-Helleday, C.; Jacobsson, H.; Linder, S.; Misra, S.; et al. Pharmacokinetics and toxicity of sodium selenite in the treatment of patients with carcinoma in a phase I clinical trial: The SECAR Study. *Nutrients* **2015**, *7*, 4978–4994. [CrossRef] [PubMed]

77. Berthier, S.; Arnaud, J.; Champelovier, P.; Col, E.; Garrel, C.; Cottet, C.; Boutonnat, J.; Laporte, F.; Faure, P.; Hazane-Puch, F. Anticancer properties of sodium selenite in human glioblastoma cell cluster spheroids. *J. Trace Elem. Med. Biol.* **2017**, *44*, 161–176. [CrossRef]

78. Lipinski, B. Sodium selenite as an anticancer agent. *Anticancer Agents Med. Chem.* **2017**, *17*, 658–661. [CrossRef]

79. Chen, W.; An, J.; Guo, J.; Wu, Y.; Yang, L.; Dai, J.; Gong, K.; Miao, S.; Xi, S.; Du, J. Sodium selenite attenuates lung adenocarcinoma progression by repressing SOX2-mediated stemness. *Cancer Chemother. Pharmacol.* **2018**, *81*, 885–895. [CrossRef] [PubMed]

80. Olm, E.; Fernandes, A.P.; Hebert, C.; Rundlöf, A.K.; Larsen, E.H.; Danielsson, O.; Björnstedt, M. Extracellular thiol-assisted selenium uptake dependent on the x_c^- cystine transporter explains the cancer-specific cytotoxicity of selenite. *PNAS* **2009**, *106*, 11400–11405. [CrossRef] [PubMed]

81. Enqvist, M.; Nilsonne, G.; Hammarfjord, O.; Wallin, R.P.A.; Bjorkstrom, N.K.; Bjornstedt, M.; Hjerpe, A.; Ljunggren, H.G.; Dobra, K.; Malmberg, K.J.; et al. Selenite induces posttranscriptional blockade of HLA-E expression and sensitizes tumor cells to CD94/NKG2A-Positive NK Cells. *J. Immunol.* **2011**, *187*, 3546–3554. [CrossRef]

82. Rigobello, M.P.; Gandin, V.; Folda, A.; Rundlof, A.K.; Fernandes, A.P.; Bindoli, A.; Marzano, C.; Bjornstedt, M. Treatment of human cancer cells with selenite or tellurite in combination with auranofin enhances cell death due to redox shift. *Free Radic. Biol. Med.* **2009**, *47*, 710–721. [CrossRef] [PubMed]

83. Zheng, X.; Xu, W.; Sun, R.; Yin, H.; Dong, C.; Zeng, H. Synergism between thioredoxin reductase inhibitor ethaselen and sodium selenite in inhibiting proliferation and inducing death of human non-small cell lung cancer cells. *Chem. Biol. Interact.* **2017**, *275*, 74–85. [CrossRef] [PubMed]

84. Li, J.; Sun, K.; Ni, L.; Wang, X.; Wang, D.; Zhang, J. Sodium selenosulfate at an innocuous dose markedly prevents cisplatin-induced gastrointestinal toxicity. *Toxicol. Appl. Pharmacol.* **2012**, *258*, 376–383. [CrossRef] [PubMed]

85. Freitas, M.; Alves, V.; Sarmento-Ribeiro, A.B.; Mota-Pinto, A. Combined effect of sodium selenite and docetaxel on PC3 metastatic prostate cancer cell line. *Biochem. Biophys. Res. Commun.* **2011**, *408*, 713–719. [CrossRef] [PubMed]

86. Schroeder, C.P.; Goeldner, E.M.; Schulze-Forster, K.; Eickhoff, C.A.; Holtermann, P.; Heidecke, H. Effect of selenite combined with chemotherapeutic agents on the proliferation of human carcinoma cell lines. *Biol. Trace Elem. Res.* **2004**, *99*, 17–25. [CrossRef]

87. Björkhem-Bergman, L.; Jönsson, K.; Eriksson, L.C.; Olsson, J.M.; Lehmann, S.; Paul, C.; Björnstedt, M. Drug-resistant human lung cancer cells are more sensitive to selenium cytotoxicity. Effects on thioredoxin reductase and glutathione reductase. *Biochem. Pharmacol.* **2002**, *63*, 1875–1884. [CrossRef]

88. Jönsson-Videsäter, K.; Björkhem-Bergman, L.; Hossain, A.; Söderberg, A.; Eriksson, L.C.; Paul, C.; Rosén, A.; Björnstedt, M. Selenite-induced apoptosis in doxorubicin-resistant cells and effects on the thioredoxin system. *Biochem. Pharmacol.* **2004**, *67*, 513–522. [CrossRef]

89. Hinrichsen, S.; Planer-Friedrich, B. Cytotoxic activity of selenosulfate versus selenite in tumor cells depends on cell line and presence of amino acids. *Environ. Sci. Pollut. Res. Int.* **2016**, *23*, 8349–8357. [CrossRef]

90. Hu, B.; Cheng, R.; Gao, X.; Pan, X.; Kong, F.; Liu, X.; Xu, K.; Tang, B. Targetable mesoporous Silica nanoprobes for mapping the subcellular distribution of H_2Se in cancer cells. *ACS Appl. Mater. Interfaces* **2018**, *10*, 17345–17351. [CrossRef]

91. Storkey, C.; Davies, M.J.; White, J.M.; Schiesser, C.H. Synthesis and antioxidant capacity of 5-selenopyranose derivatives. *Chem. Commun.* **2011**, *47*, 9693–9695. [CrossRef]

92. Jariwalla, R.J.; Gangapurkar, B.; Nakamura, D. Differential sensitivity of various human tumour-derived cell types to apoptosis by organic derivatives of selenium. *Br. J. Nutr.* **2009**, *101*, 182–189. [CrossRef] [PubMed]

93. Tung, Y.C.; Tsai, M.L.; Kuo, F.L.; Lai, C.S.; Badmaev, V.; Ho, C.T.; Pan, M.H. Se-Methyl-L-selenocysteine induces apoptosis via endoplasmic reticulum stress and the death receptor pathway in human colon adenocarcinoma COLO 205 Cells. *J. Agricult. Food Chem.* **2015**, *63*, 5008–5016. [CrossRef] [PubMed]

94. Domracheva, I.; Kanepe-Lapsa, I.; Jackevica, L.; Vasiljeva, J.; Arsenyan, P. Selenopheno quinolinones and coumarins promote cancer cell apoptosis by ROS depletion and caspase-7 activation. *Life Sci.* **2017**, *186*, 92–101. [CrossRef] [PubMed]

95. Fan, C.; Zheng, W.; Fu, X.; Li, X.; Wong, Y.S.; Chen, T. Strategy to enhance the therapeutic effect of doxorubicin in human hepatocellular carcinoma by selenocystine, a synergistic agent that regulates the ROS-mediated signaling. *Oncotarget* **2014**, *5*, 2853–2863. [CrossRef] [PubMed]

96. Deepagan, V.G.; Kwon, S.; You, D.G.; Nguyen, V.Q.; Um, W.; Ko, H.; Lee, H.; Jo, D.G.; Kang, Y.M.; Park, J.H. In situ diselenide-crosslinked polymeric micelles for ROS-mediated anticancer drug delivery. *Biomaterials* **2016**, *103*, 56–66. [CrossRef] [PubMed]

97. Çetin, E.S.; Nazıroğlu, M.; Çiğ, B.; Övey, I.S.; Koşar, P.A. Selenium potentiates the anticancer effect of cisplatin against oxidative stress and calcium ion signaling-induced intracellular toxicity in MCF-7 breast cancer cells: Involvement of the TRPV1 channel. *J. Recept. Signal. Transduct. Res.* **2017**, *37*, 84–93. [CrossRef] [PubMed]

98. Gandin, V.; Khalkar, P.; Braude, J.; Fernandes, A.P. Organic selenium compounds as potential chemotherapeutic agents for improved cancer treatment. *Free Radic. Biol. Med.* **2018**, *127*, 80–97. [CrossRef]

99. Álvarez-Pérez, M.; Ali, W.; Marć, M.A.; Handzlik, J.; Domínguez-Álvarez, E. Selenides and diselenides: A review of their anticancer and chemopreventive activity. *Molecules* **2018**, *23*, 628. [CrossRef]

100. Ali, W.; Álvarez-Pérez, M.; Marć, M.A.; Salardón-Jiménez, N.; Handzlik, J.; Domínguez-Álvarez, E. The anticancer and chemopreventive activity of selenocyanate-containing compounds. *Curr. Pharmacol. Rep.* **2018**, *4*, 468–481. [CrossRef]

101. de Bruin, E.C.; Medema, J.P. Apoptosis and non-apoptotic deaths in cancer development and treatment response. *Cancer Treat. Rev.* **2008**, *34*, 737–749. [CrossRef]

102. Poerschke, R.L.; Franklin, M.R.; Moos, P.J. Modulation of redox status in human lung cell lines by organoselenocompounds: Selenazolidines, selenomethionine, and methylseleninic acid. *Toxicol. In Vitro* **2008**, *22*, 1761–1767. [CrossRef] [PubMed]

103. Gao, R.; Zhao, L.; Liu, X.; Rowan, B.G.; Wabitsch, M.; Edwards, D.P.; Nishi, Y.; Yanase, T.; Yu, Q.; Dong, Y. Methylseleninic acid is a novel suppressor of aromatase expression. *J. Endocrinol.* **2012**, *212*, 199–205. [CrossRef] [PubMed]

104. Qi, Y.; Fu, X.; Xiong, Z.; Zhang, H.; Hill, S.M.; Rowan, B.G.; Dong, Y. Methylseleninic acid enhances paclitaxel efficacy for the treatment of triple-negative breast cancer. *PLoS ONE* **2012**, *7*, e31539. [CrossRef] [PubMed]

105. Lennicke, C.; Rahn, J.; Bukur, J.; Hochgräfe, F.; Wessjohann, L.A.; Lichtenfels, R.; Seliger, B. Modulation of MHC class I surface expression in B16F10 melanoma cells by methylseleninic acid. *Oncoimmunology* **2017**, *6*, e1259049. [CrossRef] [PubMed]

106. Li, G.X.; Lee, H.J.; Wang, Z.; Hu, H.; Liao, J.D.; Watts, J.C.; Combs, G.F.; Lü, J. Superior in vivo inhibitory efficacy of methylseleninic acid against human prostate cancer over selenomethionine or selenite. *Carcinogenesis* **2008**, *29*, 1005–1012. [CrossRef] [PubMed]

107. Wang, L.; Bonorden, M.J.L.; Li, G.; Lee, H.J.; Hu, H.; Zhang, Y.; Liao, J.D.; Cleary, M.P.; Lü, J. Methyl-selenium compounds inhibit prostate carcinogenesis in the transgenic adenocarcinoma of mouse prostate model with survival benefit. *Cancer Prev. Res.* **2009**, *2*, 484–495. [CrossRef] [PubMed]

108. Lee, S.O.; Chun, J.Y.; Nadiminty, N.; Trump, D.L.; Ip, C.; Dong, Y.; Gao, A.C. Monomethylated selenium inhibits growth of LNCaP human prostate cancer xenograft accompanied by a decrease in the expression of androgen receptor and prostate-specific antigen (PSA). *Prostate* **2006**, *66*, 1070–1075. [CrossRef] [PubMed]

109. Lobb, R.J.; Jacobson, G.M.; Cursons, R.T.; Jameson, M.B. The interaction of selenium with chemotherapy and radiation on normal and malignant human mononuclear blood cells. *Int. J. Mol. Sci.* **2018**, *19*, 3167. [CrossRef] [PubMed]

110. Sun, L.; Zhang, J.; Yang, Q.; Si, Y.; Liu, Y.; Wang, Q.; Han, F.; Huang, Z. Synergistic effects of SAM and selenium compounds on proliferation, migration and adhesion of HeLa cells. *Anticancer Res.* **2017**, *37*, 4433–4441. [PubMed]

111. Li, Z.; Carrier, L.; Rowan, B.G. Methylseleninic acid synergizes with tamoxifen to induce caspase-mediated apoptosis in breast cancer cells. *Mol. Cancer Ther.* **2008**, *7*, 3056–3063. [CrossRef]

112. Khalkar, P.; Diaz-Argelich, N.; Antonio Palop, J.; Sanmartín, C.; Fernandes, A.P. Novel methylselenoesters induce programed cell death via entosis in pancreatic cancer cells. *Int. J. Mol. Sci.* **2018**, *19*, 2849. [CrossRef] [PubMed]

113. Shi, C.; Yu, L.; Yang, F.; Yan, J.; Zeng, H. A novel organoselenium compound induces cell cycle arrest and apoptosis in prostate cancer cell lines. *Biochem. Biophys. Res. Commun.* **2003**, *309*, 578–583. [CrossRef]

114. Sharma, V.; Tewari, R.; Sk, U.H.; Joseph, C.; Sen, E. Ebselen sensitizes glioblastoma cells to Tumor Necrosis Factor (TNFα)-induced apoptosis through two distinct pathways involving NF-κβ downregulation and Fas-mediated formation of death inducing signaling complex. *Int. J. Cancer* **2008**, *123*, 2204–2212. [CrossRef] [PubMed]

115. Shiah, H.S.; Lee, W.S.; Juang, S.H.; Hong, P.C.; Lung, C.C.; Chang, C.J.; Chou, K.M.; Chang, J.Y. Mitochondria-mediated and p53-associated apoptosis induced in human cancer cells by a novel selenophene derivative, D-501036. *Biochem. Pharmacol.* **2007**, *73*, 610–619. [CrossRef] [PubMed]

116. Juang, S.H.; Lung, C.C.; Hsu, P.C.; Hsu, K.S.; Li, Y.C.; Hong, P.C.; Shiah, H.S.; Kuo, C.C.; Huang, C.W.; Wang, Y.C.; et al. D-501036, a novel selenophene-based triheterocycle derivative, exhibits potent in vitro and in vivo antitumoral activity which involves DNA damage and ataxia telangiectasia-mutated nuclear protein kinase activation. *Mol. Cancer Ther.* **2007**, *6*, 193–202. [CrossRef] [PubMed]

117. Yang, Y.N.; Chou, K.M.; Pan, W.Y.; Chen, Y.W.; Tsou, T.C.; Yeh, S.C.; Cheung, C.H.; Chen, L.T.; Chang, J.Y. Enhancement of non-homologous end joining DNA repair capacity confers cancer cells resistance to the novel selenophene compound, D-501036. *Cancer Lett.* **2011**, *309*, 110–118. [CrossRef]

118. Gumulec, J.; Balvan, J.; Sztalmachova, M.; Raudenska, M.; Dvorakova, V.; Knopfova, L.; Polanska, H.; Hudcova, K.; Ruttkay-Nedecky, B.; Babula, P.; et al. Cisplatin-resistant prostate cancer model: Differences in antioxidant system, apoptosis and cell cycle. *Int. J. Oncol.* **2014**, *44*, 923–933. [CrossRef]

119. Bartolini, D.; Sancineto, L.; Fabro de Bem, A.; Tew, K.D.; Santi, C.; Radi, R.; Toquato, P.; Galli, F. Selenocompounds in cancer therapy: An overview. *Adv. Cancer Res.* **2017**, *136*, 259–302.

120. Ho, B.N.; Pfeffer, C.M.; Singh, A.T.K. Update on nanotechnology-based drug delivery systems in cancer treatment. *Anticancer Res.* **2017**, *37*, 5975–5981.

121. Gao, F.; Yuan, Q.; Gao, L.; Cai, P.; Zhu, H.; Liu, R.; Wang, Y.; Wei, Y.; Huang, G.; Liang, J.; et al. Cytotoxicity and therapeutic effect of irinotecan combined with selenium nanoparticles. *Biomaterials* **2014**, *35*, 8854–8866. [CrossRef]

122. Shamsi, M.M.; Chekachak, S.; Soudi, S.; Gharakhanlou, R.; Quinn, L.S.; Ranjbar, K.; Rezaei, S.; Shirazi, F.J.; Allahmoradi, B.; Yazdi, M.H.; et al. Effects of exercise training and supplementation with selenium nanoparticle on T-helper 1 and 2 and cytokine levels in tumor tissue of mice bearing the 4 T1 mammary carcinoma. *Nutrition* **2018**, *57*, 141–147. [CrossRef] [PubMed]

123. Skalickova, S.; Milosavljevic, V.; Cihalova, K.; Horky, P.; Richtera, L.; Adam, V. Selenium nanoparticles as a nutritional supplement. *Nutrition* **2017**, *33*, 83–90. [CrossRef] [PubMed]

124. Zhang, J.; Teng, Z.; Yuan, Y.; Zeng, Q.Z.; Lou, Z.; Lee, S.H.; Wang, Q. Development, physicochemical characterization and cytotoxicity of selenium nanoparticles stabilized by beta-lactoglobulin. *Int. J. Biol. Macromol.* **2018**, *107*, 1406–1413. [CrossRef] [PubMed]

125. Menon, S.; KS, S.D.; Santhiya, R.; Rajeshkumar, S.; Kumar, V. Selenium nanoparticles: A potent chemotherapeutic agent and an elucidation of its mechanism. *Colloids Surf B Biointerfaces* **2018**, *170*, 280–292. [CrossRef] [PubMed]

126. Wadhwani, S.A.; Gorain, M.; Banerjee, P.; Shedbalkar, U.U.; Singh, R.; Kundu, G.C.; Chopade, B.A. Green synthesis of selenium nanoparticles using *Acinetobacter* sp. SW30: Optimization, characterization and its anticancer activity in breast cancer cells. *Int. J. Nanomed.* **2017**, *12*, 6841–6855. [CrossRef] [PubMed]

127. Sonkusre, P.; Cameotra, S.S. Biogenic selenium nanoparticles induce ROS-mediated necroptosis in PC-3 cancer cells through TNF activation. *J. Nanobiotechnol.* **2017**, *15*, 1–12. [CrossRef] [PubMed]

128. Cui, D.; Yan, C.; Miao, J.; Zhang, X.; Chen, J.; Sun, L.; Meng, L.; Liang, T.; Li, Q. Synthesis, characterization and antitumor properties of selenium nanoparticles coupling with ferulic acid. *Mater. Sci. Eng. C* **2018**, *90*, 104–112. [CrossRef] [PubMed]

129. Shahverdi, A.R.; Shahverdi, F.; Faghfuri, E.; Reza Khoshayand, M.; Mavandadnejad, F.; Yazdi, M.H.; Amini, M. Characterization of folic acid surface-coated selenium nanoparticles and corresponding in vitro and in vivo effects against breast cancer. *Arch. Med. Res.* **2018**, *49*, 10–17. [CrossRef]

130. Yang, Y.; Xie, Q.; Zhao, Z.; He, L.; Chan, L.; Liu, Y.; Chen, Y.; Bai, M.; Pan, T.; Qu, Y.; et al. Functionalized selenium nanosystem as radiation sensitizer of ^{125}I seeds for precise cancer therapy. *ACS Appl. Mater. Interfaces* **2017**, *9*, 25857–25869. [CrossRef]

131. Qiu, W.Y.; Wang, Y.Y.; Wang, M.; Yan, J.K. Construction, stability, and enhanced antioxidant activity of pectin-decorated selenium nanoparticles. *Colloids Surf. B Biointerfaces* **2018**, *170*, 692–700. [CrossRef]

132. Huang, G.; Liu, Z.; He, L.; Luk, K.H.; Cheung, S.T.; Wong, K.H.; Chen, T. Autophagy is an important action mode for functionalized selenium nanoparticles to exhibit anti-colorectal cancer activity. *Biomater. Sci.* **2018**, *6*, 2508–2517. [CrossRef] [PubMed]

133. Gautam, P.K.; Kumar, S.; Tomar, M.S.; Singh, R.K.; Acharya, A.; Kumar, S.; Ram, B. Selenium nanoparticles induce suppressed function of tumor associated macrophages and inhibit Dalton's lymphoma proliferation. *Biochem. Biophys. Rep.* **2017**, *12*, 172–184. [PubMed]

134. Zhao, G.; Wu, X.; Chen, P.; Zhang, L.; Yang, C.S.; Zhang, J. Selenium nanoparticles are more efficient than sodium selenite in producing reactive oxygen species and hyper-accumulation of selenium nanoparticles in cancer cells generates potent therapeutic effects. *Free Radic. Biol. Med.* **2018**, *126*, 55–66. [CrossRef] [PubMed]

135. Zhai, S.; Hu, X.; Hu, Y.; Wu, B.; Xing, D. Visible light-induced crosslinking and physiological stabilization of diselenide-rich nanoparticles for redox-responsive drug release and combination chemotherapy. *Biomaterials* **2017**, *121*, 41–54. [CrossRef] [PubMed]

136. Yan, Q.; Chen, X.; Gong, H.; Qiu, P.; Xiao, X.; Dang, S.; Hong, A.; Ma, Y. Delivery of a TNF-α–derived peptide by nanoparticles enhances its antitumor activity by inducing cell-cycle arrest and caspase-dependent apoptosis. *FASEB J.* **2018**. [CrossRef] [PubMed]

137. Luesakul, U.; Puthong, S.; Neamati, N.; Muangsin, N. pH-responsive selenium nanoparticles stabilized by folate-chitosan delivering doxorubicin for overcoming drug-resistant cancer cells. *Carbohydr. Polym.* **2018**, *181*, 841–850. [CrossRef] [PubMed]

138. Jalalian, S.H.; Ramezani, M.; Abnous, K.; Taghdisi, S.M. Targeted co-delivery of epirubicin and NAS-24 aptamer to cancer cells using selenium nanoparticles for enhancing tumor response in vitro and in vivo. *Cancer Lett.* **2018**, *416*, 87–93. [CrossRef] [PubMed]

139. Nonsuwan, P.; Puthong, S.; Palaga, T.; Muangsin, N. Novel organic/inorganic hybrid flower-like structure of selenium nanoparticles stabilized by pullulan derivatives. *Carbohydr. Polym.* **2018**, *184*, 9–19. [CrossRef]

140. Xia, Y.; Guo, M.; Xu, T.; Li, Y.; Wang, C.; Lin, Z.; Zhao, M.; Zhu, B. siRNA-loaded selenium nanoparticle modified with hyaluronic acid for enhanced hepatocellular carcinoma therapy. *Int. J. Nanomed.* **2018**, *13*, 1539–1552. [CrossRef]

141. Abd-Rabou, A.A.; Shalby, A.B.; Ahmed, H.H. Selenium nanoparticles induce the chemo-sensitivity of fluorouracil nanoparticles in breast and colon cancer cells. *Biol. Trace Elem. Res.* **2018**. [CrossRef]

142. Zhang, S.; Zhao, L.; Qiu, N.; Liu, Y.; Xu, B.; Zhu, H. On the hypoxic tumor targeting ability of two chitosan micelles loaded with oil-soluble CdSe quantum dots. *Pharm. Dev. Technol.* **2018**, *23*, 87–95. [CrossRef] [PubMed]

143. Pi, J.; Jiang, J.; Cai, H.; Yang, F.; Jin, H.; Yang, P.; Cai, J.; Chen, Z.W. GE11 peptide conjugated selenium nanoparticles for EGFR targeted oridonin delivery to achieve enhanced anticancer efficacy by inhibiting EGFR-mediated PI3K/AKT and Ras/Raf/MEK/ERK pathways. *Drug Deliv.* **2017**, *24*, 1549–1564. [CrossRef]

144. Chen, Q.; Xu, M.; Zheng, W.; Xu, T.; Deng, H.; Liu, J. Se/Ru-decorated porous metal-organic framework nanoparticles for the delivery of pooled siRNAs to reversing multidrug resistance in taxol-resistant breast cancer cells. *ACS Appl. Mater. Interfaces* **2017**, *9*, 6712–6724. [CrossRef] [PubMed]

145. Cadkova, M.; Kovarova, A.; Dvorakova, V.; Metelka, R.; Bilkova, Z.; Korecka, L. Electrochemical quantum dots-based magneto-immunoassay for detection of HE4 protein on metal film-modified screen-printed carbon electrodes. *Talanta* **2018**, *182*, 111–115. [CrossRef]

146. Liu, M.L.; Zou, H.Y.; Li, C.M.; Li, R.S.; Huang, C.Z. Aptamer-modified selenium nanoparticles for dark-field microscopy imaging of nucleolin. *Chem. Commun.* **2017**, *53*, 13047–13050. [CrossRef]

147. Moulick, A.; Milosavljevic, V.; Vlachova, J.; Podgajny, R.; Hynek, D.; Kopel, P.; Adam, V. Using CdTe/ZnSe core/shell quantum dots to detect DNA and damage to DNA. *Int. J. Nanomed.* **2017**, *12*, 1277–1291. [CrossRef] [PubMed]

148. Madhankumar, A.B.; Mrowczynski, O.D.; Patel, S.R.; Weston, C.L.; Zacharia, B.E.; Glantz, M.J.; Siedlecki, C.A.; Xu, L.C.; Connor, J.R. Interleukin-13 conjugated quantum dots for identification of glioma initiating cells and their extracellular vesicles. *Acta Biomater.* **2017**, *58*, 205–213. [CrossRef] [PubMed]

149. Zhao, Y.; Sun, Q.; Zhang, X.; Baeyens, J.; Su, H. Self-assembled selenium nanoparticles and their application in the rapid diagnostic detection of small cell lung cancer biomarkers. *Soft Matter* **2018**, *14*, 481–489. [CrossRef]

150. Zhu, C.N.; Chen, G.; Tian, Z.Q.; Wang, W.; Zhong, W.Q.; Li, Z.; Zhang, Z.L.; Pang, D.W. Near-infrared fluorescent Ag_2Se–cetuximab nanoprobes for targeted imaging and therapy of cancer. *Small* **2017**, *13*. [CrossRef]

151. Purohit, M.P.; Verma, N.K.; Kar, A.K.; Singh, A.; Ghosh, D.; Patnaik, S. Inhibition of thioredoxin reductase by targeted selenopolymeric nanocarriers synergizes the therapeutic efficacy of doxorubicin in MCF7 human breast cancer cells. *ACS Appl. Mater. Interfaces* **2017**, *9*, 36493–36512. [CrossRef]

152. Zhao, S.; Yu, Q.; Pan, J.; Zhou, Y.; Cao, C.; Ouyang, J.M.; Liu, J. Redox-responsive mesoporous selenium delivery of doxorubicin targets MCF-7 cells and synergistically enhances its anti-tumor activity. *Acta Biomater.* **2017**, *54*, 294–306. [CrossRef] [PubMed]

153. Pan, J.; Zhu, X.; Chen, X.; Zhao, Y.; Liu, J. Gd^{3+}-Doped $MoSe_2$ nanosheets used as a theranostic agent for bimodal imaging and highly efficient photothermal cancer therapy. *Biomater. Sci.* **2018**, *6*, 372–387. [CrossRef] [PubMed]

154. Bidkar, A.P.; Sanpui, P.; Ghosh, S.S. Efficient induction of apoptosis in cancer cells by paclitaxel-loaded selenium nanoparticles. *Nanomedicine* **2017**, *12*, 2641–2651. [CrossRef] [PubMed]

155. Hauksdóttir, H.L.; Webster, T.J. Selenium and iron oxide nanocomposites for magnetically-targeted anti-cancer applications. *J. Biomed. Nanotechnol.* **2018**, *14*, 510–525. [CrossRef] [PubMed]

156. Kolmas, J.; Pajor, K.; Pajchel, L.; Przekora, A.; Ginalska, G.; Oledzka, E.; Sobczak, M. Fabrication and physicochemical characterization of porous composite microgranules with selenium oxyanions and risedronate sodium for potential applications in bone tumors. *Int. J. Nanomed.* **2017**, *12*, 5633–5642. [CrossRef] [PubMed]

157. Fang, X.; Li, C.; Zheng, L.; Yang, F.; Chen, T. Dual-targeted selenium nanoparticles for synergistic photothermal therapy and chemotherapy of tumors. *Chem. Asian J.* **2018**, *13*, 996–1004. [CrossRef]

158. Chen, F.; Zhang, X.H.; Hu, X.D.; Liu, P.D.; Zhang, H.Q. The effects of combined selenium nanoparticles and radiation therapy on breast cancer cells in vitro. *Artif. Cells Nanomed. Biotechnol.* **2018**, *46*, 937–948. [CrossRef]

159. Maiyo, F.; Singh, M. Selenium nanoparticles: Potential in cancer gene and drug delivery. *Nanomedicine* **2017**, *12*, 1075–1089. [CrossRef]

160. Zhai, X.; Zhang, C.; Zhao, G.; Stoll, S.; Ren, F.; Leng, X. Antioxidant capacities of the selenium nanoparticles stabilized by chitosan. *J. Nanobiotechnol.* **2017**, *15*, 1–12. [CrossRef]

161. Kumari, M.; Ray, L.; Purohit, M.P.; Patnaik, S.; Pant, A.B.; Shukla, Y.; Kumar, P.; Gupta, K.C. Curcumin loading potentiates the chemotherapeutic efficacy of selenium nanoparticles in HCT116 cells and Ehrlich's ascites carcinoma bearing mice. *Eur. J. Pharm. Biopharm.* **2017**, *117*, 346–362. [CrossRef]

162. Wadhwani, S.A.; Shedbalkar, U.U.; Singh, R.; Chopade, B.A. Biogenic selenium nanoparticles: Current status and future prospects. *Appl. Microbiol. Biotechnol.* **2016**, *100*, 2555–2566. [CrossRef] [PubMed]

163. Vetchinkina, E.; Loshchinina, E.; Kursky, V.; Nikitina, V. Reduction of organic and inorganic selenium compounds by the edible medicinal basidiomycete *Lentinula edodes* and the accumulation of elemental selenium nanoparticles in its mycelium. *J. Microbiol.* **2013**, *51*, 829–835. [CrossRef] [PubMed]

164. Tugarova, A.V.; Kamnev, A.A. Proteins in microbial synthesis of selenium nanoparticles. *Talanta* **2017**, *174*, 539–547. [CrossRef] [PubMed]

165. Prasad, K.S.; Patel, H.; Patel, T.; Patel, K.; Selvaraj, K. Biosynthesis of Se nanoparticles and its effect on UV-induced DNA damage. *Colloids Surf. B Biointerfaces* **2013**, *103*, 261–266. [CrossRef] [PubMed]

166. Zonaro, E.; Piacenza, E.; Presentato, A.; Monti, F.; Dell'Anna, R.; Lampis, S.; Vallini, G. *Ochrobactrum* sp. MPV1 from a dump of roasted pyrites can be exploited as bacterial catalyst for the biogenesis of selenium and tellurium nanoparticles. *Microb. Cell Fact.* **2017**, *16*, 215. [CrossRef] [PubMed]

167. Ramya, S.; Shanmugasundaram, T.; Balagurunathan, R. Biomedical potential of actinobacterially synthesized selenium nanoparticles with special reference to anti-biofilm, anti-oxidant, wound healing, cytotoxic and anti-viral activities. *J. Trace Elem. Med. Biol.* **2015**, *32*, 30–39. [CrossRef] [PubMed]

168. Saratale, R.G.; Karuppusamy, I.; Saratale, G.D.; Pugazhendhi, A.; Kumar, G.; Park, Y.; Ghodake, G.S.; Bharagava, R.N.; Banu, J.R.; Shin, H.S. A comprehensive review on green nanomaterials using biological systems: Recent perception and their future applications. *Colloids Surf. B Biointerfaces* **2018**, *170*, 20–35. [CrossRef]

169. Eswayah, A.S.; Smith, T.J.; Scheinost, A.C.; Hondow, N.; Gardiner, P.H.E. Microbial transformations of selenite by methane-oxidizing bacteria. *Appl. Microbiol. Biotechnol.* **2017**, *101*, 6713–6724. [CrossRef]

170. Song, D.; Li, X.; Cheng, Y.; Xiao, X.; Lu, Z.; Wang, Y.; Wang, F. Aerobic biogenesis of selenium nanoparticles by Enterobacter cloacae Z0206 as a consequence of fumarate reductase mediated selenite reduction. *Sci. Rep.* **2017**, *7*, 3239. [CrossRef]

171. Fernandez-Llamosas, H.; Castro, L.; Blazquez, M.L.; Diaz, E.; Carmona, M. Speeding up bioproduction of selenium nanoparticles by using *Vibrio natriegens* as microbial factory. *Sci. Rep.* **2017**, *7*, 16046. [CrossRef]

172. Quintana, M.; Haro-Poniatowski, E.; Morales, J.; Batina, N. Synthesis of selenium nanoparticles by pulsed laser ablation. *Appl. Surf. Sci.* **2002**, *195*, 175–186. [CrossRef]

173. Chang, S.Q.; Dai, Y.D.; Kang, B.; Han, W.; Chen, D. Gamma-radiation synthesis of silk fibroin coated CdSe quantum dots and their biocompatibility and photostability in living cells. *J. Nanosci. Nanotechnol.* **2009**, *9*, 5693–5700. [CrossRef] [PubMed]

174. Guisbiers, G.; Lara, H.H.; Mendoza-Cruz, R.; Naranjo, G.; Vincent, B.A.; Peralta, X.G.; Nash, K.L. Inhibition of Candida albicans biofilm by pure selenium nanoparticles synthesized by pulsed laser ablation in liquids. *Nanomedicine* **2017**, *13*, 1095–1103. [CrossRef] [PubMed]

175. Kasinski, A.L.; Slack, F.J. Epigenetics and genetics. MicroRNAs en route to the clinic: Progress in validating and targeting microRNAs for cancer therapy. *Nat. Rev. Cancer* **2011**, *11*, 849–864. [CrossRef] [PubMed]

176. Berndsen, R.H.; Abdul, U.K.; Weiss, A.; Zoetemelk, M.; Te Winkel, M.T.; Dyson, P.J.; Griffioen, A.W.; Nowak-Sliwinska, P. Epigenetic approach for angiostatic therapy: Promising combinations for cancer treatment. *Angiogenesis* **2017**, *20*, 245–267. [CrossRef] [PubMed]

177. Speckmann, B.; Grune, T. Epigenetic effects of selenium and their implications for health. *Epigenetics* **2015**, *10*, 179–190. [CrossRef] [PubMed]

178. Jablonska, E.; Reszka, E. Selenium and epigenetics in cancer: Focus on DNA methylation. *Adv. Cancer Res.* **2017**, *136*, 193–234.

179. Xiang, N.; Zhao, R.; Song, G.; Zhong, W. Selenite reactivates silenced genes by modifying DNA methylation and histones in prostate cancer cells. *Carcinogenesis* **2008**, *29*, 2175–2181. [CrossRef]

180. de Miranda, J.X.; Andrade Fde, O.; Conti, A.; Dagli, M.L.; Moreno, F.S.; Ong, T.P. Effects of selenium compounds on proliferation and epigenetic marks of breast cancer cells. *J. Trace Elem. Med. Biol.* **2014**, *28*, 486–491. [CrossRef]

181. Hu, C.; Liu, M.; Zhang, W.; Xu, Q.; Ma, K.; Chen, L.; Wang, Z.; He, S.; Zhu, H.; Xu, N. Upregulation of KLF4 by methylseleninic acid in human esophageal squamous cell carcinoma cells: Modification of histone H3 acetylation through HAT/HDAC interplay. *Mol. Carcinog.* **2015**, *54*, 1051–1059. [CrossRef]

182. Wu, J.C.; Wang, F.Z.; Tsai, M.L.; Lo, C.Y.; Badmaev, V.; Ho, C.T.; Wang, Y.J.; Pan, M.H. *Se*-Allylselenocysteine induces autophagy by modulating the AMPK/mTOR signaling pathway and epigenetic regulation of PCDH17 in human colorectal adenocarcinoma cells. *Mol. Nutr. Food Res.* **2015**, *59*, 2511–2522. [CrossRef] [PubMed]

183. Silvers, A.L.; Lin, L.; Bass, A.J.; Chen, G.; Wang, Z.; Thomas, D.G.; Lin, J.; Giordano, T.J.; Orringer, M.B.; Beer, D.G.; et al. Decreased selenium-binding protein 1 in esophageal adenocarcinoma results from posttranscriptional and epigenetic regulation and affects chemosensitivity. *Clin. Cancer Res.* **2010**, *16*, 2009–2021. [CrossRef] [PubMed]

184. Khalkar, P.; Ali, H.A.; Codo, P.; Argelich, N.D.; Martikainen, A.; Arzenani, M.K.; Lehmann, S.; Walfridsson, J.; Ungerstedt, J.; Fernandes, A.P. Selenite and methylseleninic acid epigenetically affects distinct gene sets in myeloid leukemia: A genome wide epigenetic analysis. *Free Radic. Biol. Med.* **2018**, *117*, 247–257. [CrossRef] [PubMed]

185. Hosnedlova, B.; Kepinska, M.; Skalickova, S.; Fernandez, C.; Ruttkay-Nedecky, B.; Peng, Q.; Baron, M.; Melcova, M.; Opatrilova, R.; Zidkova, J.; et al. Nano-selenium and its nanomedicine applications: A critical review. *Int. J. Nanomed.* **2018**, *13*, 2107–2128. [CrossRef] [PubMed]

186. Sakr, T.M.; Korany, M.; Katti, K.V. Selenium nanomaterials in biomedicine—An overview of new opportunities in nanomedicine of selenium. *J. Drug Deliv. Sci. Tec.* **2018**, *46*, 223–233. [CrossRef]

International Journal of
Molecular Sciences

MDPI

Article

Methylseleninic Acid Induces Lipid Peroxidation and Radiation Sensitivity in Head and Neck Cancer Cells

John T. Lafin [1], Ehab H. Sarsour [2], Amanda L. Kalen [2], Brett A. Wagner [2], Garry R. Buettner [2] and Prabhat C. Goswami [2,*]

[1] Department of Urology, University of Texas Southwestern Medical Center, Dallas, TX 75235, USA; john.lafin@utsouthwestern.edu
[2] The University of Iowa Free Radical and Radiation Biology Program, Department of Radiation Oncology, University of Iowa, Iowa City, IA 52242, USA; ehab-sarsour@uiowa.edu (E.H.S.); amanda-kalen@uiowa.edu (A.L.K.); brett-wagner@uiowa.edu (B.A.W.); garry-buettner@uiowa.edu (G.R.B.)
* Correspondence: prabhat-goswami@uiowa.edu; Tel.: +1-319-384-4666

Received: 10 November 2018; Accepted: 4 January 2019; Published: 8 January 2019

Abstract: Combination radiation and chemotherapy are commonly used to treat locoregionally advanced head and neck squamous cell carcinoma (HNSCC). Aggressive dosing of these therapies is significantly hampered by side effects due to normal tissue toxicity. Selenium represents an adjuvant that selectively sensitizes cancer cells to these treatments modalities, potentially by inducing lipid peroxidation (LPO). This study investigated whether one such selenium compound, methylseleninic acid (MSA), induces LPO and radiation sensitivity in HNSCC cells. Results from 4,4-difluoro-4-bora-3a,4a-diaza-*S*-indacene (BODIPY) C11 oxidation and ferric thiocyanate assays revealed that MSA induced LPO in cells rapidly and persistently. Propidium iodide (PI) exclusion assay found that MSA was more toxic to cancer cells than other related selenium compounds; this toxicity was abrogated by treatment with α-tocopherol, an LPO inhibitor. MSA exhibited no toxicity to normal fibroblasts at similar doses. MSA also sensitized HNSCC cells to radiation as determined by clonogenic assay. Intracellular glutathione in cancer cells was depleted following MSA treatment, and supplementation of the intracellular glutathione pool with *N*-acetylcysteine sensitized cells to MSA. The addition of MSA to a cell-free solution of glutathione resulted in an increase in oxygen consumption, which was abrogated by catalase, suggesting the formation of H_2O_2. Results from this study identify MSA as an inducer of LPO, and reveal its capability to sensitize HNSCC to radiation. MSA may represent a potent adjuvant to radiation therapy in HNSCC.

Keywords: head and neck cancer; selenium; methylseleninic acid; radiation; lipid peroxidation; glutathione; tocopherol

---

1. Introduction

Head and neck squamous cell carcinoma (HNSCC) is a diverse group of cancers that originate from the mouth, nose, throat, or other nearby areas. Over 50,000 new cases of HNSCC are anticipated to arise in the US in 2018; the five-year survival rate is ~64% [1]. Locoregionally advanced HNSCC is often treated with combination radio- and chemotherapy. However, side effects from primary therapy can be debilitating. Radiation and chemotherapy can result in oral mucositis, which can significantly reduce tolerable doses [2]. Even with these aggressive therapy options, about 40% of HNSCC deaths will occur due to the development of therapy resistance [3]. Additional options to sensitize HNSCC cells to current therapies are sorely needed.

Selenium administration shows great promise as a sensitizer to radio- and chemotherapy. Sodium selenite, an inorganic selenium derivative, induces toxicity and radiation sensitization in various cancer cell types with limited effects on normal fibroblasts [4–6]. A recent metastudy reported that selenite supplementation in patients receiving radiation therapy reduced deleterious side effects with no protective effects noted in tumors, supporting the use of selenium as an adjuvant to therapy [7]. These studies and others suggest that selenium may sensitize tumors to intervention, while potentially protecting normal tissue. Unfortunately, sodium selenite exhibits toxicity at relatively low doses, with a reported maximum tolerable dose of 10.2 mg m^{-2} [8]. Organoselenium derivatives, such as selenomethionine (SLM) and methylselenocysteine (MSC), are much less toxic than their inorganic counterparts while maintaining the selective effects noted with selenite [9,10]. Organoselenium derivatives exert their anticancer activities through the formation of a common active metabolite, methylselenol [11,12]. SLM and MSC require the action of specific lyase enzymes, such as methionine gamma-lyase (MGL), to release methylselenol [13]. MGL expression is reported to decline in a number of cancer types, suggesting that the formation of methylselenol in tumor tissue by SLM may be slow, limiting its efficacy [14–17].

Methylseleninic acid (MSA) is an organoselenium derivative that generates methylselenol through its spontaneous reaction with free thiols, such as glutathione [18]. Because the activity of MSA is not reliant upon the expression of lyases, such as MGL, it may represent a more effective antitumor agent than other organoselenium compounds. Previous studies have reported that MSA more effectively reduces TM2H and TM12 hyperplastic mammary cell accumulation than MSC, even at 10-fold lower doses [12]. Additionally, oral administration of MSA reduced the size of PC-3 xenografts in mice by approximately 40%, while administration of SLM, MSC, or sodium selenite exhibited no effects, indicating that MSA is more effective both in vitro and in vivo than other selenium compounds [19]. A combination of MSA and paclitaxel reduced the size of MDA-MB-231 xenografts in mice by about 50%, compared to paclitaxel alone, suggesting that MSA may be an effective adjuvant to current therapies [20]. In these studies, no change in body weight was observed, suggesting that MSA was well tolerated.

Although methylselenol has been identified as the active antitumor metabolite of organoselenium compounds, the mechanism of action following its generation is poorly understood. The combination of MSA and glutathione has been demonstrated to increase lucigenin-based chemiluminescence, which was abrogated by the presence of superoxide dismutase, suggesting the generation of superoxide $(O_2^{\bullet-})$ [18]. Superoxide may be protonated to form hydroperoxyl radical $(HO_2^{\bullet})$ or dismutated to form hydrogen peroxide (H_2O_2), both of which may contribute to the initiation of lipid peroxidation (LPO) [21,22]. Selenium administration has been associated with elevated markers of LPO, suggesting that these initiating effects occur in vivo [23–25]. Because end products of LPO can be highly toxic, MSA-generated methylselenol could, therefore, exhibit toxicity through the superoxide-mediated initiation of LPO. Furthermore, HNSCC patients exhibit higher levels of plasma markers of LPO than matched healthy subjects, suggesting that HNSCC may be particularly susceptible to LPO [26,27].

Results presented herein reveal that MSA exhibits toxicity and radiation sensitization of Cal27 and SCC25 HNSCC cells. Cal27 cells were found to be much more sensitive to MSA compared to SLM or MSC, while normal human fibroblasts were resistant to MSA-induced toxicity. Initiation of oxidative distress via lipid peroxidation appears to be the underlying mechanism for toxicity. Our data suggest that the toxic effects of MSA are mediated by a glutathione-dependent formation of an initiator of LPO. MSA may be a useful adjuvant to radiation therapy.

2. Results

2.1. MSA is More Toxic to HNSCC Cells than Other Organoselenium Derivatives, and Causes Cell Death in a Dose- and Time-Dependent Manner

Organoselenium derivatives SLM and MSC require enzymatic action to generate methylselenol, the common active metabolite. Therefore, methylselenol generation from SLM and MSC will be less in situations where these enzymes are poorly expressed. MSA requires no enzymatic activity to generate methylselenol [18]. To determine if MSA exhibits greater toxicity than other organoselenium compounds, Cal27 cells were treated with MSA, MSC, or SLM for 72 h, and viability was assessed with propidium iodide (PI) exclusion assay. MSC and SLM exhibited no toxicity at doses up to 10 µM, Figure 1A. Treatment with 1 µM MSA resulted in a small but significant increase in PI positive cells, while 10 µM MSA resulted in about 30% PI-positive cells. To ensure that MSA toxicity was not cell-line dependent, the effects of MSA treatment on SCC25 cells were also examined. A dose-dependent increase in PI positive SCC25 cells was observed with MSA treatment, with approximately 30% of the cells staining positive at 10 µM, Figure 1B. These results indicate that Cal27 cells are more sensitive to MSA than other organoselenium derivatives and that MSA-induced toxicity is dose-dependent in both Cal27 and SCC25 cell lines.

The sensitizing effects of organoselenium compounds on cancer cells have been reported to occur as early as 24 h following the beginning of treatment [28]. To determine the temporal aspects of the toxicity resulting from exposure to MSA, Cal27 and SCC25 cells were treated with MSA for varying durations and the toxicity was examined by measuring changes in cell numbers as well as flow cytometry measurements of the percentage of PI-positive (non-viable) and PI-negative (viable) cell populations. SCC25 and Cal27 cells both showed a marked decline in cell number as early as 24 h after initiation of treatment with MSA, with a reduction in cell number of about 75% and 95% in Cal27 and SCC25, respectively, Figure 1C. Cal27 cell numbers continued to decline up to 72 h, while SCC25 cell number appeared to begin to recover at 72 h. The rapid onset of a reduction in cell number correlated with an increase in the percentage of PI-positive Cal 27 cells at 48 h of treatment, Figure 1D. SCC25 cells exhibited significant toxicity as early as 24 h. SCC25 maximal toxicity (45% PI-positive cells) was reached by 48 h in the period examined, while Cal27 reached similar levels at 72 h. Together, these results indicate that the MSA treatment exhibits greater toxicity to HNSCC than treatments with MSC and SLM and that this toxicity is dose- and time-dependent. Furthermore, treatment with MSA appears to be more toxic to SCC25 compared to Cal27 cells.

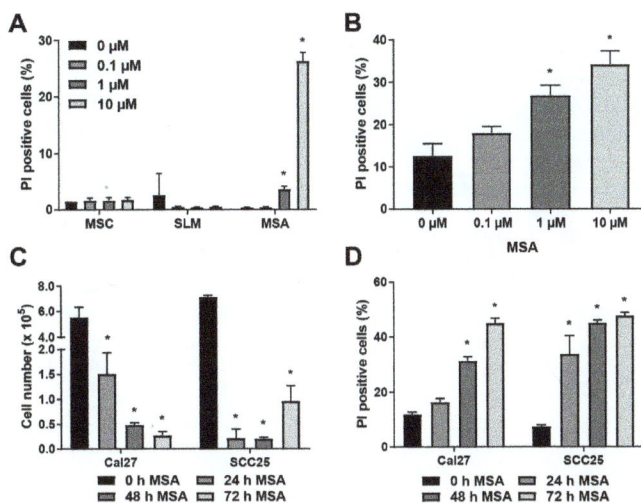

Figure 1. Methylseleninic acid (MSA) is toxic to Cal27 and SCC25 HNSCC cells in a dose- and time-dependent manner. (**A**) PI exclusion assay of Cal27 cells treated with the shown concentrations of Se-methylselenocysteine (MSC), seleno-L-methionine (SLM), or MSA for 72 h. (**B**) Propidium iodide (PI) exclusion assay of SCC25 cells treated with 0 to 10 µM MSA for 72 h. (**C**) Cell counts of Cal27 and SCC25 cells following treatment with 10 µM MSA for 0 to 72 h. (**D**) PI exclusion assay of Cal27 and SCC25 cells after treatment with 10 µM MSA for 0 to 72 h. *, statistical significance relative to 0 µM MSA controls; $p < 0.05$, $n = 3$.

2.2. MSA Treatment Sensitizes HNSCC Cells to Radiation

Selenium compounds, such as sodium selenite and seleno-L-methionine, sensitize cancer cells to radiation [4,5,10,29]. Furthermore, this sensitization is frequently noted to be selective for cancer cells [29]. Fibroblasts are often thought to make up the majority of the non-cancer cellular fraction in the tumor stroma [30,31]. To determine if normal human fibroblasts (NHF) were resistant to MSA toxicity, a PI exclusion assay was utilized. PI-positive (non-viable) NHF population did not increase following MSA treatment, Figure 2A. MSA (1 µM) treatment more than doubled non-viable Cal27 and SCC25 populations, Figure 1A,B, demonstrating the selective effects of MSA to HNSCC over NHF. To determine if MSA sensitizes HNSCC to radiation, Cal27 cells were treated with MSA for 48 h before 2 or 4 Gy irradiation, and toxicity was analyzed by using a clonogenic assay. Irradiated cells without MSA treatment showed a surviving fraction of 0.75 and 0.28 at 2 and 4 Gy, respectively, Figure 2B. Treatment with 0.1 µM MSA did not significantly alter surviving fraction of Cal27 cells: 0.66 and 0.22 at 2 and 4 Gy, respectively. Interestingly, prior treatment with 1 µM MSA significantly reduced the surviving fraction to 0.3 and 0.03 at 2 and 4 Gy compared to a surviving fraction of 0.75 and 0.28 without MSA treatment.

Radiation response is frequently dependent upon the support of the tumor stroma. To determine if the tumor stroma impacts the ability of MSA to sensitize Cal27 cells to radiation, a co-culture clonogenic assay was utilized. Cal27 cells were plated on lawns of quiescent normal human fibroblasts (NHF), and co-cultures were treated with 1 µM MSA for 48 h before irradiation. Even with NHF present, MSA treatment resulted in a 40% decline of surviving fraction following 2 Gy radiation, Figure 2D. Additionally, the lawn of NHF was not disturbed by MSA, further indicating that MSA was not toxic to NHF even in combination with radiation, Figure 2C. These results indicate that MSA treatment potently and selectively sensitizes Cal27 cells to radiation in co-cultures of NHF.

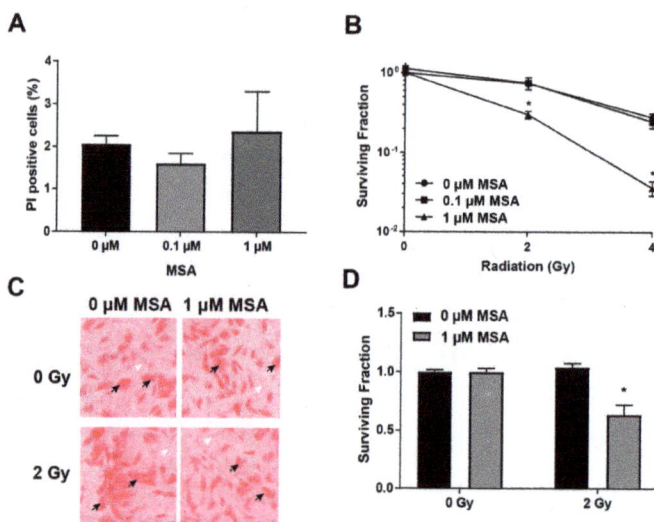

Figure 2. MSA selectively sensitizes head and neck squamous cell carcinoma (HNSCC) cells to radiation. (**A**) PI exclusion assay of normal human fibroblasts (NHF) treated with MSA 24 h. (**B**) Clonogenic assay of Cal27 cells treated with MSA 48 h before irradiation with γ-rays. (**C**) Representative images of Cal27 cells in co-cultures with NHF that were treated with MSA 48 h before irradiation with γ-rays. Black arrows: Cal27 colonies; white arrows: quiescent NHF. (**D**) Quantitation of Cal27 clonogenic survival in co-cultures of Cal27 and NHF that were treated with MSA 48 h before irradiation with γ-rays. *, statistical significance relative to 0 µM MSA controls; $p < 0.05$, $n = 3$.

2.3. MSA Treatment Induces Lipid Peroxidation in HNSCC Cells

Organoselenium compounds are theorized to be metabolized through a multitude of pathways to a central active metabolite, methylselenol, which exerts toxicity. Due to the highly reactive nature of methylselenol, studies concerning its mechanism of toxicity are sorely lacking. However, markers of lipid peroxidation have been found to rise in patients treated with selenium [24], suggesting that high dose selenium may induce lipid peroxidation. To determine if MSA treatment induces lipid peroxidation in HNSCC cells, MSA-treated Cal27 cells were labeled with the dye BODIPY C-11. This dye integrates into membranes and emits maximally at 590 nm. Upon oxidation by an initiator or propagator of lipid peroxidation, the maximal emission shifts to 510 nm. By reading both channels simultaneously, a ratio of oxidized to reduced dye can be calculated, providing a snapshot of lipid peroxidation initiation and propagation. Lipid peroxidation was found to be up to 30% elevated in Cal27 cells treated with MSA for 72 h, Figure 3A. An elevation was noted at a dose as low as 0.1 μM MSA, suggesting a powerful potential for initiation. Examination of lipid peroxidation at very early time intervals indicated that lipid peroxidation was initiated as early as 2 h, and maintained at a relatively stable level up to 72 h, Figure 3B. To determine if these increases in initiation and propagation events resulted in elevated lipid hydroperoxides, the Cayman Lipid Hydroperoxide Assay kit was utilized. Cal27 cells treated with 10 μM MSA for 72 h were found to have 1.16 fmol lipid hydroperoxide per cell, nearly 40 times as much as untreated cells, Figure 3C. These data indicate that treatment with MSA induces lipid peroxidation potently and persistently and that this induction results in a significant accumulation of lipid hydroperoxides.

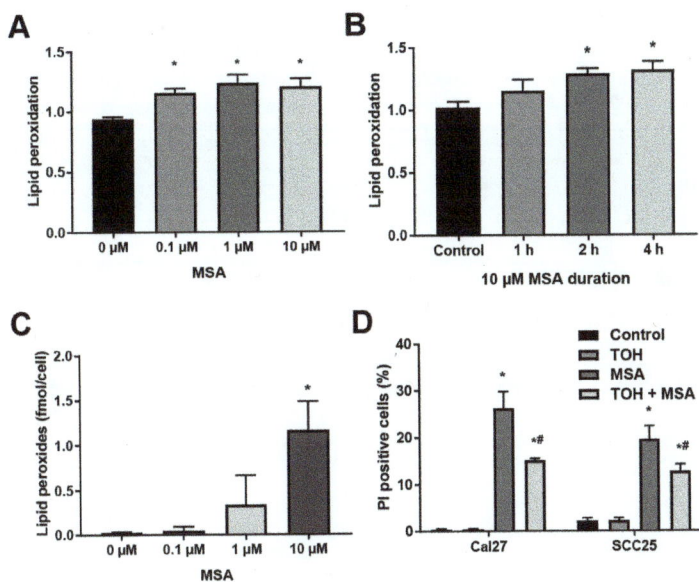

Figure 3. MSA induces lipid peroxidation in HNSCC cells. (**A**) Lipid peroxidation in Cal27 cells assessed by 4,4-difluoro-4-bora-3a,4a-diaza-S-indacene (BODIPY) C-11 staining following 72 h treatment with 0 to 10 μM MSA. (**B**) Lipid peroxidation in Cal27 cells following treatment with 10 μM MSA. (**C**) Lipid peroxides in Cal27 cells as assessed by the Cayman Chemical LPO Kit following treatment with 0 to 10 μM MSA for 72 h. (**D**) PI exclusion assay of Cal27 and SCC25 cells treated with 20 μM α-tocopherol acetate (TOH) for 24 h, 10 μM MSA for 72 h, or pre-treatment with TOH for 24 h followed by treatment with MSA. *, statistical significance relative to 0 μM MSA controls; #, statistical significance relative to MSA alone; $p < 0.05$, $n = 3$.

Lipid peroxidation is a deleterious oxidative chain reaction that can form toxic products, such as MDA or 4-HNE. Induction of uncontrolled lipid peroxidation can damage critical biomolecules resulting in cell death. To determine if MSA-induced toxicity is caused by lipid peroxidation, Cal27 and SCC25 cells were pre-treated with a lipid peroxidation chain terminator, α-tocopherol acetate (TOH), before MSA treatment. Treatment with TOH alone did not impact the percentage of PI-positive cell populations in either cell line, Figure 3D. However, treatment with TOH before MSA treatment reduced the percentage of the PI-positive cell population in Cal27 from 26% to 15%, and in SCC25 from 20% to 12%, an approximate 40% decline in the PI-positive populations. These results show that lipid peroxidation is an essential step in MSA-induced toxicity of HNSCC cells.

2.4. N-Acetyl-L-Cysteine Exacerbates MSA Toxicity in HNSCC

MSA spontaneously reacts with glutathione (GSH) to form its active metabolite, methylselenol, and GSH disulfide (GSSG), see below [18]. As the principal intracellular redox buffer, GSH is critical to normal cellular function. GSSG may be cytotoxic; it can be recycled to GSH by glutathione reductase, exported from the cell, or form mixed protein disulfides. To determine if MSA treatment influences GSH levels in HNSCC cells, the intracellular GSH content was measured in MSA-treated Cal27 cells using a biochemical assay [32]. Results indicate a dose-dependent decrease of total GSH in MSA-treated Cal27 cells, Figure 4A. Untreated Cal27 cells exhibited a total GSH concentration of about 17 nmol (mg protein)$^{-1}$. Treatment with MSA lowered this to below 10 nmol (mg protein)$^{-1}$, a 40% decline. Despite this marked decline, intracellular GSSG was not found to increase following MSA treatment, but rather also declined, Figure 4B. The decline in GSH levels suggests that GSH may have a significant role in the MSA-induced toxicity of HNSCC. This premise is further supported by results showing N-Acetyl-L-Cysteine (NAC) treatments exacerbating MSA-induced toxicity in Cal27 cells, Figure 4C. NAC is a membrane permeable precursor to GSH, stimulating its production; its effects are detectable within 4 h [33,34]. Cal27 cells were treated with 5 mM NAC for 24 h, washed, and treated with 10 μM MSA for 72 h. Cell number declined from 3×10^5 in cultures treated with MSA alone, to 0.5×10^5 cells, Figure 4C. A PI exclusion assay revealed that the combination of NAC and MSA resulted in a 30% non-viable cell population, while MSA alone resulted in only 25%, Figure 4D. Although NAC alone reduced cell number from 11×10^5 to 0.6×10^5, it had no effect on viability. This is consistent with prior reports from our lab that NAC treatment induces a cell cycle arrest [33,34]. These results indicate that GSH facilitates MSA-induced free radical chemistry (see below), leading to lipid peroxidation and HNSCC cytotoxicity.

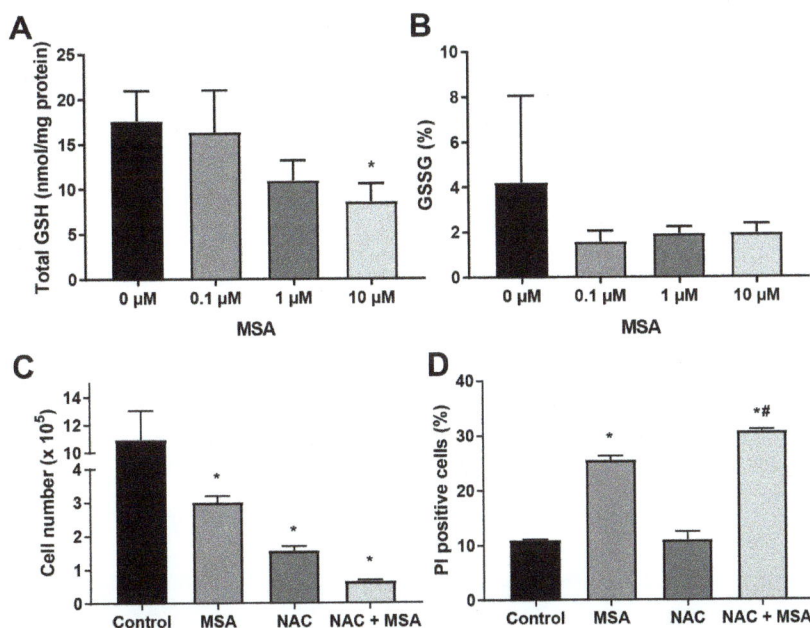

Figure 4. MSA exhibits toxicity in a glutathione (GSH)-dependent manner. (**A**) Total GSH in Cal27 cells treated with 0 to 10 μM MSA for 72 h. (**B**) Percent of GSH existing as glutathione disulfide (GSSG) in Cal27 cells treated with 0 to 10 μM MSA for 72 h. (**C**) Cell counts of Cal27 cells following treatment with 5 mM N-Acetyl-L-Cysteine (NAC) for 24 h, and/or 10 μM MSA for 72 h. The NAC + MSA group received 5 mM NAC for 24 h before treatment with MSA. (**D**) PI exclusion assay of Cal27 cells treated with 5 mM NAC for 24 h, and/or 10 μM MSA for 72 h. The NAC + MSA group received 5 mM NAC for 24 h before treatment with MSA. *, statistical significance relative to 0 μM MSA controls; #, statistical significance relative to NAC alone; $p < 0.05$, $n = 3$.

2.5. MSA Treatment Enhances GSH-Dependent O_2 Consumption

MSA can be reduced by free thiols, such as GSH, to form its active metabolite, methylselenol, Figure 5C. Because the pK_a of the selenohydryl group is 5.2, it primarily exists in biological systems as its conjugate base, the highly reactive methylselenolate anion (MeSe$^-$) [35]. This reactive species may initiate a cyclic reaction with molecular oxygen and GSH to cycle between a methylselenyl radical intermediate and methylselenolate anion, forming $O_2^{\bullet-}$ (which is rapidly dismuted to H_2O_2) and GSH to GSSG as products. This cyclic reaction is anticipated to account for the toxicity of organoselenium compounds [18,36]. In support of this chemistry, a previous study demonstrated the involvement of oxygen to the cyclic reaction between a related compound, selenocystamine, and GSH [36]. To determine if MSA may also undergo a similar reaction, O_2 consumption was monitored in a cell-free system containing MSA and GSH, Figure 5A. This system was held at pH 9.2, as this is the reported optimum pH for selenium-catalyzed $O_2^{\bullet-}$ generation [18]. O_2 was observed to disappear from the buffer at a rate of approximately 3 nM s^{-1}. Addition of MSA to buffer did not change the rate of consumption of O_2 (data not shown). Addition of GSH in the absence of MSA resulted in the disappearance of O_2 at a rate of approximately 15 nM s^{-1}, Figure 5B. Interestingly, the addition of MSA (250 μM) following GSH doubled the rate of O_2 consumption to approximately 32 nM s^{-1}. Furthermore, the addition of 500 IU catalase caused the rate of O_2 consumption to return to 13 nM s^{-1}, approximately the same as GSH alone. These results indicate an O_2-dependent reaction occurring between MSA and GSH that may result in the formation, but not accumulation, of H_2O_2. H_2O_2

may also contribute to another O_2-dependent reaction between GSH and MSA. The resulting flux of H_2O_2 may facilitate lipid peroxidation through the iron-dependent generation of $HO^\bullet$, resulting in MSA-induced toxicity of HNSCC. Previous reports indicate elevated markers of LPO in HNSCC patients [26,27], suggesting that MSA-induced selective cytotoxicity may be due to higher baseline levels of LPO in HNSCC compared to normal cells.

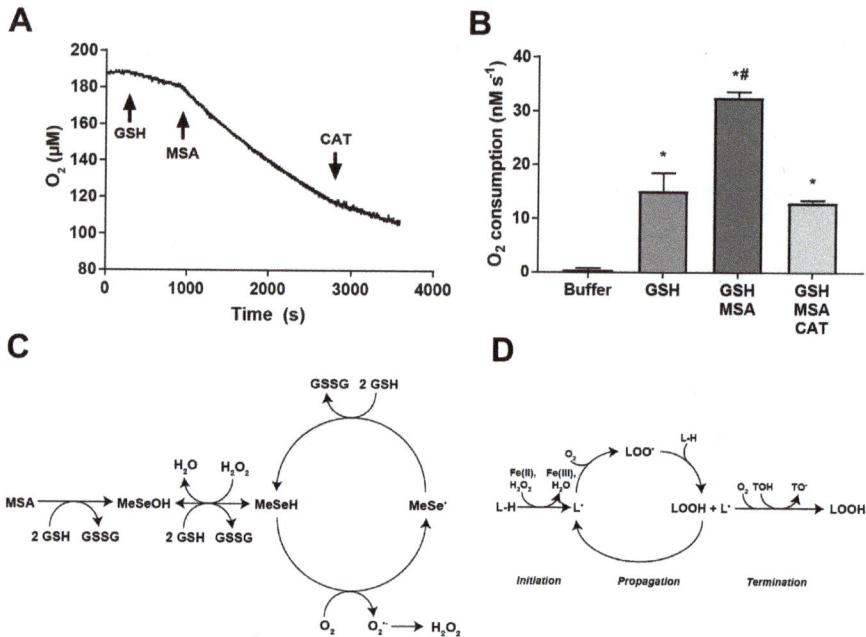

Figure 5. MSA enhances the consumption of O_2 by glutathione. (**A**) Representative trace of O_2 concentration in 3.00 mL pH 9.2 50 mM borate buffer at 37 °C. GSH and MSA added to a final concentration of 2 mM and 250 μM, respectively, at indicated time points. (**B**) Rate of O_2 consumption in 3.00 mL pH 9.2, 50 mM borate buffer at 37 °C. (**C**) Schematic of metabolism of MSA. MSA is reduced by GSH to methylselenol (MeSeH) through a methylselenenic acid (MeSeOH) intermediate. MeSeH may cycle with O_2 and GSH through a methylselenyl radical intermediate (MeSe$^\bullet$) to generate H_2O_2, potentially through an $O_2^{\bullet-}$ intermediate. MeSeH may also exhibit peroxidase activity, consuming H_2O_2. (**D**) Schematic of the process of lipid peroxidation. The process is initiated by abstraction of a hydrogen atom from a lipid, forming a carbon-centered lipid radical (L$^\bullet$). The reaction is propagated by the addition of O_2, followed by abstraction of another hydrogen atom from a neighboring lipid, forming LOOH and a new L$^\bullet$. The chain can be terminated by a donor antioxidant, such as tocopherol (TOH). The resulting tocopheroxyl radical (TO$^\bullet$) radical does not efficiently further oxidize lipids. *, statistical significance relative to buffer alone; #, statistical significance to GSH alone; $p < 0.05$, $n = 3$.

3. Discussion

Results presented here show that MSA treatment results in toxicity and enhanced radiation sensitivity in HNSCC cells and that this toxicity may be facilitated by glutathione and oxygen-mediated reactions resulting in toxicity. BODIPY C-11 oxidation data indicate that lipid peroxidation is associated with this oxidative process, can be detected within 2 h of the initiation of MSA treatment, and persists for at least 72 h. Lipid hydroperoxides were detected in a dose-dependent manner following treatment with MSA, consistent with a role for lipid peroxidation and corroborating BODIPY C-11 oxidation experiments. In a cell-free system, MSA doubled the rate of GSH-dependent O_2 consumption.

Consistent with results from the cell-free experiments, intracellular GSH levels decline in Cal27 cells following exposure to MSA. Supplementation of the intracellular GSH pool by pre-treatment with NAC further sensitized, rather than protected, the cells from MSA-induced toxicity. These results suggest that MSA reacts with intracellular GSH, yielding reactive oxygen species capable of inducing lipid peroxidation and cell death.

MSA exhibited significant toxicity to Cal27 and SCC25 HNSCC cells, Figure 1. Similar organoselenium compounds SLM and MSC did not show any toxicity at doses up to 10 µM, while 10 µM MSA resulted in a nearly 30% PI-positive (non-viable) population of cells, Figure 1A. The active metabolite of all three of the examined compounds is methylselenol [11,12]. SLM and MSC release methylselenol following processing by lyase enzymes, such as MGL [13]. Several reports have identified MGL as a tumor suppressor gene, suggesting that its expression is reduced in tumors [14–17]. Overexpression of MGL in ovarian cancer cells resulted in up to 1000-fold sensitization to SLM [37]. Furthermore, hepatoma xenografts treated with the combination of adenovirus-delivered MGL and SLM (1 µmol d^{-1}, IP) exhibited a drastic decline in tumor size compared to SLM alone, indicating the necessity of MGL for SLM to generate methylselenol [37]. MSA generates methylselenol by a direct and spontaneous reaction with GSH, obviating the need for MGL, Figure 5C [13,18]. Previous studies have identified MSA as more effective than MSC at inducing apoptosis and inhibiting cell growth in murine mammary cell cultures [12]. MSA also more effectively reduced prostate cancer xenograft size than SLM or MSC with no change in body weight [19]. Our results indicate that HNSCC is also more sensitive to MSA than SLM or MSC. MSA was also found to render Cal27 HNSCC cells sensitive to radiation, Figure 2B. Although other selenium compounds have been reported to induce sensitivity of cancer cells to radiation, the ability of MSA to do so has not yet been reported [4,10]. The increased toxicity of MSA compared to other selenium compounds suggests that it may also more effectively sensitize cancer cells to radiation. Additionally, the cytotoxic and sensitizing effects appear to be selective to cancer cells, as NHF were relatively resistant to MSA-induced toxicity, Figure 2A, and quiescent lawns of NHF were undisturbed by a combination of MSA and radiation, Figure 2C.

MSA treatment increased BODIPY C11 oxidation in Cal27 cells, Figure 3A,B. BODIPY C11 may be oxidized by LPO initiators, such as hydroperoxyl radical, and propagators, such as lipid peroxyl radical [38]. The dye is insensitive to LOOH and aldehydic end products of LPO, such as malondialdehyde (MDA) or 4-hydroxynonenal (4-HNE). A dose-dependent increase of BODIPY C11 oxidation was not observed at 72 h of MSA treatment. However, MSA treatment did result in a dose-dependent increase in the accumulation of LOOH, as determined by the Cayman LPO Kit, Figure 3C. Additionally, BODIPY C11 dye oxidation was found to stabilize as early as 2 h of MSA treatment, Figure 3B. These results suggest that MSA induces LPO rapidly and persistently and that this results in an accumulation of toxic lipid hydroperoxides. Furthermore, LPO appears essential to the toxicity of MSA, as pre-treatment with α-tocopherol acetate, an inhibitor of LPO, protected the cells from MSA-induced toxicity, Figure 3D. Previous studies have mainly focused on selenium administration as an inhibitor of LPO, presumably due to the induction of the glutathione peroxidase system [39]. However, evidence exists to suggest that selenium may initiate rather than inhibit LPO. Pre-treatment with sodium selenite (single dose 2 mg kg^{-1}, IP) in a murine model of heavy metal poisoning found a 4-fold increase of liver MDA content over metal alone [23]. Administration of ebselen, a synthetic organoselenium compound, increased liver MDA content in rats by approximately 20% [25]. Serum MDA levels were elevated in ovarian cancer patients following the administration of 50 µg of selenium as selenized yeast (4.8 µM compared to 3.9 µM) [24]. These reports and others suggest that in some circumstances, selenium administration may induce LPO, although the mechanisms are yet unclear. Because MSA more effectively generates methylselenol, the active metabolite, than similar selenium compounds, it may also more potently induce LPO.

MSA treatment of Cal27 cells resulted in a depletion of total intracellular GSH content, Figure 4A,B. MSA has been reported to deplete GSH in A549 lung cancer and HepG2 hepatoma cells [40,41]. In HepG2 cells, a biphasic response was noted: A 10 µM treatment of MSA for 24 h caused intracellular

GSH to increase by about 75%, while 25 µM MSA resulted in an approximate 20% decline in GSH, with no change in GSSG. This biphasic response was not noted in Cal27 cells; total GSH content was unchanged or declined at all doses tested, Figure 3A. Furthermore, treatment with MSA (10 µM) for 24 h decreased cell numbers of Cal27 and SCC25; a significant increase in the non-viable population, Figure 1C,D. Apoptosis of HepG2 cells as detected by LDH was not noted at 10 µM MSA for 24 h, suggesting that HepG2 cells are more resistant to MSA than HNSCC [41]. This concentration coincides with elevated GSH in HepG2, which suggests that the biphasic response may play a role in resistance of MSA. Interestingly, despite a decline in intracellular GSH levels following MSA treatment, no change was noted in GSSG, Figure 4B. This may be due to increased GSSG efflux through membrane transporters, such as MRP1 [42]. GSH may also be consumed without GSSG formation through conjugation, suggesting that MSA treatment may induce increased GSH conjugation forming mixed disulfides. Supplementation of the intracellular thiol pool by pre-treatment with NAC sensitized Cal27 cells to MSA, Figure 4C,D. Similar results were reported with MSA treatment of HepG2 cells [41]. These results suggest that a reaction between MSA and a thiol, such as GSH, is essential for MSA-induced cytotoxicity.

Oxygen consumption by GSH was doubled in the presence of MSA in a cell-free system, Figure 5A,B. Furthermore, the addition of catalase to this system returned the rate of O_2 consumption to those similar to GSH alone. These results suggest the formation, but not accumulation, of H_2O_2. Had H_2O_2 accumulated, the addition of catalase would have returned O_2 to the system, i.e., an increase in the concentration of O_2 would have been observed. Selenol species may exhibit peroxidase activity, as evidenced by the active site selenol of selenium-containing glutathione peroxidases [36]. Methylselenol may, therefore, exhibit peroxidase activity, yielding a methylselenenic acid, which may again react with GSH to regenerate methylselenol, Figure 5C. Previous reports indicate the potential for the formation of $O_2^{\bullet-}$ in systems containing selenium and GSH. The addition of MSA to a solution of GSH is reported to increase lucigenin-based chemiluminescence [13,18]. This effect was abrogated by the presence of superoxide dismutase, suggesting the formation of $O_2^{\bullet-}$. Additionally, an examination of the reaction kinetics of GSH and selenocystamine, an organoselenium compound similar to MSA, suggests a cyclic reaction yielding superoxide [36]. Following $O_2^{\bullet-}$ generation, methylselenol may be regenerated by an additional reduction by GSH, yielding a cyclic reaction capable of generating large amounts of $O_2^{\bullet-}$ [18,36]. These reactions are summarized in Figure 5C. Following its formation, $O_2^{\bullet-}$ can dismute to H_2O_2, either spontaneously or through the action of superoxide dismutase [43]. H_2O_2 may contribute to LPO initiation through the generation of hydroxyl radical by iron-mediated Fenton chemistry [44,45]. Superoxide can also be protonated to form hydroperoxyl radical, which is a powerful LPO initiator [21,22]. The contribution of $O_2^{\bullet-}$ to selenium-based cytotoxicity has been further demonstrated by the protective effects of a superoxide dismutase mimetic [46]. Methylselenol may, therefore, initiate lipid peroxidation chain reactions through superoxide-mediated products.

Overall, results from this study indicate that MSA sensitizes HNSCC cell to radiation and exhibits toxicity through a GSH-dependent induction of LPO. LPO occurs more readily in cells with higher polyunsaturated fatty acid content [47]. Many types of cancer exhibit greater lipid content than their respective normal counterparts, including colon, prostate, pancreatic, and clear cell renal carcinoma [48–52]. MSA may, therefore, exhibit selective cytotoxicity in cancer cells on the basis of altered lipid content. A pre-clinical study examining the efficacy of MSA (4 mg kg^{-1} d^{-1}, PO) in treating prostate cancer xenografts found a nearly 40% reduction in tumor size following MSA treatment, with no change in body weight [19]. Similarly, MSA (4.5 mg kg^{-1} d^{-1}, PO) combined with paclitaxel (10 mg kg^{-1} week^{-1}, IP) reduced breast cancer xenograft tumor size with no change in body weight [20]. These studies suggest that MSA is selectively toxic to cancer cells. The work presented herein suggests that this selectivity may be in part due to a differential sensitivity to LPO in cancer compared to normal cells.

4. Materials and Methods

4.1. Cell Culture and Reagents

Head and neck squamous carcinoma cell lines Cal27 (tongue origin, CRL-2095) and SCC25 (tongue origin, CRL-1628) were purchased from ATCC (Manassas, VA, USA). Both lines have mutated p53, are epidermal growth factor receptor (EGFR) positive, and human papillomavirus (HPV) negative. Normal human fibroblasts (NHF) were obtained from the Coriell cell repository (AG01522D). Cells were cultured in Dulbecco's Modified Eagle's Medium (DMEM) (Thermo Fisher Scientific, Waltham, MA, USA), supplemented with antibiotics and 10% bovine calf serum (HNSCC cells, Thermo Fisher Scientific, Waltham, MA, USA) or 10% fetal bovine serum (NHF, Sigma-Aldrich, St. Louis, MO, USA). All cells were grown in humidified incubators set to 37 °C, 5% CO_2, and atmospheric oxygen.

Seleno-L-methionine (SLM, S3132), N-acetylcysteine (NAC, A9165), glutathione (GSH, G4251), glutathione disulfide (GSSG, G4376), glutathione reductase (GR, G3664), 2-vinylpyridine (2-VP, 132292), 5,5-dithio-bis-(2-nitrobenzoic acid) (DTNB, D8130), and 5-sulfosalicylic acid (SSA, S2130) were purchased from Sigma-Aldrich. Reduced nicotinamide adenine dinucleotide phosphate (NADPH) (481973) was purchased from EMD Millipore (Burlington, MA, USA). Methyl-Se-selenocysteine (MSC) and methylseleninic acid (MSA) were generous gifts of the laboratory of Youcef Rustum (Roswell Park Cancer Institute, Buffalo, NY, USA). BODIPY 581/591 C-11 (D3861) and CellTracker Green CMFDA (C7025) were purchased from Thermo Fisher Scientific (Waltham, MA, USA). Lipid Hydroperoxide (LPO) Assay Kit (705002) was purchased from Cayman Chemical Company (Ann Arbor, MI, USA).

4.2. Irradiation

Exponentially growing cells were irradiated at the Free Radical and Radiation Research Core Facility at The University of Iowa. All irradiated cells received a single dose of γ-rays from a cesium-137 irradiator (JL Shephard, San Fernando, CA, USA) at a dose rate of 0.65 Gy min^{-1}. Cell survival was measured using a clonogenic assay following a previously published method [53].

4.3. Propidium Iodide Exclusion Assay

Following treatment, cultures were trypsinized, washed, and resuspended in cold phosphate buffered saline (PBS). The suspended cultures were filtered and labeled with 1 μg mL^{-1} propidium iodide for 5 min on ice. Flow cytometry was completed on a Becton-Dickinson FACScan at the University of Iowa Flow Cytometry Core. Data from 10,000 events were collected in list mode. The population of PI-positive (non-viable) and negative (viable) cells were calculated with FlowJo software (FlowJo, LLC, Ashland, Oregon, USA).

4.4. BODIPY C-11 Assay

Following MSA treatment, adherent cells were washed and labeled with 5 μM BODIPY C-11 in DMEM lacking serum and antibiotics for 15 min at 37 °C. Following labeling, cultures were collected by trypsinization, washed, resuspended in cold PBS and filtered. Samples were read on a Becton-Dickinson LSR II flow cytometer using channels for Texas Red (reduced dye) and fluorescein isothiocyanate (FITC, oxidized dye) simultaneously at The University of Iowa Flow Cytometry Core. Populations were gated and analyzed with FlowJo software (version 7.6.5), and ratios of oxidized:reduced dye were calculated.

4.5. Total Lipid Hydroperoxide Determination

A Cayman Lipid Hydroperoxide Assay Kit was used to determine total lipid hydroperoxides in cell samples. Cal27 cells were treated with MSA for 72 h. Following treatment, cells were collected by trypsinization and counted. Total lipid extracts were obtained and analyzed as recommended by the manufacturer in glass cuvettes on a Beckman DU650 spectrophotometer. Lipid hydroperoxides per

Int. J. Mol. Sci. **2019**, *20*, 225

cell was quantified by construction of an appropriate standard curve and normalized to cell number, as determined by a Z1 Coulter Counter (Beckman-Coulter, Brea, CA, USA).

4.6. Glutathione Determination

Following treatment, cells were collected by trypsinization, washed, and pellets were lysed in ice cold 5% sulfosalicylic acid. Extracts were stored at $-80\ ^\circ$C until analysis. Following centrifugation, the supernatant was removed and used for the glutathione assay; protein precipitate was dissolved in 1% SDS, 0.1 M NaOH for protein determination. Total glutathione was determined as described previously on a Beckman DU-650 spectrophotometer [54]. Glutathione disulfide (GSSG) was determined using the method of Griffith and Anderson [32]. Rates of reaction were compared to glutathione or glutathione disulfide standard curves and normalized to protein content as determined by bicinchoninic acid (BCA) (Thermo Fisher Scientific, Manassas, VA, USA).

4.7. Oxygen Consumption

Involvement of oxygen in the reactions of glutathione and MSA was investigated by recording oxygen consumption during reaction progress in a cell-free system with an ESA BioStat Multi Electrode System and YSI Oxygen Probe (Yellow Springs Instrument Co., Yellow Springs, OH, USA) at the Free Radical and Radiation Research Core Facility at The University of Iowa. Measurements were conducted in 3.00 mL of 50 mM pH 9.2 borate buffer at 37 $^\circ$C. Initial oxygen concentration was assumed to be 188 μM, as reported previously [55]. The reactants glutathione, MSA, and catalase were introduced sequentially to final concentrations of 2 mM, 250 μM, and 167 IU mL^{-1}, respectively.

4.8. Statistical Analysis

Statistical analysis was completed using Prism (GraphPad Software, San Diego, CA, USA). Two-way ANOVA with post-hoc analysis was completed to determine statistical significance. Homogeneity of variance was assumed at a 95% confidence interval. Results from at least three biological replicates with $p < 0.05$ were considered significant.

Author Contributions: Conceptualization, P.C.G., G.R.B., J.T.L., E.H.S., B.A.W.; methodology, J.T.L., E.H.S., B.A.W., G.R.B.; validation, J.T.L. and A.L.K.; formal analysis, J.T.L., B.A.W., E.H.S.; investigation, J.T.L., A.L.K.; resources, P.C.G., G.R.B.; writing—original draft preparation, J.T.L.; writing—review and editing, P.C.G., G.R.B.; visualization, J.T.L.; supervision, P.C.G., G.R.B.; project administration, P.C.G.; funding acquisition, P.C.G.

Funding: This research was supported by funding from NIH 2R01 CA111365 (P.C.G.), R01 CA169046 (G.R.B.) and T32 CA078586.

Acknowledgments: Youcef Rustum provided methylseleninic acid (MSA) and methylselenocysteine (MSC) as a generous gift for the completion of these studies. We thank Justin Fishbaugh and The University of Iowa Flow Cytometry Core for their aid in completing the flow cytometry measurements. We also thank Michael McCormick and The University of Iowa Free Radical and Radiation Research Core Facility for their expertise in biochemical measurements and irradiation services.

Conflicts of Interest: The authors declare no conflict of interest.

References

1. Howlader, N.A.; Krapcho, M.; Miller, D.; Bishop, K.; Altekruse, S.F.; Kosary, C.L.; Yu, M.; Ruhl, J.; Tatalovich, Z.; Mariotto, A.; et al. *SEER Cancer Statistics Review, 1975–2013*; National Cancer Institute: Bethesda, MD, USA, 2016.
2. Orlandi, E.; Iacovelli, N.A.; Rancati, T.; Cicchetti, A.; Bossi, P.; Pignoli, E.; Bergamini, C.; Licitra, L.; Fallai, C.; Valdagni, R.; et al. Multivariable model for predicting acute oral mucositis during combined IMRT and chemotherapy for locally advanced nasopharyngeal cancer patients. *Oral Oncol.* **2018**, *86*, 266–272. [CrossRef]
3. Pulte, D.; Brenner, H. Changes in survival in head and neck cancers in the late 20th and early 21st century: A period analysis. *Oncologist* **2010**, *15*, 994–1001. [CrossRef]
4. Schueller, P.; Puettmann, S.; Micke, O.; Senner, V.; Schaefer, U.; Willich, N. Selenium influences the radiation sensitivity of C6 rat glioma cells. *Anticancer Res.* **2004**, *24*, 2913–2917.

5. Husbeck, B.; Peehl, D.M.; Knox, S.J. Redox modulation of human prostate carcinoma cells by selenite increases radiation-induced cell killing. *Free Radic. Biol. Med.* **2005**, *38*, 50–57. [CrossRef]
6. Eckers, J.C.; Kalen, A.L.; Xiao, W.; Sarsour, E.H.; Goswami, P.C. Selenoprotein P inhibits radiation-induced late reactive oxygen species accumulation and normal cell injury. *Int. J. Radiat. Oncol. Biol. Phys.* **2013**, *87*, 619–625. [CrossRef]
7. Puspitasari, I.M.; Abdulah, R.; Yamazaki, C.; Kameo, S.; Nakano, T.; Koyama, H. Updates on clinical studies of selenium supplementation in radiotherapy. *Radiat. Oncol.* **2014**, *9*, 125. [CrossRef]
8. Brodin, O.; Eksborg, S.; Wallenberg, M.; Asker-Hagelberg, C.; Larsen, E.H.; Mohlkert, D.; Lenneby-Helleday, C.; Jacobsson, H.; Linder, S.; Misra, S.; et al. Pharmacokinetics and Toxicity of Sodium Selenite in the Treatment of Patients with Carcinoma in a Phase I Clinical Trial: The SECAR Study. *Nutrients* **2015**, *7*, 4978–4994. [CrossRef]
9. Ammar, E.M.; Couri, D. Acute toxicity of sodium selenite and selenomethionine in mice after ICV or IV administration. *Neurotoxicology* **1981**, *2*, 383–386.
10. Shin, S.H.; Yoon, M.J.; Kim, M.; Kim, J.I.; Lee, S.J.; Lee, Y.S.; Bae, S. Enhanced lung cancer cell killing by the combination of selenium and ionizing radiation. *Oncol. Rep.* **2007**, *17*, 209–216. [CrossRef]
11. Ip, C. Lessons from basic research in selenium and cancer prevention. *J. Nutr.* **1998**, *128*, 1845–1854. [CrossRef]
12. Ip, C.; Thompson, H.J.; Zhu, Z.; Ganther, H.E. In vitro and in vivo studies of methylseleninic acid: Evidence that a monomethylated selenium metabolite is critical for cancer chemoprevention. *Cancer Res.* **2000**, *60*, 2882–2886.
13. Spallholz, J.E.; Palace, V.P.; Reid, T.W. Methioninase and selenomethionine but not Se-methylselenocysteine generate methylselenol and superoxide in an in vitro chemiluminescent assay: Implications for the nutritional carcinostatic activity of selenoamino acids. *Biochem. Pharmacol.* **2004**, *67*, 547–554. [CrossRef]
14. Tan, Y.; Sun, X.; Xu, M.; Tan, X.; Sasson, A.; Rashidi, B.; Han, Q.; Tan, X.; Wang, X.; An, Z.; et al. Efficacy of recombinant methioninase in combination with cisplatin on human colon tumors in nude mice. *Clin. Cancer Res. Off. J. Am. Assoc. Cancer Res.* **1999**, *5*, 2157–2163.
15. Yoshioka, T.; Wada, T.; Uchida, N.; Maki, H.; Yoshida, H.; Ide, N.; Kasai, H.; Hojo, K.; Shono, K.; Maekawa, R.; et al. Anticancer efficacy in vivo and in vitro, synergy with 5-fluorouracil, and safety of recombinant methioninase. *Cancer Res.* **1998**, *58*, 2583–2587.
16. Kokkinakis, D.M.; Hoffman, R.M.; Frenkel, E.P.; Wick, J.B.; Han, Q.; Xu, M.; Tan, Y.; Schold, S.C. Synergy between methionine stress and chemotherapy in the treatment of brain tumor xenografts in athymic mice. *Cancer Res.* **2001**, *61*, 4017–4023.
17. Hu, J.; Cheung, N.K. Methionine depletion with recombinant methioninase: In vitro and in vivo efficacy against neuroblastoma and its synergism with chemotherapeutic drugs. *Int. J. Cancer* **2009**, *124*, 1700–1706. [CrossRef]
18. Spallholz, J.E.; Shriver, B.J.; Reid, T.W. Dimethyldiselenide and methylseleninic acid generate superoxide in an in vitro chemiluminescence assay in the presence of glutathione: Implications for the anticarcinogenic activity of L-selenomethionine and L-Se-methylselenocysteine. *Nutr. Cancer* **2001**, *40*, 34–41. [CrossRef]
19. Li, G.-X.; Lee, H.-J.; Wang, Z.; Hu, H.; Liao, J.D.; Watts, J.C.; Combs, G.F.; Lü, J. Superior in vivo inhibitory efficacy of methylseleninic acid against human prostate cancer over selenomethionine or selenite. *Carcinogenesis* **2008**, *29*, 1005–1012. [CrossRef]
20. Qi, Y.; Fu, X.; Xiong, Z.; Zhang, H.; Hill, S.M.; Rowan, B.G.; Dong, Y. Methylseleninic acid enhances paclitaxel efficacy for the treatment of triple-negative breast cancer. *PLoS ONE* **2012**, *7*, e31539. [CrossRef]
21. Aikens, J.; Dix, T.A. Perhydroxyl radical (HOO.) initiated lipid peroxidation. The role of fatty acid hydroperoxides. *J. Biol. Chem.* **1991**, *266*, 15091–15098.
22. Bielski, B.H.; Arudi, R.L.; Sutherland, M.W. A study of the reactivity of HO2/O2- with unsaturated fatty acids. *J. Biol. Chem.* **1983**, *258*, 4759–4761.
23. Rungby, J. Silver-induced lipid peroxidation in mice: Interactions with selenium and nickel. *Toxicology* **1987**, *45*, 135–142. [CrossRef]
24. Sieja, K.; Talerczyk, M. Selenium as an element in the treatment of ovarian cancer in women receiving chemotherapy. *Gynecol. Oncol.* **2004**, *93*, 320–327. [CrossRef]
25. Farina, M.; Soares, F.A.; Zeni, G.; Souza, D.O.; Rocha, J.B. Additive pro-oxidative effects of methylmercury and ebselen in liver from suckling rat pups. *Toxicol. Lett.* **2004**, *146*, 227–235. [CrossRef]

26. Rasheed, M.H.; Beevi, S.S.; Geetha, A. Enhanced lipid peroxidation and nitric oxide products with deranged antioxidant status in patients with head and neck squamous cell carcinoma. *Oral Oncol.* **2007**, *43*, 333–338. [CrossRef]

27. Gupta, A.; Bhatt, M.L.; Misra, M.K. Lipid peroxidation and antioxidant status in head and neck squamous cell carcinoma patients. *Oxid. Med. Cell. Longev.* **2009**, *2*, 68–72. [CrossRef]

28. Chintala, S.; Toth, K.; Cao, S.; Durrani, F.A.; Vaughan, M.M.; Jensen, R.L.; Rustum, Y.M. Se-methylselenocysteine sensitizes hypoxic tumor cells to irinotecan by targeting hypoxia-inducible factor 1alpha. *Cancer Chemother. Pharmacol.* **2010**, *66*, 899–911. [CrossRef]

29. Büntzel, J.; Micke, O.; Glatzel, M.; Schafer, U.; Riesenbeck, D.; Kisters, K.; Bruns, F.; Schänekaes, K.G.; Dawczynski, H.; Mücke, R. Selenium substitution during radiotherapy in head and neck cancer. *Trace Elem. Electrolytes* **2010**, *27*, 235–239. [CrossRef]

30. Orimo, A.; Weinberg, R.A. Stromal fibroblasts in cancer: A novel tumor-promoting cell type. *Cell Cycle* **2006**, *5*, 1597–1601. [CrossRef]

31. Arina, A.; Idel, C.; Hyjek, E.M.; Alegre, M.L.; Wang, Y.; Bindokas, V.P.; Weichselbaum, R.R.; Schreiber, H. Tumor-associated fibroblasts predominantly come from local and not circulating precursors. *Proc. Natl. Acad. Sci. USA* **2016**, *113*, 7551–7556. [CrossRef]

32. Griffith, O.W. Determination of glutathione and glutathione disulfide using glutathione reductase and 2-vinylpyridine. *Anal. Biochem.* **1980**, *106*, 207–212. [CrossRef]

33. Menon, S.G.; Sarsour, E.H.; Spitz, D.R.; Higashikubo, R.; Sturm, M.; Zhang, H.; Goswami, P.C. Redox regulation of the G1 to S phase transition in the mouse embryo fibroblast cell cycle. *Cancer Res.* **2003**, *63*, 2109–2117.

34. Menon, S.G.; Sarsour, E.H.; Kalen, A.L.; Venkataraman, S.; Hitchler, M.J.; Domann, F.E.; Oberley, L.W.; Goswami, P.C. Superoxide signaling mediates *N*-acetyl-L-cysteine-induced G_1 arrest: Regulatory role of cyclin D1 and manganese superoxide dismutase. *Cancer Res.* **2007**, *67*, 6392–6399. [CrossRef]

35. Huber, R.E.; Criddle, R.S. Comparison of the chemical properties of selenocysteine and selenocystine with their sulfur analogs. *Arch. Biochem. Biophys.* **1967**, *122*, 164–173. [CrossRef]

36. Chaudiere, J.; Courtin, O.; Leclaire, J. Glutathione oxidase activity of selenocystamine: A mechanistic study. *Arch. Biochem. Biophys.* **1992**, *296*, 328–336. [CrossRef]

37. Miki, K.; Xu, M.; Gupta, A.; Ba, Y.; Tan, Y.; Al-Refaie, W.; Bouvet, M.; Makuuchi, M.; Moossa, A.R.; Hoffman, R.M. Methioninase cancer gene therapy with selenomethionine as suicide prodrug substrate. *Cancer Res.* **2001**, *61*, 6805–6810.

38. Drummen, G.P.; van Liebergen, L.C.; Op den Kamp, J.A.; Post, J.A. C11-BODIPY(581/591), an oxidation-sensitive fluorescent lipid peroxidation probe: (Micro)spectroscopic characterization and validation of methodology. *Free Radic. Biol. Med.* **2002**, *33*, 473–490. [CrossRef]

39. Rotruck, J.T.; Pope, A.L.; Ganther, H.E.; Swanson, A.B.; Hafeman, D.G.; Hoekstra, W.G. Selenium: Biochemical Role as a Component of Glutathione Peroxidase. *Science* **1973**, *179*, 588–590. [CrossRef]

40. Okuno, T.; Honda, E.; Arakawa, T.; Ogino, H.; Ueno, H. Glutathione-dependent cell cycle G1 arrest and apoptosis induction in human lung cancer A549 cells caused by methylseleninic acid: Comparison with sodium selenite. *Biol. Pharm. Bull.* **2014**, *37*, 1831–1837. [CrossRef]

41. Shen, H.M.; Ding, W.X.; Ong, C.N. Intracellular glutathione is a cofactor in methylseleninic acid-induced apoptotic cell death of human hepatoma HEPG(2) cells. *Free Radic. Biol. Med.* **2002**, *33*, 552–561. [CrossRef]

42. Gordillo, G.M.; Biswas, A.; Khanna, S.; Spieldenner, J.M.; Pan, X.; Sen, C.K. Multidrug Resistance-associated Protein-1 (MRP-1)-dependent Glutathione Disulfide (GSSG) Efflux as a Critical Survival Factor for Oxidant-enriched Tumorigenic Endothelial Cells. *J. Biol. Chem.* **2016**, *291*, 10089–10103. [CrossRef]

43. McCord, J.M.; Fridovich, I. Superoxide dismutase. An enzymic function for erythrocuprein (hemocuprein). *J. Biol. Chem.* **1969**, *244*, 6049–6055.

44. Liochev, S.I.; Fridovich, I. The role of $O_2{}^{\cdot-}$ in the production of $HO^{\cdot}$: In vitro and in vivo. *Free Radic. Biol. Med.* **1994**, *16*, 29–33. [CrossRef]

45. Agil, A.; Fuller, C.J.; Jialal, I. Susceptibility of plasma to ferrous iron/hydrogen peroxide-mediated oxidation: Demonstration of a possible Fenton reaction. *Clin. Chem.* **1995**, *41*, 220–225.

46. Zhong, W.; Oberley, T.D. Redox-mediated effects of selenium on apoptosis and cell cycle in the LNCaP human prostate cancer cell line. *Cancer Res.* **2001**, *61*, 7071–7078.

47. Wagner, B.A.; Buettner, G.R.; Burns, C.P. Free radical-mediated lipid peroxidation in cells: Oxidizability is a function of cell lipid bis-allylic hydrogen content. *Biochemistry* **1994**, *33*, 4449–4453. [CrossRef]

48. Accioly, M.T.; Pacheco, P.; Maya-Monteiro, C.M.; Carrossini, N.; Robbs, B.K.; Oliveira, S.S.; Kaufmann, C.; Morgado-Diaz, J.A.; Bozza, P.T.; Viola, J.P. Lipid bodies are reservoirs of cyclooxygenase-2 and sites of prostaglandin-E2 synthesis in colon cancer cells. *Cancer Res.* **2008**, *68*, 1732–1740. [CrossRef]

49. Yue, S.; Li, J.; Lee, S.Y.; Lee, H.J.; Shao, T.; Song, B.; Cheng, L.; Masterson, T.A.; Liu, X.; Ratliff, T.L.; et al. Cholesteryl ester accumulation induced by PTEN loss and PI3K/AKT activation underlies human prostate cancer aggressiveness. *Cell Metab.* **2014**, *19*, 393–406. [CrossRef]

50. Guillaumond, F.; Bidaut, G.; Ouaissi, M.; Servais, S.; Gouirand, V.; Olivares, O.; Lac, S.; Borge, L.; Roques, J.; Gayet, O.; et al. Cholesterol uptake disruption, in association with chemotherapy, is a promising combined metabolic therapy for pancreatic adenocarcinoma. *Proc. Natl. Acad. Sci. USA* **2015**, *112*, 2473–2478. [CrossRef]

51. Toth, K.; Peter, I.; Kremmer, T.; Sugar, J. Lipid-rich cell thyroid adenoma: Histopathology with comparative lipid analysis. *Virchows Arch. APathol. Anat. Histopathol.* **1990**, *417*, 273–276. [CrossRef]

52. Ericsson, J.L.E.; Seljelid, R.; Orrenius, S. Comparative light and electron microscopic observations of the cytoplasmic matrix in renal carcinomas. *Virchows Archiv für Pathologische Anatomie und Physiologie und für Klinische Medizin* **1966**, *341*, 204–223. [CrossRef]

53. Kalen, A.L.; Sarsour, E.H.; Venkataraman, S.; Goswami, P.C. Mn-superoxide dismutase overexpression enhances G2 accumulation and radioresistance in human oral squamous carcinoma cells. *Antioxid. Redox Signal.* **2006**, *8*, 1273–1281. [CrossRef]

54. Tietze, F. Enzymic method for quantitative determination of nanogram amounts of total and oxidized glutathione: Applications to mammalian blood and other tissues. *Anal. Biochem.* **1969**, *27*, 502–522. [CrossRef]

55. Wagner, B.A.; Venkataraman, S.; Buettner, G.R. The rate of oxygen utilization by cells. *Free Radic. Biol. Med.* **2011**, *51*, 700–712. [CrossRef]

International Journal of
Molecular Sciences

MDPI

Article

Identification of a Novel Quinoxaline-Isoselenourea Targeting the STAT3 Pathway as a Potential Melanoma Therapeutic

Verónica Alcolea [1,2], Deepkamal N. Karelia [3,4], Manoj K. Pandey [3,4], Daniel Plano [1,2], Parvesh Singh [5], Juan Antonio Palop [1,2], Shantu Amin [3,4], Carmen Sanmartín [1,2] and Arun K. Sharma [3,4,*]

[1] Department of Pharmaceutical Technology and Chemistry, School of Pharmacy and Nutrition, University of Navarra, Irunlarrea 1, E-31008 Pamplona, Spain; valcolea@alumni.unav.es (V.A.); dplano@unav.es (D.P.); jpalop@unav.es (J.A.P.); sanmartin@unav.es (C.S.)
[2] Instituto de Investigación Sanitaria de Navarra (IdiSNA), Irunlarrea 3, E-31008 Pamplona, Spain
[3] Department of Pharmacology, Penn State Cancer Institute, CH72, Penn State College of Medicine, 500 University Drive, Hershey, PA 17033, USA; dkarelia@pennstatehealth.psu.edu (D.N.K.); pandey@rowan.edu (M.K.P.); sga3@psu.edu (S.A.)
[4] Department of Biomedical Sciences, Cooper Medical School of Rowan University, Camden, NJ 08103, USA
[5] School of Chemistry and Physics, University of Kwa-Zulu Natal (UKZN), Westville Campus, Durban 4000, South Africa; singhp4@ukzn.ac.za
* Correspondence: aks14@psu.edu; Tel.: +1-717-531-4563

Received: 3 January 2019; Accepted: 24 January 2019; Published: 26 January 2019

Abstract: The prognosis for patients with metastatic melanoma remains very poor. Constitutive signal transducer and activator of transcription 3 (STAT3) activation has been correlated to metastasis, poor patient survival, larger tumor size, and acquired resistance against vemurafenib (PLX-4032), suggesting its potential as a molecular target. We recently designed a series of isoseleno- and isothio-urea derivatives of several biologically active heterocyclic scaffolds. The cytotoxic effects of lead isoseleno- and isothio-urea derivatives (compounds 1 and 3) were studied in a panel of five melanoma cell lines, including B-RAF^{V600E}-mutant and wild-type (WT) cells. Compound 1 (IC$_{50}$ range 0.8–3.8 µM) showed lower IC$_{50}$ values than compound 3 (IC$_{50}$ range 8.1–38.7 µM) and the mutant B-RAF specific inhibitor PLX-4032 (IC$_{50}$ ranging from 0.4 to >50 µM), especially at a short treatment time (24 h). These effects were long-lasting, since melanoma cells did not recover their proliferative potential after 14 days of treatment. In addition, we confirmed that compound 1 induced cell death by apoptosis using Live-and-Dead, Annexin V, and Caspase3/7 apoptosis assays. Furthermore, compound 1 reduced the protein levels of STAT3 and its phosphorylation, as well as decreased the expression of STAT3-regulated genes involved in metastasis and survival, such as survivin and c-myc. Compound 1 also upregulated the cell cycle inhibitor p21. Docking studies further revealed the favorable binding of compound 1 with the SH2 domain of STAT3, suggesting it acts through STAT3 inhibition. Taken together, our results suggest that compound 1 induces apoptosis by means of the inhibition of the STAT3 pathway, non-specifically targeting both B-RAF-mutant and WT melanoma cells, with much higher cytotoxicity than the current therapeutic drug PLX-4032.

Keywords: selenium; isoselenourea; melanoma; STAT3; apoptosis

1. Introduction

Melanoma is originated from the malignant transformation of melanocytes. Its incidence has been rapidly increasing, with a 5-fold increase in the last three decades [1]. The American Cancer Society's estimates for melanoma in the United States for 2018 are: About 91,270 new melanomas will be diagnosed (about 55,150 in men and 36,120 in women) and about 9320 people are expected to die of melanoma (about 5990 men and 3330 women). Nearly 1,330,300 new cases of melanoma and about 126,000 deaths are estimated to occur worldwide in 2018. The regions more affected are those with white population, highest incidences being in Australia, New Zealand, Northern America, and Northern and Western Europe [2]. If diagnosed in early stages, melanoma can be easily removed by surgical excision but in metastatic form, it is one of the most aggressive malignancies with low survival rates [3].

The mitogen-activated protein kinase (MAPK) signaling cascade is a key pathway in melanoma survival and proliferation. Constitutive activation of B-RAF kinase due to B-RAF^{V600E} mutation occurs in nearly 50% of cutaneous melanomas [4]. Moreover, constitutive activation of signal transducer and activator of transcription 3 (STAT3) has been implicated in many oncogenic features, as it regulates the transcription of a variety of genes involved in cell proliferation, apoptosis, angiogenesis, and metastasis. The constitutive activation of STAT3 has been reported in melanoma patients [5]. The high expression of STAT3 has been correlated to large tumor diameter and depth, lymph node metastasis, high expression of MMP-2 and -9, and poor patient survival [6–8].

Traditional treatments for melanoma have shown low response rates and are associated with severe adverse effects. The introduction of agents that specifically target the MAPK pathway, such as vemurafenib (PLX-4032), dabrafenib, and trametinib, has significantly improved the overall survival of patients bearing the B-RAF^{V600E} mutation [9]. Although treatment with these inhibitors induces a quick initial response, the effect is not durable because of the rapid development of drug resistance [10]. The acquired resistance against vemurafenib is implicated in the activation of STAT3 and its signaling pathways [11]. Importantly, suppression of STAT3 activity disrupts B-RAF^{V600E}-mediated induction of anti-apoptotic proteins and reduces melanoma cell survival [12]. In addition, these treatments are ineffective in patients without the B-RAF^{V600E} mutation [10]. Therefore, there is an urgent need to develop novel agents for melanoma treatment.

In the last decade, selenium-containing compounds have emerged as promising anticancer agents due to their efficacy and selectivity [13]. Several studies, including those from our laboratories, have reported the role of different selenium-containing compounds in the prevention and treatment of melanoma, including its metastasis [14–20]. Of these, *Se,Se'*-1,4-phenylenebis(1,2-ethanediyl)bisisoselenourea (PBISe), an isoselenourea derivative, has been found to inhibit proliferation and promote apoptosis in melanoma cells in cell culture models [19]. It should be noted that PBISe was over 10-fold more effective than its isosteric sulfur analog *S,S'*-1,4-phenylenebis(1,2-ethanediyl)bisisothiourea (PBIT). Furthermore, the topical application of PBISe significantly delayed xenografted melanoma tumors growth [21].

In continuation of our pursuits towards developing novel organoselenium compounds as anticancer agents, we recently reported the synthesis and screening of a series of novel isoselenourea and corresponding isothiourea analogs [22]. This series consisted of new hybrid compounds containing an isoselenourea or an isothiourea group and different carbo- and hetero-cyclic scaffolds which have been traditionally included in anticancer agents [22]. Considering the encouraging results of PBISe in melanoma, we decided to evaluate the potential of the novel analogs as potential melanoma therapeutics. For this purpose, we selected an isoselenourea and an isothiourea derivative which showed the highest activity against a melanoma cell line (1205Lu) in our previous screening (compounds 1 and 3, respectively) [22]. The corresponding sulfur analog of compound 1, i.e., isothiourea compound 2, was also tested in order to determine if the compound selenium-containing 1 was more active than the sulfur compound 2 against melanoma cells, similar to what observed in the case of PBISe. This study examines the effects of these compounds on various melanoma cell lines and further evaluates the underlying mechanisms of action of compound 1.

2. Results

2.1. Synthesis of Compounds 1, 2, and 3

The structures of quinoxaline-2,3-diylbis(methylene)dicarbamimidoselenoate dihydrobromide (compound 1), its sulfur analog quinoxaline-2,3-diylbis(methylene)dicarbamimidothioate dihydrobromide (compound 2), and (9,10-dioxo-9,10-dihydroanthracen-2-yl)methyl carbamimidothioate hydrochloride (compound 3) are depicted in Figure 1A. These derivatives were synthesized as previously reported [22]. Briefly, compound 1 was synthesized by treating 2,3-bis(bromomethyl)quinoxaline with selenourea (molar ratio 1:1.1) in absolute ethanol at room temperature for 2 h, and compound 2 by treating 2,3-bis(bromomethyl)quinoxaline with thiourea in absolute ethanol for 3.5 h at reflux. Compound 3 was obtained from 2-(chloromethyl)anthraquinone and thiourea (1:1.1 molar ratio) in absolute ethanol, stirring for 3 h at reflux. The purity of compounds was ≥ 99%.

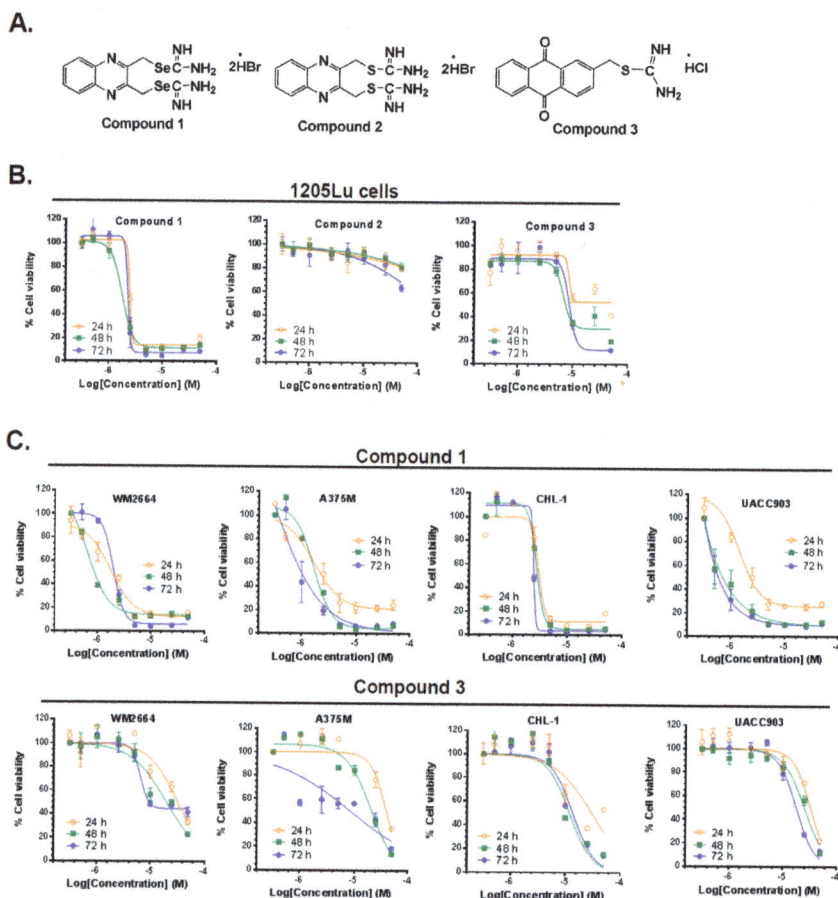

Figure 1. Compounds 1 and 3 were effective at reducing cell viability of different melanoma cancer cell lines. (**A**) Structures of the compounds, (**B**) Screening of the three agents in melanoma cells (1205Lu) for 24, 48, and 72 h using the MTT (cell viability) assay (**C**) Cell viability (MTT) results for compounds 1 and 3 in four different melanoma cancer cell lines at three different time points. Graphs were obtained by performing non-linear regression analysis using variable slope. Error bars represent mean ± SD.

2.2. Compounds 1 and 3 Reduced the Viability of Different Melanoma Cancer Cells

All the compounds were tested against a melanoma cell line (1205Lu) for their effect on cancer cell viability. The MTT (3-(4,5-dimethylthiazol-2-yl)-2,5-diphenyltetrazolium bromide) assay was performed in order to measure cancer cell viability, as previously described [22]. The anti-cancer effect of each agent was tested at seven different concentrations between 0.1 and 50 μM and three time points (24, 48, and 72 h). As shown in Figure 1B, compounds 1 and 3 potently inhibited the cancer cell viability of melanoma cells, while the isosteric sulfur analog (compound 2) of 1 was essentially ineffective up to the maximum concentration (50 μM) used. These results were in accordance with many of our previous reports [14,16,22–24] where isosteric replacement of sulfur by selenium in a small molecule significantly enhanced the anticancer activity. Therefore, compounds 1 and 3 were further screened against a panel of four melanoma cancer cell lines (WM2664, A375M, UACC903, and CHL-1) using the MTT assay. We compared the results with those obtained with PLX-4032, a B-RAF kinase inhibitor used currently in the clinic for melanoma treatment. The results are summarized in Table 1, expressed as IC_{50}, the concentration that produces 50 % of growth inhibition.

Table 1. Effects on cell viability of compounds 1 and 3 and PLX-4032 in five different melanoma cell lines. The results are expressed as mean values for IC_{50} with their respective standard deviation ($\pm$ SD).

Comp.	Time (h)	Cell Lines				
		WM2664	1205Lu	A375M	UACC903	CHL-1
1	24	1.8 ± 0.4	2.6 ± 0.3	2.6 ± 1.0	3.8 ± 1.6	2.9 ± 0.4
	48	1.1 ± 0.3	1.9 ± 0.3	1.9 ± 0.1	0.8 ± 0.2	3.1 ± 0.4
	72	2.0 ± 0.2	2.0 ± 0.6	1.1 ± 0.2	0.3 ± 0.6	2.5 ± 0.2
3	24	32.7 ± 3.6	33.7 ± 10.1	38.7 ± 1.0	32.6 ± 1.9	29.3 ± 10.6
	48	21.0 ± 2.9	11.3 ± 3.0	21.3 ± 0.8	24.7 ± 2.3	12.3 ± 2.2
	72	22.8 ± 7.0	8.3 ± 0.9	8.1 ± 4.2	22.3 ± 5.6	13.6 ± 1.7
PLX-4032	24	>50.0	>50.0	>50.0	38.5 ± 7.2	>50.0
	48	32.1 ± 14.2	31.4 ± 8.0	3.5 ± 1.1	10.3 ± 2.6	21.2 ± 3.4
	72	3.6 ± 1.5	9.3 ± 3.4	0.4 ± 0.3	17.9 ± 1.3	12.2 ± 1.5

Compound 1 was more cytotoxic across all cell lines compared to compound 3 and PLX-4032 (Figure 1, Table 1). Interestingly, while PLX-4032 was more effective when the B-RAF^{V600E} mutation was present, compounds 1 and 3 also reduced cell viability in the B-RAF wild-type cell line CHL-1, compound 1 being the most potent. It should be noted that PLX-4032 showed no effect at 24 h in four out of the five lines tested, whereas both compounds 1 and 3 produced a reduction of cell viability in the five cell lines at that time point (Figure 1, Table 1). Compound 1 exhibited an IC_{50} lower than 4 μM in all the cell lines tested at 24 h. Furthermore, this compound was more potent than the reference drug in all cell lines at the three time points, with the exception of A375M cell line after 72 h of treatment. Overall, the isoselenourea derivative (compound 1) exhibited much higher potency than the isothiourea analog (compound 3) and, therefore, we selected compound 1 for further in vitro efficacy and mechanism of action elucidation.

2.3. Compound 1 Suppressed the Proliferative Ability of Melanoma Cells (1205Lu and UACC903)

In order to evaluate the long-term effects of compound 1 on melanoma cells, a colony formation assay was employed. For this assay, the cells were treated with either compound 1 or dimethyl sulfoxide (DMSO, control) for 24 h, and then 500 live cells were counted and re-seeded in new plates in the absence of compound 1 or DMSO and allowed to form colonies over a period of 14 days. As shown in Figure 2, compound 1 effectively inhibited the colony formation ability of both 1205Lu and UACC903 melanoma cell lines. At the dose of 1 μM, the ability of melanoma cells to form colonies was reduced to less than 50%, and when cells were treated with 2.5 μM, this percentage was dramatically reduced at levels lower than 20% and 5% for 1205Lu and UACC903 cells, respectively. These studies clearly indicate that the effects of compound 1 on the inhibition of melanoma cells replication are long-lasting.

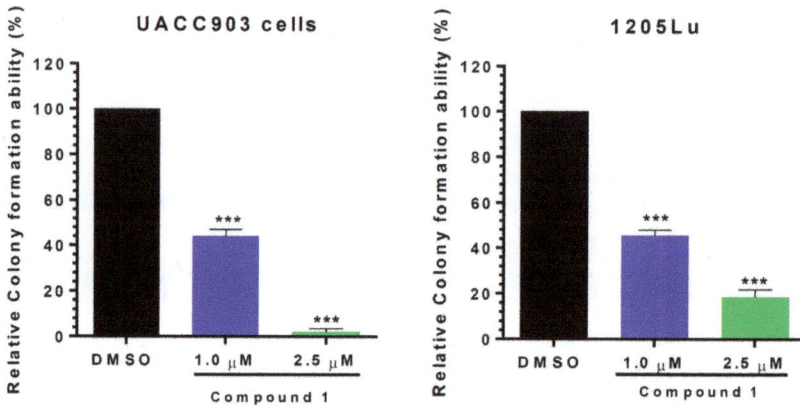

Figure 2. Compound 1 dramatically reduced the colony formation ability of melanoma cells. 1205Lu (**A**) and UACC903 (**B**) cells were treated with 1 or 2.5 µM of compound 1 or dimethyl sulfoxide (DMSO) (control) for 24 h. Subsequently, 500 cells were re-seeded with free compound 1 or DMSO medium for 14 days and stained with 0.5% alcoholic crystal violet. The results are expressed as mean ± SD. *** $p < 0.001$.

2.4. Compound 1 Increased Melanoma Cell Death in Vitro

In order to study whether the reduction of cell viability caused by compound 1 was due to cell death and not cell growth inhibition, 1205Lu cells were subjected to the Live-and-Dead assay. As shown in Figure 3, compound 1 increased the number of cells positive for ethidium homodimer staining (dead cells, upper left quadrant) and reduced the cells stained with calcein AM (live cells, lower right quadrant) compared to control cells. After treatment with 1 µM compound 1, no difference was observed between control and treated cells. However, when the dose of compound 1 was increased to 5 µM, the percentage of dead cells increased dramatically up to 25 %. These results suggest that compound 1 was able to induce cell death in vitro in melanoma cells.

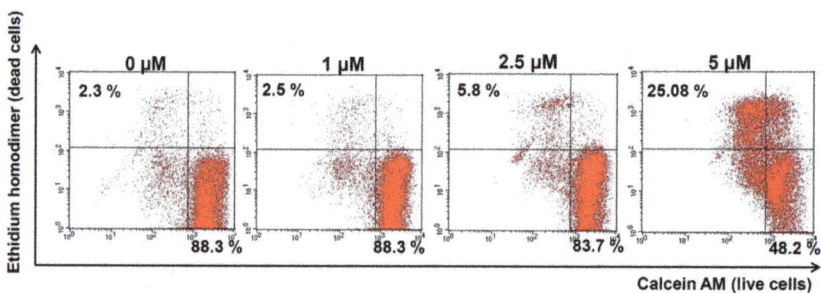

Figure 3. Compound 1 induced cell death in melanoma cells. 1205Lu cells were incubated with 1, 2.5, or 5 µM of compound 1 or DMSO (control) for 24 h and stained with ethidium homodimer and calcein AM. Live and dead cells were quantified by flow cytometry.

2.5. Compound 1 Induced Apoptosis in Melanoma Cells

With the aim of investigating whether the increase in cell death induced by compound 1 was due to apoptosis induction, the Muse™ Annexin V & Dead Cell assay was carried out. Annexin V was employed in this assay to detect the externalization of phosphatidylserine to the cell surface, a process occurring in apoptosis but not in necrosis [25]. A dead cell marker (7-ADD) was also included in the kit as an indicator of cell membrane structural integrity. Therefore, cells negative for both markers (lower left quadrant) were healthy cells, cells positive for Annexin V only (lower right quadrant) were in early apoptosis, and cells positive for both Annexin V and 7-ADD were undergoing apoptotic death (upper right quadrant). Cells positive for 7-ADD only were undergoing necrosis (upper left quadrant).

Compound 1 was tested at three concentrations: 1.75, 2.5, and 5 µM. The dose of 1 µM was not tested because we observed no significant effect at this dose in the previous assay. As shown in Figure 4A, after the treatment with compound 1 at 1.75 µM concentration, 15% of cells were in early apoptosis (lower right quadrant). At 5 µM of compound 1, less than 50% of cells were healthy cells and 25% of cells died by apoptosis (upper right quadrant). Less than 1% of cells died without externalization of phosphatidylserine (upper left quadrant), indicating that compound 1 induced cell death through apoptosis.

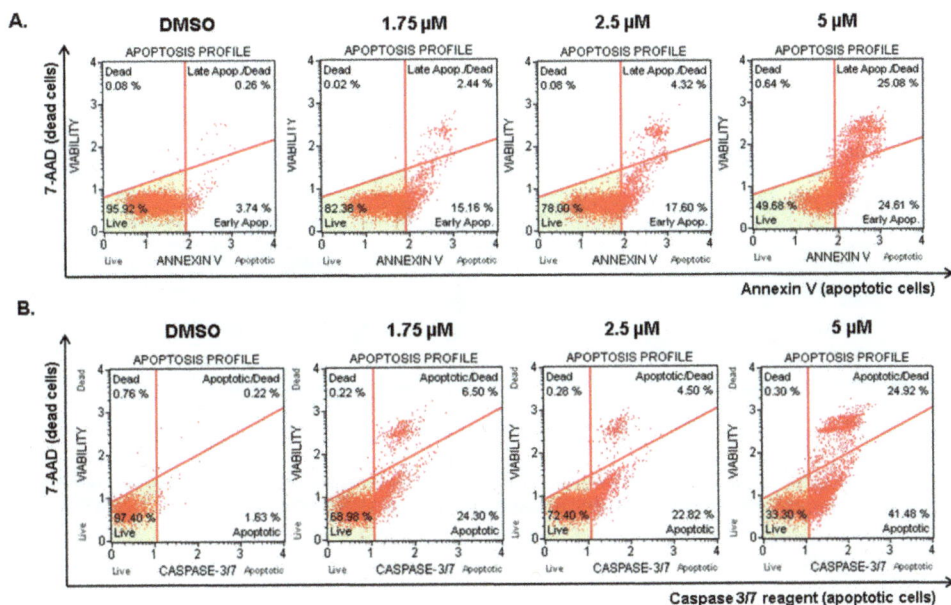

Figure 4. Compound 1 induced apoptotic cell death. After 24 h of incubation with the indicated concentration of compound 1 or DMSO (control), the apoptotic status of 1205Lu cells was analyzed using the Muse™ Annexin V & Dead Cell Kit according to the manufacturer's instructions. (**A**) Analogous independent experiments were analyzed with Muse™ Caspase 3/7 Kit to confirm the results. (**B**) The results of both experiments were analyzed by flow cytometry.

In order to confirm these results, the Muse™ Caspase-3/7 kit was also employed. The kit includes a reagent with a DNA binding dye. In non-apoptotic cells, this reagent is linked to an effector caspase recognition sequence which does not bind to DNA. However, when caspases are active, the dye is released by caspase cleavage and translocated to the nucleus, where it binds to DNA, producing high fluorescence. This kit also includes the dead cell marker 7-ADD. Hence, cells negative for both dyes were healthy cells (lower left quadrant), cells only positive for Caspase 3/7 reagent (lower right quadrant) were supposed to be in early apoptosis, cells positive for both dyes (upper right quadrant) were undergoing apoptotic death, and cells only positive for 7-ADD (upper left quadrant) were undergoing caspase-independent death. The results (Figure 4B) showed the same tendency observed in the Annexin V assay. The doses of 1.75 and 2.5 µM increased the amount of early apoptotic cells (lower right quadrant), and, after treatment with 5 µM compound 1, more than 60% of the cells were positive for caspase 3/7 activity (lower and upper right quadrants). Less than 0.5% of cells died through caspase-independent mechanisms. Overall, our results confirmed that compound 1 is a potent apoptosis inducer.

2.6. Compound 1 did not Inhibit the Phosphorylation of Akt and ERK1/2

The dysregulated expression of Akt and ERK1/2 is associated with cell proliferation and melanoma cell survival [26,27]. Hence, we decided to evaluate whether compound 1 affected the expression and phosphorylation status of these proteins. Interestingly, as shown in Figure 5A, compound 1 did not affect the expression or phosphorylation of Akt or ERK1/2. These observations suggest that the cell inhibitory response to compound 1 is not mediated through the inhibition of Akt and ERK1/2; therefore, other pathways might be implicated.

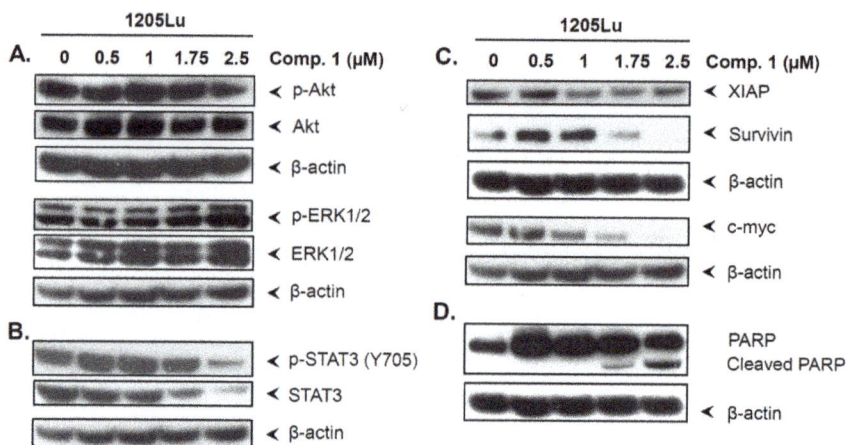

Figure 5. Compound 1 downregulated STAT3 and its downstream target proteins expression. 1205Lu cells were incubated with either DMSO (control) or compound 1 for 24 h. Whole cell lysates were subjected to western blot analysis. Expression of different proteins related to STAT3 and its downstream targets (**A–D**) were monitored. ß-actin was used as a loading control.

2.7. Compound 1 Inhibited STAT3 and Related Proteins Expression

To further examine the mechanism of action of compound 1, its effects on proteins implicated in cancer survival and metastasis were studied. STAT3 is highly upregulated in melanoma, and its activation by phosphorylation contributes to cancer progression and survival [12]. Hence, we decided to establish whether compound 1 affected the expression and phosphorylation status of the transcription factor STAT3. As shown in Figure 5B, compound 1 reduced protein levels and phosphorylation of STAT3. At the dose of 2.5 µM, the expression and phosphorylation of STAT3 were dramatically reduced.

STAT3 controls the expression of proteins involved in proliferation, survival, and metastasis formation, such as XIAP, survivin, and c-myc [28–30]. As observed in Figure 5C, compound 1 also downregulated these proteins at the doses at which STAT3 inhibition was observed. In addition, the inhibition of these anti-apoptotic proteins was accompanied by PARP (poly(ADP-ribose) polymerase) cleavage (Figure 5D), which is traditionally employed as a marker of apoptosis [31]. These results suggest that compound 1 has the ability to inhibit STAT3 signaling pathway. Moreover, the induction of apoptosis and cell proliferation inhibitory response of compound 1 may be mediated by STAT3 inhibition.

2.8. Compound 1 Induced the Cell Cycle Inhibitor p21

Recent studies have shown a correlation between the cell cycle inhibitor p21 and transcription factor STAT3 [32]. These studies revealed that p21 is part of a feedback network controlling the down-modulation of STAT activity [32]. Thus, we sought to investigate whether compound 1 activated the expression of this cell cycle inhibitor. As shown in Figure 6, compound 1 induced the expression of p21. These results suggest that the cell inhibitory activities of compound 1 may be mediated by activation of p21.

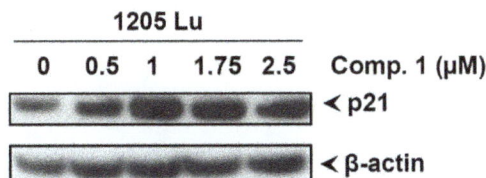

Figure 6. Compound 1 induced p21 expression. After 24 h of treatment with either compound 1 or DMSO (control), 1205Lu cells were collected, and the expression of p21 was analyzed by western blot analysis. ß-actin was used as a loading control.

2.9. Compound 1 Inhibited the Phosphorylation of STAT3, Increased the Expression of p21, and Induced Apoptotic Cell Death in Different Melanoma Cell Lines

The data above demonstrate that compound 1 effectively decreased p-STAT3 levels and increased the expression of p21 and apoptotic cell death in 1205LU cells. However, to rule out cell line-dependent effects of compound 1, we tested all the above biological activities of compound 1 in two additional melanoma cell lines (UACC903 and SK-Mel-8). As shown in Figure 7A, compound 1 inhibited cell proliferation of both melanoma cell lines, with UACC903 cells being more sensitive than SK-Mel-8. As suggested by our Annexin V and Caspase 3/7 activity assay (Figure 7B,C), compound 1 induced apoptotic cell death. Additionally, compound 1 also decreased the phosphorylation of STAT3 (Figure 7D,E) and induced p21 expression similar in both cell lines, to what observed in 1205Lu cells. Hence, the biological activities of compound 1 were consistent across a variety of human melanoma cell lines.

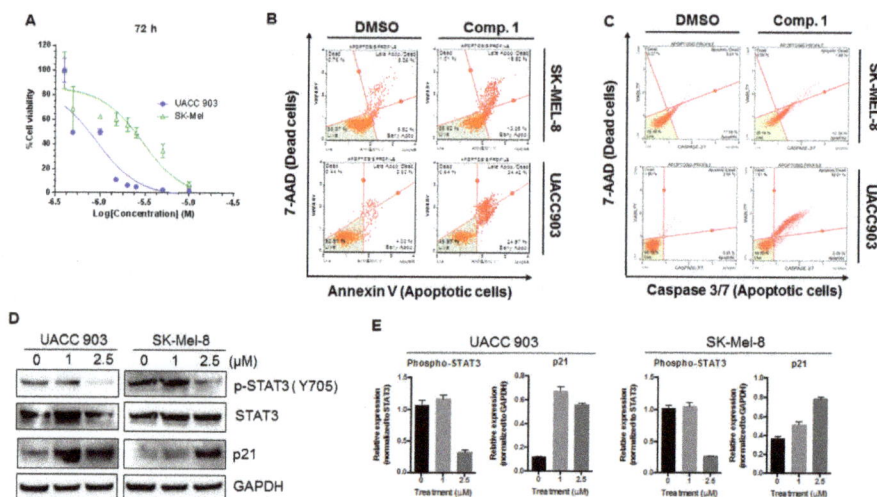

Figure 7. Compound 1 inhibited the phosphorylation of STAT3, increased the expression of p21, and induced apoptotic cell death in UACC 903 and Sk-Mel-8 melanoma cell lines. (**A**) Cell viability (MTT) results for compound 1 in two different melanoma cell lines at 72 h. The graphs were obtained by performing non-linear regression analysis using variable slope. Error bars represent mean ± SD. (**B**,**C**) Human melanoma cells UACC 903 and SK-Mel-8 were treated with compound 1 (2.5 μM) for 24 h, and an apoptotic assay was performed using the Muse[TM] Annexin V & Dead Cell (**B**) and Caspase 3/7 Kit (**C**) according to the manufacturer's instructions. (**D**) Melanoma cells (UACC 903 and SK-Mel-8) cells were treated with the mentioned concentrations of Compound 1. After 24 h, cells were collected and lysed, and the expression of p-STAT3, STAT3, and p21 was analyzed by western blot. GAPDH was used as a loading control; (**E**) Quantification of protein expression was performed using Image J software, and graphs represent the relative expressions of proteins. The relative expressions were determined using either STAT3 or GAPDH as mentioned.

2.10. Docking

In order to substantiate our experimental findings and to investigate the binding propensities of compound 1 to the SH2 domain of STAT3, docking simulations were employed. The molecular structure [33] of STAT3 (pdb id: 1BG1, resolution 2.25 Å) was retrieved from the protein data bank (www.rcsb.org). CDocker [34], a CHARMm force field-based algorithm embedded in DS version 4.0 (Accelrys; San Diego, CA, USA), was used to flexibly dock compound 1 within the conservative SH2 domain of STAT3. The docking analysis revealed favorable binding of compound 1 with STAT3, based on the computed scoring function (CDocker energy = −23.8 kcal/mol), where the most negative value of CDocker energy indicates good binding affinity of a ligand for a protein. The visualization of the complex (Figure 8A) further revealed that compound 1 penetrated deep inside the STAT3 cavity and settled well by establishing a network of hydrogen bonding and electrostatic forces with the protein. Specifically, compound 1 utilized its nitrogen atom (proton acceptor) of the quinoxaline moiety to form two strong concurrent hydrogen bonds (2.2 Å, 2.8 Å) with the amine (proton donor) functionalities of Arg609 (Figure 8B). This amino acid contributed significantly to STAT3 and SH2 peptide binding, as the mutation of Arg609 has been reported to abolish the peptide-binding ability of this domain [35,36]. Several anticancer agents targeting the same amino acid residue (Arg609) in STAT3 SH2 domain have already been documented in the literature [37,38]. Additionally, a hydrophobic interaction and two electrostatic forces (cation-type) between the –NH₃ group of Lys591 and the aromatic network of compound 1 also facilitated locking its conformation in the binding domain of STAT3. Overall, the docking results revealed compound 1 as a good inhibitor of STAT3 protein and supported our experimental observations.

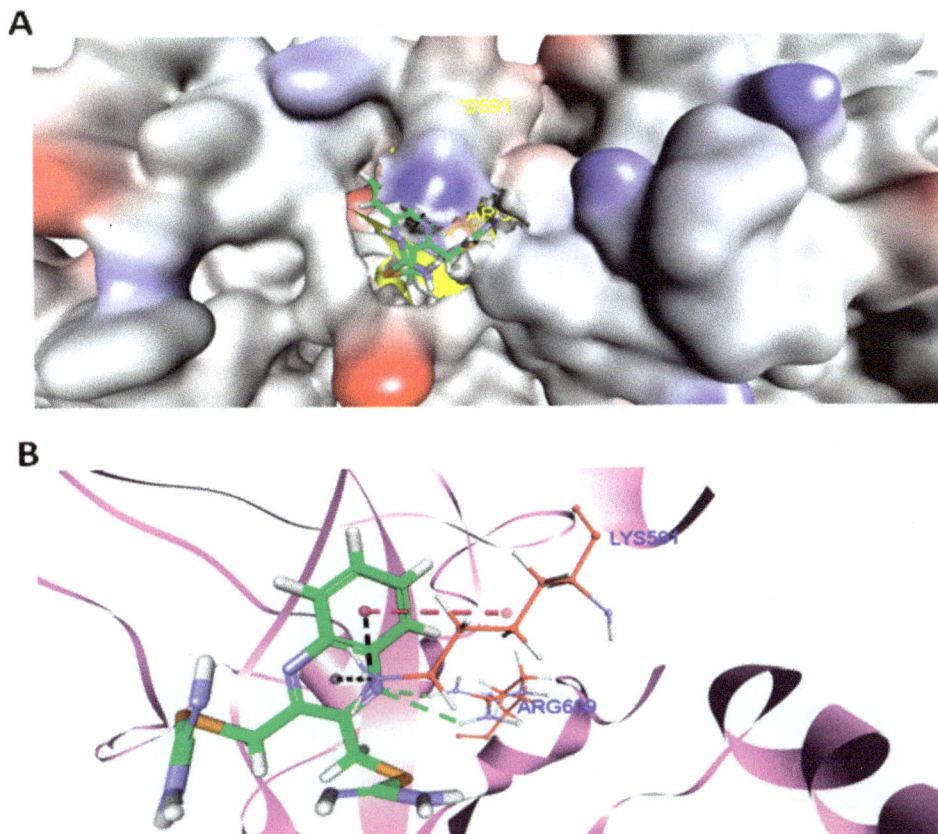

Figure 8. Docked complex of compound 1 with STAT3. (**A**) The surface representation of STAT3 showing deep penetration of compound 1 into the SH2 domain of STAT3 (pdb id: 1BG1), generated by DS (Accelrys). (**B**) Docked conformation of compound 1 in the SH2 domain of STAT3 (pdb id: 1BG1), viewed by DS (Accelrys). Carbon atoms of compound 1 (in sticks format) are colored green. The interacting amino acids of the SH2 domain of STAT3 are colored in red. All other amino acids of STAT3 are depicted in violet (flat ribbon format). Hydrogen bonds are shown in green, electrostatic interactions in black, and hydrophobic interactions are shown as magenta dotted lines.

3. Discussion

In this work, we tested the effects of an isoselenourea derivative and an isothiourea compound on cell viability in a panel of five melanoma cell lines bearing different mutations. The sulfur analog of compound 1 (compound 2) was inactive at the studied doses. This was in accordance with a previous report showing the isoselenourea derivative PBISe, an isosteric selenium analog of PBIT, to be over 10-fold more effective in inhibiting the viability of melanoma cells [19]. However, our recent report [22] also indicates that replacing the isothiourea functionality by an isoselenourea may not always lead to a more potent compound, suggesting that the potency depends on the overall structure of the molecule. For example, the isothiourea analog compound 3 was more effective than its corresponding isoselenourea analog [22]. Compounds 1 and 3 were active against both B-RAF mutant and wild-type

(CHL-1) cells lines, whereas PLX-4032 is only effective against the altered form [39]. In addition, our compounds were also more effective in reducing cell viability at 24 h in all the tested cell lines, compound 1 being at least 10-fold more potent than PLX-4032. In general, as reported in previous works [14,22,23], the selenium-containing derivative was much more active than the isothiourea one in reducing cancer cell growth.

The results from Annexin V and Caspase3/7 assays, as well as the observation of cleaved PARP in western blot, indicated that compound 1 induced apoptotic cell death. Apoptosis is a typical mechanism for selenium compounds to induce cell death [13]. However, the underlying mechanism of apoptosis induction could be remarkably different between distinct selenium-containing small molecules. Activation of caspases, modulation of anti-apoptotic proteins, alteration of oxidative stress-related proteins, cell cycle arrest, or kinases regulation are some of the described mechanisms associated with apoptosis [40]. This fact could explain the different mechanism of action between PBISe and compound 1. Although PBISe inhibited the phosphorylation and total protein levels of Akt in melanoma cells [41], compound 1 did not regulate this protein. We conclude that compound 1 is not effective against either Akt or ERK1/2 and indeed we found that compound 1 inhibits the expression of STAT3. In silico docking simulations conducted on compound 1 using STAT3 as a molecular target also suggested the compound to be a good inhibitor for this target protein. Hydrogen bonding and electrostatic and hydrophobic forces were found to be accountable for their host–guest relationship. Moreover, just like other known STAT3 inhibitors, compound 1 also interacted with Arg609, an essential amino acid for STAT3 function, suggesting its anti-melanoma activity may well be through inhibition of STAT3, in agreement with the experimental results.

STAT3 is a member of the signal-transducer-and-activators-of-transcription family. When phosphorylated, STAT3 dimerizes and translocates to the nucleus, where it modulates the transcription of genes involved in cell survival and proliferation [42]. Constitutive activation of STAT3 has been reported in several tumor types and has been proved to be involved in proliferation, survival, inflammation, invasion, metastasis, and angiogenesis [7]. Several in vitro and in vivo studies have demonstrated that the inhibition of STAT3 leads to tumor growth inhibition [5]. Thus, STAT3 is considered a valid therapeutic target for cancer therapy. STAT3 also has an important role in melanoma development and survival. This protein was found to be constitutively activated in numerous melanoma cell lines and tumor specimens [5]. Furthermore, inhibition of STAT3 signaling led to apoptosis of melanoma cells. Besides, STAT3 upregulation has been associated with acquired resistances to PLX-4032. Melanoma cells resistant to PLX-4032 showed increased STAT3 pathway activity [43], and in vitro silencing of this signaling inhibited the growth of cells resistant to PLX-4032 [11]. Moreover, a combination treatment with WP1066 (a STAT3 inhibitor) and PLX-4032 resulted in more significant growth inhibition of both resistant and sensitive cells to PLX-4032. The downregulation of other proliferative and anti-apoptotic proteins which are under transcriptional regulation of STAT3, such as XIAP, survivin, and c-myc, supports that compound 1 induces apoptosis through STAT3 inhibition. A previous study also reported a relationship between selenium-containing compounds and STAT3 inhibition [44]. The studies by Zuazo et al. demonstrated that imidoselenocarbamate derivatives blocked hypoxia-induced STAT3 phosphorylation [44].

4. Materials and Methods

4.1. Chemistry

All the chemicals were obtained from Sigma Aldrich (Alcobendas, Madrid, Spain) and Acros Organics (Janssen Pharmaceuticalaan, Geel, Belgium). Quinoxaline-2,3-diylbis(methylene)dicarbamimidoselenoate dihydrobromide (compound 1), quinoxaline-2,3-diylbis(methylene)dicarbamimidothioate dihydrobromide (compound 2), and (9,10-dioxo-9,10-dihydroanthracen-2-yl)methyl carbamimidothioate hydrochloride (compound 3) were synthesized and characterized as previously described [22]. Briefly, the corresponding alkyl halide (1 mmol) was added to a mixture of selenourea (2.2 mmol; compound 1) or thiourea (2.2 mmol for compound 2 and 1.1. mmol for 3) in absolute ethanol (20 mL). The mixture was stirred 2 h at r.t. for compound 1 or 3 hat reflux for 2 and 3. The precipitate was filtered and washed with 50 mL of ether (1 and 2) or with dichloromethane and ether (3).

4.2. Reagents and Antibodies

All the chemicals were obtained from Sigma Aldrich (Alcobendas, Madrid, Spain) and Acros Organics (Janssen Pharmaceuticalaan, Geel, Belgium). Quinoxaline-2,3-diylbis(methylene)dicarbamimidoselenoate dihydrobromide (compound 1), quinoxaline-2,3-diylbis(methylene)dicarbamimidothioate dihydrobromide (compound 2) and (9,10-dioxo-9,10-dihydroanthracen-2-yl)methyl carbamimidothioate hydrochloride (compound 3) were synthesized and characterized as previously described [22]. Briefly, the corresponding alkyl halide (1 mmol) was added to a mixture of selenourea (2.2 mmol; compound 1) or thiourea (2.2 mmol for compound 2 and 1.1. mmol for 3) in absolute ethanol (20 mL). The mixture was stirred 2 h at r.t. for compound 1, or 3 h at reflux for compounds 2 and 3. The precipitate was filtered and washed with 50 mL of ether (1 and 2) or with dichloromethane and ether (3). Rabbit anti-phospho-Akt (Cat # 4060), rabbit anti-Akt (Cat # 4685), rabbit anti-phospho-STAT3 (Cat # 9145), rabbit anti-STAT3 (Cat # 12640), rabbit anti-phospho-p44/42 MAPK (Erk1/2) (Cat # 4370), rabbit anti-p44/42 MAPK (Erk1/2) (Cat # 4695), rabbit anti-XIAP (Cat # 14334), rabbit anti-survivin (Cat # 2803), rabbit anti-c-Myc (Cat #9402), rabbit anti-p21wafl/Cip1 (Cat # 2947), and rabbit anti-PARP (Cat # 9542) were obtained from Cell Signaling (Danvers, MA). Antibodies against GAPDH, β-actin, goat anti-rabbit, and goat anti-mouse horseradish peroxidase conjugates, and MTT were purchased from Sigma-Aldrich (St. Louis, MO, USA).

4.3. Cell Culture

Human melanoma cell lines were grown in DMEM medium supplemented with 10 % fetal bovine serum (FBS) and 100 units/mL of penicillin and streptomycin (Corning; Corning, NY, USA). The cells were maintained at 37 °C and 5% CO_2.

4.4. Cell Viability

A total of 3000 cells/well were grown in 96-well plates for 12 h and then treated with either DMSO (control) or increasing concentrations (0.5–50 μM) of compound 1, 3, or PLX-4032 for 24, 48, and 72 h. Three hours before the termination point, 20 μL of MTT were added to measure cellular viability. The resultant formazan crystals were dissolved in 50 μL of DMSO, and absorbance was measured at 570 nm and 630 nm wavelengths. IC_{50} values were calculated using GraphPad Prism version 6.01.

4.5. Colony Formation Assay

Melanoma cells (UACC903 and 1205Lu) were treated with DMSO or compound 1 (1 and 2.5 μM) for 24 h. The cells were then trypsinized and counted, and 500 live cells were seeded in new 10 cm tissue culture plates separated by treatments. The cells were allowed to form colonies for 14 days in DMEM without compound 1 or DMSO in a 5% CO_2 incubator. After 14 days, the medium was removed, and the formed colonies were stained with 0.5% alcoholic crystal violet. The percentage plating efficiency (PE) was calculated by using the following formula:

$$\%PE = \frac{number\ of\ colonies}{number\ of\ cells\ plated} \times 100 \tag{1}$$

The results correspond to the mean ± SD of three independent experiments. Differences between control and treated cells were determined by one-way ANOVA tests, using GraphPad Prism version 6.01.

4.6. Live-and-Dead Assay

The LIVE/DEAD viability/cytotoxicity kit for mammalian cells (Molecular Probes, Invitrogen; Carlsbad, CA, USA) was employed for this experiment. A total of 7.5×10^5 1205Lu cells/well were plated in a 6-well plate and treated with DMSO (control) or increasing amounts of compound 1. After 24 h of incubation, both floating and attached cells were collected and re-suspended in PBS. Further, cells were stained with a calcein AM and ethidium homodimer solution according to the manufacturer's instructions. The number of live and dead cells was determined by flow cytometry (BD FACSCalibur; Heidelberg, Germany).

4.7. Annexin V Assay

1205Lu, UACC903, and SK-Mel-8 (4×10^5 cells/well) were pleated and treated with DMSO (control) or compound 1 at the given concentrations. After 24 h of treatment, both floating and attached cells were collected and stained with the Muse™ Annexin V & Dead Cell Reagent (Millipore; Bedford, MA, USA) for 20 min at room temperature in the dark. The results were collected by Muse™ Cell Analyzer (Millipore).

4.8. Caspase-3/7 Assay

The Muse™ Caspase 3/7 kit (Millipore) was employed for this experiment according to the manufacturer's instructions and our previously published method [22]. 1205Lu, UACC903, and Sk-Mel-8 cells were seeded at a density of 4×10^5 cells/well and treated with DMSO or the indicated amounts of compounds 1 and 3 for 24 h. Next, both floating and attached cells were collected, out of which 50 µL of cells were stained with 5 µL of Muse™ Caspase-3/7 working solution and incubated for 30 min at 37 °C with 5 % CO_2. After incubation, 150 µL of Muse™ Caspase 7-AAD working solution were added to each sample, and the samples were incubated in the dark for 5 min more. The samples were analyzed by Muse™ Cell Analyzer.

4.9. Western Blotting

1205Lu, UACC903, and SK-Mel-8 cells (7×10^5 cells/well) were plated in 6-well plates and treated with compound 1 or DMSO for 24 h. The cells were then collected, and whole cell lysates were prepared in RIPA buffer (Thermo Scientific #89900; Rockford, IL, USA) supplemented with 1% phosphatase inhibitor cocktail 2 (Sigma #P5726-5ML), 1% protease inhibitor (Complete mini, Roche #11836170001; Branchburg, NJ, USA), and 0.5 % of 200 mM phenylmethanesulfonyl fluoride (PMSF) (Sigma #P7626-250 mg). The cell lysates were spun at $15,000 \times g$ for 10 min to remove any insoluble cell debris. The resultant supernatants were collected and stored at -80 °C until use. The whole cell lysates were resolved by SDS-PAGE. The proteins were transferred to Immobilon®-P PVDF membranes (Millipore # IPVH304F0) and blotted with the indicated antibody overnight at 4 °C. The dilution for primary antibodies was 1:1000, with the exception of p-STAT3 (1:2000), STAT3 (1:200), and β-actin (1:3000). Further, the membranes were incubated with the corresponding peroxidase-linked secondary antibodies (dilution 1:3000) for 1–4 h at room temperature. The antibodies were detected by an enhanced chemiluminescence reagent (Thermo Scientific #1856135 and #1856136).

4.10. Docking Methodology

Different 3D conformations of compound 1 were generated and energetically minimized using the "Generate Conformations" tool in Discovery Studio (DS) 4.0 client (Accelrys). The lowest energetic conformation thus obtained was subjected to the "Prepare Ligands" module to generate its isomers at physiological pH. The CHARMm force field was employed to develop the partial atomic charges on each atom of the isomer. The isomer with the lowest CHARMm energy was used for the docking study.

The X-ray co-ordinates of STAT3 (pdb id: 1BG1, resolution 2.25 Å) were retrieved from the protein data bank (www.rcsb.org). The "Prepare Protein" tool in DS was used to add missing atoms/chains and remove water molecules in the protein structure. The "Prepare Protein" algorithm was employed to protonate amino acid residues according to the physiological conditions. Prior to docking, a binding sphere covering the SH2 domain of STAT3 was generated. CDOCKER [34], a grid-based docking program, was used to dock compound 1 in the SH2 domain, considering the default parameters. The most favorable pose of compound 1 was identified based on the CDOCKER energy (-CDE).

5. Conclusions

In conclusion, compound 1, a quinoxaline-isoselenourea, showed promising efficacy in vitro against melanoma cells. Compound 1 was more effective than the isothiourea derivative 3 and the reference drug PLX-4032, especially at short times of treatment (24 h). As demonstrated by the colony formation assay, the effects of compound 1 in inhibiting cell proliferation were long-lasting. Overall, we demonstrated for the first time that the cell growth inhibitory response of compound 1 may be mediated through the STAT3 pathway as it does not inhibit other pro-survival signaling pathways such as Akt and ERK1/2. On account of the fact that compound 1 inhibits STAT3, it could also be employed in combination therapy with PLX-4032 to overcome acquired resistance to the latter. Interestingly, the response to compound 1 was not dependent on the mutation status of BRAF. Only about half of all melanomas have a mutation in the BRAF gene, and, therefore, compound 1, targeting both BRAF-mutant and WT melanoma cells, may have a broader clinical impact. However, more preclinical studies examining in vivo efficacy, Absorption, Distribution, Metabolism and Excretion (ADME), toxicity, and mechanism of action are required to determine the therapeutic properties and future clinical potential of compound 1.

Author Contributions: Conceptualization, D.P., J.A.P., S.A., C.S., A.K.S.; Methodology, V.A., D.N.K., M.K.P., D.P., P.S.; Writing – Original Draft Preparation, V.A., D.N.K.; Writing – Review & Editing, M.J.P., D.P., J.A.P., S.A., C.S., A.K.S.

Funding: This research was funded by Caixa Foundation-UNED-Caja Navarra.

Acknowledgments: The authors thank the Department of Pharmacology and Penn State Cancer Institute, Pennsylvania State University College of Medicine for financial support, and the Flow Cytometry Core at of the Pennsylvania State University, College of Medicine. The melanoma cell lines were graciously provided by Gavin Robertson, Penn State College of Medicine, Hershey, PA. V. Alcolea wishes to express her gratitude to the Asociación de Amigos de la Universidad de Navarra for the pre-doctoral fellowship. The Authors are also grateful to the Centre for High-Performance Computing, an initiative supported by the Department of Science and Technology of South Africa for usage of software and cluster support.

Conflicts of Interest: The authors declare no conflict of interest.

Abbreviations

AKT	Protein kinase B
c-myc	Avian myelocytomatosis virus oncogene cellular homolog
ERK1/2	Extracellular signal-regulated kinases 1 and 2
FBS	Fetal bovine serum
MAPK	Mitogen-activated protein kinase
MMP	Matrix metalloproteinase
MTT	3-(4,5-dimethylthiazol-2-yl)-2,5-diphenyltetrazolium bromide
PARP	Poly(ADP-ribose)polymerase
PBISe	*Se,Se′*-1,4-phenylenebis(1,2-ethanediyl)bisisoselenourea
PBIT	*S,S′*-1,4-phenylenebis(1,2-ethanediyl)bisisothiourea
PMSF	Phenylmethanesulfonyl fluoride
STAT3	Signal transducer and activator of transcription 3
XIAP	X-linked inhibitor of apoptosis protein

References

1. Leiter, U.; Eigentler, T.; Garbe, C. Epidemiology of skin cancer. *Adv. Exp. Med. Biol.* **2014**, *810*, 120–140. [PubMed]
2. Bray, F.; Ferlay, J.; Soerjomataram, I.; Siegel, R.L.; Torre, L.A.; Jemal, A. Global cancer statistics 2018: GLOBOCAN estimates of incidence and mortality worldwide for 36 cancers in 185 countries. *CA Cancer J. Clin.* **2018**, *68*, 394–424. [CrossRef] [PubMed]
3. Di Trolio, R.; Simeone, E.; Di Lorenzo, G.; Buonerba, C.; Ascierto, P.A. The use of interferon in melanoma patients: A systematic review. *Cytokine Growth Factor Rev.* **2015**, *26*, 203–212. [CrossRef] [PubMed]
4. Marzuka, A.; Huang, L.; Theodosakis, N.; Bosenberg, M. Melanoma treatments: Advances and mechanisms. *J. Cell. Physiol.* **2015**, *230*, 2626–2633. [CrossRef] [PubMed]
5. Niu, G.; Bowman, T.; Huang, M.; Shivers, S.; Reintgen, D.; Daud, A.; Chang, A.; Kraker, A.; Jove, R.; Yu, H. Roles of activated Src and Stat3 signaling in melanoma tumor cell growth. *Oncogene* **2002**, *21*, 7001–7010. [CrossRef] [PubMed]
6. Xie, T.X.; Wei, D.; Liu, M.; Gao, A.C.; Ali-Osman, F.; Sawaya, R.; Huang, S. Stat3 activation regulates the expression of matrix metalloproteinase-2 and tumor invasion and metastasis. *Oncogene* **2004**, *23*, 3550–3560. [CrossRef]
7. Siveen, K.S.; Sikka, S.; Surana, R.; Dai, X.; Zhang, J.; Kumar, A.P.; Tan, B.K.; Sethi, G.; Bishayee, A. Targeting the STAT3 signaling pathway in cancer: Role of synthetic and natural inhibitors. *Biochim. Biophys. Acta* **2014**, *1845*, 136–154. [CrossRef]
8. Kusaba, T.; Nakayama, T.; Yamazumi, K.; Yakata, Y.; Yoshizaki, A.; Inoue, K.; Nagayasu, T.; Sekine, I. Activation of STAT3 is a marker of poor prognosis in human colorectal cancer. *Oncol. Rep.* **2006**, *15*, 1445–1451. [CrossRef]
9. Strickland, L.R.; Pal, H.C.; Elmets, C.A.; Afaq, F. Targeting drivers of melanoma with synthetic small molecules and phytochemicals. *Cancer Lett.* **2015**, *359*, 20–35. [CrossRef]
10. Mukherjee, N.; Schwan, J.V.; Fujita, M.; Norris, D.A.; Shellman, Y.G. Alternative treatments for melanoma: Targeting Bcl-2 family members to de-bulk and kill cancer stem cells. *J. Investig. Dermatol.* **2015**, *135*, 2155–2161. [CrossRef]
11. Liu, F.; Cao, J.; Wu, J.; Sullivan, K.; Shen, J.; Ryu, B.; Xu, Z.; Wei, W.; Cui, R. Stat3-targeted therapies overcome the acquired resistance to vemurafenib in melanomas. *J. Investig. Dermatol.* **2013**, *133*, 2041–2049. [CrossRef] [PubMed]
12. Becker, T.M.; Boyd, S.C.; Mijatov, B.; Gowrishankar, K.; Snoyman, S.; Pupo, G.M.; Scolyer, R.A.; Mann, G.J.; Kefford, R.F.; Zhang, X.D.; et al. Mutant B-RAF-Mcl-1 survival signaling depends on the STAT3 transcription factor. *Oncogene* **2014**, *33*, 1158–1166. [CrossRef] [PubMed]
13. Fernandes, A.P.; Gandin, V. Selenium compounds as therapeutic agents in cancer. *Biochim. Biophys. Acta* **2015**, *1850*, 1642–1660. [CrossRef] [PubMed]
14. Sharma, A.K.; Sharma, A.; Desai, D.; Madhunapantula, S.V.; Huh, S.J.; Robertson, G.P.; Amin, S. Synthesis and anticancer activity comparison of phenylalkyl isoselenocyanates with corresponding naturally occurring and synthetic isothiocyanates. *J. Med. Chem.* **2008**, *51*, 7820–7826. [CrossRef] [PubMed]
15. Chen, Y.C.; Prabhu, K.S.; Mastro, A.M. Is selenium a potential treatment for cancer metastasis? *Nutrients* **2013**, *5*, 1149–1168. [CrossRef] [PubMed]
16. Sharma, A.; Sharma, A.K.; Madhunapantula, S.V.; Desai, D.; Huh, S.J.; Mosca, P.; Amin, S.; Robertson, G.P. Targeting Akt3 signaling in malignant melanoma using isoselenocyanates. *Clin. Cancer Res.* **2009**, *15*, 1674–1685. [CrossRef]
17. Nguyen, N.; Sharma, A.; Nguyen, N.; Sharma, A.K.; Desai, D.; Huh, S.J.; Amin, S.; Meyers, C.; Robertson, G.P. Melanoma chemoprevention in skin reconstructs and mouse xenografts using isoselenocyanate-4. *Cancer Prev. Res.* **2011**, *4*, 248–258. [CrossRef]
18. Cassidy, P.B.; Fain, H.D.; Cassidy, J.P., Jr.; Tran, S.M.; Moos, P.J.; Boucher, K.M.; Gerads, R.; Florell, S.R.; Grossman, D.; Leachman, S.A. Selenium for the prevention of cutaneous melanoma. *Nutrients* **2013**, *5*, 725–749. [CrossRef]
19. Madhunapantula, S.V.; Desai, D.; Sharma, A.; Huh, S.J.; Amin, S.; Robertson, G.P. PBISe, a novel selenium-containing drug for the treatment of malignant melanoma. *Mol. Cancer Ther.* **2008**, *7*, 1297–1308. [CrossRef]

20. Karelia, D.N.; Sk, U.H.; Singh, P.; Gowda, A.S.P.; Pandey, M.K.; Ramisetti, S.R.; Amin, S.; Sharma, A.K. Design, synthesis, and identification of a novel napthalamide-isoselenocyanate compound NISC-6 as a dual Topoisomerase-IIalpha and Akt pathway inhibitor, and evaluation of its anti-melanoma activity. *Eur. J. Med. Chem.* **2017**, *135*, 282–295. [CrossRef]

21. Chung, C.Y.; Madhunapantula, S.V.; Desai, D.; Amin, S.; Robertson, G.P. Melanoma prevention using topical PBISe. *Cancer Prev. Res.* **2011**, *4*, 935–948. [CrossRef] [PubMed]

22. Alcolea, V.; Plano, D.; Karelia, D.N.; Palop, J.A.; Amin, S.; Sanmartin, C.; Sharma, A.K. Novel seleno- and thio-urea derivatives with potent in vitro activities against several cancer cell lines. *Eur. J. Med. Chem.* **2016**, *113*, 134–144. [CrossRef] [PubMed]

23. Ibanez, E.; Plano, D.; Font, M.; Calvo, A.; Prior, C.; Palop, J.A.; Sanmartin, C. Synthesis and antiproliferative activity of novel symmetrical alkylthio- and alkylseleno-imidocarbamates. *Eur. J. Med. Chem.* **2011**, *46*, 265–274. [CrossRef] [PubMed]

24. Alcolea, V.; Plano, D.; Encio, I.; Palop, J.A.; Sharma, A.K.; Sanmartin, C. Chalcogen containing heterocyclic scaffolds: New hybrids with antitumoral activity. *Eur. J. Med. Chem.* **2016**, *123*, 407–418. [CrossRef] [PubMed]

25. van Engeland, M.; Nieland, L.J.; Ramaekers, F.C.; Schutte, B.; Reutelingsperger, C.P. Annexin V-affinity assay: A review on an apoptosis detection system based on phosphatidylserine exposure. *Cytometry* **1998**, *31*, 1–9. [CrossRef]

26. Stahl, J.M.; Sharma, A.; Cheung, M.; Zimmerman, M.; Cheng, J.Q.; Bosenberg, M.W.; Kester, M.; Sandirasegarane, L.; Robertson, G.P. Deregulated Akt3 activity promotes development of malignant melanoma. *Cancer Res.* **2004**, *64*, 7002–7010. [CrossRef] [PubMed]

27. Zhang, X.D.; Borrow, J.M.; Zhang, X.Y.; Nguyen, T.; Hersey, P. Activation of ERK1/2 protects melanoma cells from TRAIL-induced apoptosis by inhibiting Smac/DIABLO release from mitochondria. *Oncogene* **2003**, *22*, 2869–2881. [CrossRef] [PubMed]

28. Fofaria, N.M.; Srivastava, S.K. Critical role of STAT3 in melanoma metastasis through anoikis resistance. *Oncotarget* **2014**, *5*, 7051–7064. [CrossRef] [PubMed]

29. Kiuchi, N.; Nakajima, K.; Ichiba, M.; Fukada, T.; Narimatsu, M.; Mizuno, K.; Hibi, M.; Hirano, T. STAT3 is required for the gp130-mediated full activation of the c-myc gene. *J. Exp. Med.* **1999**, *189*, 63–73. [CrossRef]

30. Ishdorj, G.; Johnston, J.B.; Gibson, S.B. Inhibition of constitutive activation of STAT3 by curcurbitacin-I (JSI-124) sensitized human B-leukemia cells to apoptosis. *Mol. Cancer Ther.* **2010**, *9*, 3302–3314. [CrossRef]

31. Soldani, C.; Scovassi, A.I. Poly(ADP-ribose) polymerase-1 cleavage during apoptosis: An update. *Apoptosis* **2002**, *7*, 321–328. [CrossRef] [PubMed]

32. Coqueret, O.; Gascan, H. Functional interaction of STAT3 transcription factor with the cell cycle inhibitor p21WAF1/CIP1/SDI1. *J. Biol. Chem.* **2000**, *275*, 18794–18800. [CrossRef] [PubMed]

33. Becker, S.; Groner, B.; Muller, C.W. Three-dimensional structure of the Stat3beta homodimer bound to DNA. *Nature* **1998**, *394*, 145–151. [CrossRef] [PubMed]

34. Wu, G.; Robertson, D.H.; Brooks, C.L., 3rd; Vieth, M. Detailed analysis of grid-based molecular docking: A case study of CDOCKER-A CHARMm-based MD docking algorithm. *J. Comput. Chem.* **2003**, *24*, 1549–1562. [CrossRef] [PubMed]

35. Wang, Y.; Ren, X.; Deng, C.; Yang, L.; Yan, E.; Guo, T.; Li, Y.; Xu, M.X. Mechanism of the inhibition of the STAT3 signaling pathway by EGCG. *Oncol. Rep.* **2013**, *30*, 2691–2696. [CrossRef] [PubMed]

36. Zhang, T.; Kee, W.H.; Seow, K.T.; Fung, W.; Cao, X. The coiled-coil domain of Stat3 is essential for its SH2 domain-mediated receptor binding and subsequent activation induced by epidermal growth factor and interleukin-6. *Mol. Cell. Biol.* **2000**, *20*, 7132–7139. [CrossRef] [PubMed]

37. Yu, W.; Xiao, H.; Lin, J.; Li, C. Discovery of novel STAT3 small molecule inhibitors via in silico site-directed fragment-based drug design. *J. Med. Chem.* **2013**, *56*, 4402–4412. [CrossRef]

38. Bhasin, D.; Cisek, K.; Pandharkar, T.; Regan, N.; Li, C.; Pandit, B.; Lin, J.; Li, P.K. Design, synthesis, and studies of small molecule STAT3 inhibitors. *Bioorg. Med. Chem. Lett.* **2008**, *18*, 391–395. [CrossRef]

39. Tsai, J.; Lee, J.T.; Wang, W.; Zhang, J.; Cho, H.; Mamo, S.; Bremer, R.; Gillette, S.; Kong, J.; Haass, N.K.; et al. Discovery of a selective inhibitor of oncogenic B-Raf kinase with potent antimelanoma activity. *Proc. Natl. Acad. Sci. USA* **2008**, *105*, 3041–3046. [CrossRef]

40. Sanmartin, C.; Plano, D.; Palop, J.A. Selenium compounds and apoptotic modulation: A new perspective in cancer therapy. *Mini Rev. Med. Chem.* **2008**, *8*, 1020–1031. [CrossRef]

41. Tagaram, H.R.; Desai, D.; Li, G.; Liu, D.; Rountree, C.B.; Gowda, K.; Berg, A.; Amin, S.; Staveley-O'Carroll, K.F.; Kimchi, E.T. A Selenium Containing Inhibitor for the Treatment of Hepatocellular Cancer. *Pharmaceuticals* **2016**, *9*, 18. [CrossRef] [PubMed]

42. Thomas, S.J.; Snowden, J.A.; Zeidler, M.P.; Danson, S.J. The role of JAK/STAT signalling in the pathogenesis, prognosis and treatment of solid tumours. *Br. J. Cancer* **2015**, *113*, 365–371. [CrossRef] [PubMed]

43. Girotti, M.R.; Pedersen, M.; Sanchez-Laorden, B.; Viros, A.; Turajlic, S.; Niculescu-Duvaz, D.; Zambon, A.; Sinclair, J.; Hayes, A.; Gore, M.; et al. Inhibiting EGF receptor or SRC family kinase signaling overcomes BRAF inhibitor resistance in melanoma. *Cancer Discov.* **2013**, *3*, 158–167. [CrossRef] [PubMed]

44. Zuazo, A.; Plano, D.; Anso, E.; Lizarraga, E.; Font, M.; Martinez Irujo, J.J. Cytotoxic and proapoptotic activities of imidoselenocarbamate derivatives are dependent on the release of methylselenol. *Chem. Res. Toxicol.* **2012**, *25*, 2479–2489. [CrossRef] [PubMed]

MDPI

St. Alban-Anlage 66

4052 Basel

Switzerland

Tel. +41 61 683 77 34

Fax +41 61 302 89 18

www.mdpi.com

International Journal of Molecular Sciences Editorial Office

E-mail: ijms@mdpi.com

www.mdpi.com/journal/ijms